Python Data Analytics

With Pandas, NumPy, and Matplotlib

Third Edition

Fabio Nelli

Apress®

Python Data Analytics: With Pandas, NumPy, and Matplotlib

Fabio Nelli
Rome, Italy

ISBN-13 (pbk): 978-1-4842-9531-1   ISBN-13 (electronic): 978-1-4842-9532-8
https://doi.org/10.1007/978-1-4842-9532-8

Managing Director, Apress Media LLC: Welmoed Spahr
Acquisitions Editor: Celestin Suresh John
Development Editor: James Markham
Coordinating Editor: Mark Powers
Copy Editor: Kezia Endsley

Cover image by Tyler B on Unsplash (www.unsplash.com)

Distributed to the book trade worldwide by Springer Science+Business Media New York, 233 Spring Street, 6th Floor, New York, NY 10013. Phone 1-800-SPRINGER, fax (201) 348-4505, e-mail orders-ny@springer-sbm.com, or visit www.springeronline.com. Apress Media, LLC is a California LLC and the sole member (owner) is Springer Science + Business Media Finance Inc (SSBM Finance Inc). SSBM Finance Inc is a **Delaware** corporation.

For information on translations, please e-mail booktranslations@springernature.com; for reprint, paperback, or audio rights, please e-mail bookpermissions@springernature.com.

Apress titles may be purchased in bulk for academic, corporate, or promotional use. eBook versions and licenses are also available for most titles. For more information, reference our Print and eBook Bulk Sales web page at http://www.apress.com/bulk-sales.

Any source code or other supplementary material referenced by the author in this book is available to readers on GitHub (github.com/apress). For more detailed information, please visit https://www.apress.com/gp/services/source-code.

Paper in this product is recyclable

"Science leads us forward in knowledge, but only analysis makes us more aware"

This book is dedicated to all those who are constantly looking for awareness

Table of Contents

About the Author

Fabio Nelli is a data scientist and Python consultant who designs and develops Python applications for data analysis and visualization. He also has experience in the scientific world, having performed various data analysis roles in pharmaceutical chemistry for private research companies and universities. He has been a computer consultant for many years at IBM, EDS, and Hewlett-Packard, along with several banks and insurance companies. He holds a master's degree in organic chemistry and a bachelor's degree in information technologies and automation systems, with many years of experience in life sciences (as a tech specialist at Beckman Coulter, Tecan, and SCIEX).

For further info and other examples, visit his page at www.meccanismocomplesso.org and the GitHub page at https://github.com/meccanismocomplesso.

About the Technical Reviewer

Akshay R. Kulkarni is an artificial intelligence (AI) and machine learning (ML) evangelist and thought leader. He has consulted with several Fortune 500 and global enterprises to drive AI and data science-led strategic transformations. He is a Google developer, an author, and a regular speaker at major AI and data science conferences, including the O'Reilly Strata Data & AI Conference and the Great International Developer Summit (GIDS). He has been a visiting faculty member at some of the top graduate institutes in India. In 2019, he was featured as one of India's "top 40 under 40" data scientists. In his spare time, Akshay enjoys reading, writing, coding, and helping aspiring data scientists. He lives in Bangalore with his family.

Preface

About five years have passed since the last edition of this book. In drafting this third edition, I made some necessary changes, both to the text and to the code. First, all the Python code has been ported to 3.8 and greater, and all references to Python 2.x versions have been dropped. Some chapters required a total rewrite because the content was no longer compatible. I'm referring to TensorFlow 3.x which, compared to TensorFlow 2.x (covered in the previous edition), has completely revamped its entire reference system. In five years, the deep learning modules and code developed with version 2.x have proven completely unusable. Keras and all its modules have been incorporated into the TensorFlow library, replacing all the classes, functions, and modules that performed similar functions. The construction of neural network models, their learning phases, and the functions they use have all completely changed. In this edition, therefore, you have the opportunity to learn the methods of TensorFlow 3.x and to acquire familiarity with the concepts and new paradigms in the new version.

Regarding data visualization, I decided to add information about the Seaborn library to the matplotlib chapter. Seaborn, although still in version 0.x, is proving to be a very useful matplotlib extension for data analysis, thanks to its statistical display of plots and its compatibility with pandas dataframes. I hope that, with this completely updated third edition, I can further entice you to study and deepen your data analysis with Python. This book will be a valuable learning tool for you now, and serve as a dependable reference in the future.

—Fabio Nelli

## CHAPTER 1

An Introduction to Data Analysis

In this chapter, you'll take your first steps in the world of data analysis, learning in detail the concepts and processes that make up this discipline. The concepts discussed in this chapter are helpful background for the following chapters, where these concepts and procedures are applied in the form of Python code, through the use of several libraries that are discussed in later chapters.

Data Analysis

In a world increasingly centralized around information technology, huge amounts of data are produced and stored each day. Often these data come from automatic detection systems, sensors, and scientific instrumentation, or you produce them daily and subconsciously every time you make a withdrawal from the bank or purchase something, when you record various blogs, or even when you post on social networks.

But what are the data? The data actually are not information, at least in terms of their form. In the formless stream of bytes, at first glance it is difficult to understand their essence, if they are not strictly numbers, words, or times. This information is actually the result of processing, which, taking into account a certain dataset, extracts conclusions that can be used in various ways. This process of extracting information from raw data is called *data analysis*.

The purpose of data analysis is to extract information that is not easily deducible but, when understood, enables you to carry out studies on the mechanisms of the systems that produced the data. This in turn allows you to forecast possible responses of these systems and their evolution in time.

Starting from a simple methodical approach to data protection, data analysis has become a real discipline, leading to the development of real methodologies that generate *models*. The model is in fact a translation of the system to a mathematical form. Once there is a mathematical or logical form that can describe system responses under different levels of precision, you can predict its development or response to certain inputs. Thus, the aim of data analysis is not the model, but the quality of its *predictive power*.

The predictive power of a model depends not only on the quality of the modeling techniques but also on the ability to choose a good dataset upon which to build the entire analysis process. So the *search for data*, their *extraction*, and their subsequent *preparation*, while representing preliminary activities of an analysis, also belong to data analysis itself, because of their importance in the success of the results.

So far I have spoken of data, their handling, and their processing through calculation procedures. In parallel to all the stages of data analysis processing, various methods of *data visualization* have also been developed. In fact, to understand the data, both individually and in terms of the role they play in the dataset, there is no better system than to develop the techniques of graphical representation. These techniques are capable of transforming information, sometimes implicitly hidden, into figures, which help you more easily understand the meaning of the data. Over the years, many display modes have been developed for different modes of data display, called *charts*.

© Fabio Nelli 2023

F. Nelli, *Python Data Analytics*, https://doi.org/10.1007/978-1-4842-9532-8_1

At the end of the data analysis process, you have a model and a set of graphical displays and you can predict the responses of the system under study; after that, you move to the test phase. The model is tested using another set of data for which you know the system response. These data do not define the predictive model. Depending on the ability of the model to replicate real, observed responses, you get an error calculation and knowledge of the validity of the model and its operating limits.

These results can be compared to any other models to understand if the newly created one is more efficient than the existing ones. Once you have assessed that, you can move to the last phase of data analysis—*deployment*. This phase consists of implementing the results produced by the analysis, namely, implementing the decisions to be made based on the predictions generated by the model and its associated risks.

Data analysis is well suited to many professional activities. So, knowledge of it and how it can be put into practice is relevant. It allows you to test hypotheses and understand the systems you've analyzed more deeply.

Knowledge Domains of the Data Analyst

Data analysis is basically a discipline suitable to the study of problems that occur in several fields of applications. Moreover, data analysis includes many tools and methodologies and requires knowledge of computing, mathematical, and statistical concepts.

A good data analyst must be able to move and act in many disciplinary areas. Many of these disciplines are the basis of the data analysis methods, and proficiency in them is almost necessary. Knowledge of other disciplines is necessary, depending on the area of application and the particular data analysis project. More generally, sufficient experience in these areas can help you better understand the issues and the type of data you need.

Often, regarding major problems of data analysis, it is necessary to have an interdisciplinary team of experts who can contribute in the best possible way to their respective fields of competence. Regarding smaller problems, a good analyst must be able to recognize problems that arise during data analysis, determine which disciplines and skills are necessary to solve these problems, study these disciplines, and maybe even ask the most knowledgeable people in the sector. In short, the analyst must be able to search not only for data, but also for information on how to treat that data.

Computer Science

Knowledge of computer science is a basic requirement for any data analyst. In fact, only when you have good knowledge of and experience in computer science can you efficiently manage the necessary tools for data analysis. In fact, every step concerning data analysis involves using calculation software (such as IDL, MATLAB, etc.) and programming languages (such as C ++, Java, and Python).

The large amount of data available today, thanks to information technology, requires specific skills in order to be managed as efficiently as possible. Indeed, data research and extraction require knowledge of these various formats. The data are structured and stored in files or database tables with particular formats. XML, JSON, or simply XLS or CSV files, are now the common formats for storing and collecting data, and many applications allow you to read and manage the data stored in them. When it comes to extracting data contained in a database, things are not so immediate, but you need to know the SQL Query language or use software specially developed for the extraction of data from a given database.

Moreover, for some specific types of data research, the data are not available in an explicit format, but are present in text files (documents and log files) or web pages, or shown as charts, measures, number of visitors, or HTML tables. This requires specific technical expertise to parse and eventually extract these data (called *web scraping*).

Knowledge of information technology is necessary for using the various tools made available by contemporary computer science, such as applications and programming languages. These tools, in turn, are needed to perform data analysis and data visualization.

The purpose of this book is to provide all the necessary knowledge, as far as possible, regarding the development of methodologies for data analysis. The book uses the Python programming language and specialized libraries that contribute to the performance of the data analysis steps, from data research to data mining, to publishing the results of the predictive model.

Mathematics and Statistics

As you will see throughout the book, data analysis requires a lot of complex math to treat and process the data. You need to be competent in all of this, at least enough to understand what you are doing. Some familiarity with the main statistical concepts is also necessary because the methods applied to the analysis and interpretation of data are based on these concepts. Just as you can say that computer science gives you the tools for data analysis, you can also say that statistics provide the concepts that form the basis of data analysis.

This discipline provides many tools to the analyst, and a good knowledge of how to best use them requires years of experience. Among the most commonly used statistical techniques in data analysis are

- Bayesian methods

- Regression

- Clustering

Having to deal with these cases, you'll discover how mathematics and statistics are closely related. Thanks to the special Python libraries covered in this book, you will be able to manage and handle them.

Machine Learning and Artificial Intelligence

One of the most advanced tools that falls in the data analysis camp is machine learning. In fact, despite the data visualization and techniques such as clustering and regression, which help you find information about the dataset, during this phase of research, you may often prefer to use special procedures that are highly specialized in searching patterns within the dataset.

Machine learning is a discipline that uses a whole series of procedures and algorithms that analyze the data in order to recognize patterns, clusters, or trends and then extracts useful information for analysis in an automated way.

This discipline is increasingly becoming a fundamental tool of data analysis, and thus knowledge of it, at least in general, is of fundamental importance to the data analyst.

Professional Fields of Application

Another very important point is the domain of data competence (its source—biology, physics, finance, materials testing, statistics on population, etc.). In fact, although analysts have had specialized preparation in the field of statistics, they must also be able to document the source of the data, with the aim of perceiving and better understanding the mechanisms that generated the data. In fact, the data are not simple strings or numbers; they are the expression, or rather the measure, of any parameter observed. Thus, a better understanding of where the data came from can improve their interpretation. Often, however, this is too costly for data analysts, even ones with the best intentions, and so it is good practice to find consultants or key figures to whom you can pose the right questions.

Understanding the Nature of the Data

The object of data analysis is basically the data. The data then will be the key player in all processes of data analysis. The data constitute the raw material to be processed, and thanks to their processing and analysis, it is possible to extract a variety of information in order to increase the level of knowledge of the system under study.

When the Data Become Information

Data are the events recorded in the world. Anything that can be measured or categorized can be converted into data. Once collected, these data can be studied and analyzed, both to understand the nature of events and very often also to make predictions or at least to make informed decisions.

When the Information Becomes Knowledge

You can speak of knowledge when the information is converted into a set of rules that helps you better understand certain mechanisms and therefore make predictions on the evolution of some events.

Types of Data

Data can be divided into two distinct categories:

- Categorical (nominal and ordinal)
- Numerical (discrete and continuous)

Categorical data are values or observations that can be divided into groups or categories. There are two types of categorical values: *nominal* and *ordinal*. A nominal variable has no intrinsic order that is identified in its category. An ordinal variable instead has a predetermined order.

Numerical data are values or observations that come from measurements. There are two types of numerical values: *discrete* and *continuous* numbers. Discrete values can be counted and are distinct and separated from each other. Continuous values, on the other hand, are values produced by measurements or observations that assume any value within a defined range.

The Data Analysis Process

Data analysis can be described as a process consisting of several steps in which the raw data are transformed and processed in order to produce data visualizations and make predictions, thanks to a mathematical model based on the collected data. Then, data analysis is nothing more than a sequence of steps, each of which plays a key role in the subsequent ones. So, data analysis is schematized as a process chain consisting of the following sequence of stages:

- Problem definition
- Data extraction
- Data preparation - data cleaning
- Data preparation - data transformation
- Data exploration and visualization

- Predictive modeling

- Model validation/testing

- Visualization and interpretation of results

- Deployment of the solution (implementation of the solution in the real world)

Figure 1-1 shows a schematic representation of all the processes involved in data analysis.

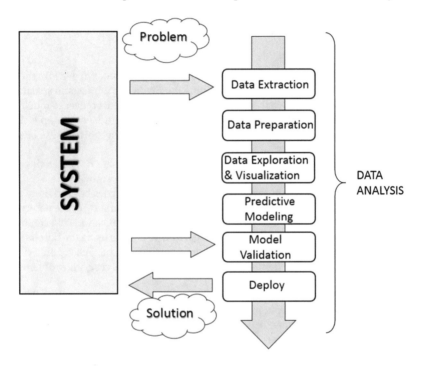

Figure 1-1. *The data analysis process*

Problem Definition

The process of data analysis actually begins long before the collection of raw data. In fact, data analysis always starts with a problem to be solved, which needs to be defined.

The problem is defined only after you have focused the system you want to study; this may be a mechanism, an application, or a process in general. Generally this study can be in order to better understand its operation, but in particular, the study is designed to understand the principles of its behavior in order to be able to make predictions or choices (defined as an informed choice).

The definition step and the corresponding documentation (*deliverables*) of the scientific problem or business are both very important in order to focus the entire analysis strictly on getting results. In fact, a comprehensive or exhaustive study of the system is sometimes complex and you do not always have enough information to start with. So the definition of the problem and especially its planning can determine the guidelines for the whole project.

Once the problem has been defined and documented, you can move to the *project planning* stage of data analysis. Planning is needed to understand which professionals and resources are necessary to meet the requirements to carry out the project as efficiently as possible. You consider the issues involving the resolution of the problem. You look for specialists in various areas of interest and install the software needed to perform data analysis.

Also during the planning phase, you choose an effective team. Generally, these teams should be cross-disciplinary in order to solve the problem by looking at the data from different perspectives. So, building a good team is certainly one of the key factors leading to success in data analysis.

Data Extraction

Once the problem has been defined, the first step is to obtain the data in order to perform the analysis. The data must be chosen with the basic purpose of building the predictive model, and so data selection is crucial for the success of the analysis as well. The sample data collected must reflect as much as possible the real world, that is, how the system responds to stimuli from the real world. For example, if you're using huge datasets of raw data and they are not collected competently, these may portray false or unbalanced situations.

Thus, poor choice of data, or even performing analysis on a dataset that's not perfectly representative of the system, will lead to models that will move away from the system under study.

The search and retrieval of data often require a form of intuition that goes beyond mere technical research and data extraction. This process also requires a careful understanding of the nature and form of the data, which only good experience and knowledge in the problem's application field can provide.

Regardless of the quality and quantity of data needed, another issue is using the best *data sources*.

If the studio environment is a laboratory (technical or scientific) and the data generated are experimental, then in this case the data source is easily identifiable. In this case, the problems will be only concerning the experimental setup.

But it is not possible for data analysis to reproduce systems in which data are gathered in a strictly experimental way in every field of application. Many fields require searching for data from the surrounding world, often relying on external experimental data, or even more often collecting them through interviews or surveys. So in these cases, finding a good data source that is able to provide all the information you need for data analysis can be quite challenging. Often it is necessary to retrieve data from multiple data sources to supplement any shortcomings, to identify any discrepancies, and to make the dataset as general as possible.

When you want to get the data, a good place to start is the web. But most of the data on the web can be difficult to capture; in fact, not all data are available in a file or database, but might be content that is inside HTML pages in many different formats. To this end, a methodology called *web scraping* allows the collection of data through the recognition of specific occurrence of HTML tags within web pages. There is software specifically designed for this purpose, and once an occurrence is found, it extracts the desired data. Once the search is complete, you will get a list of data ready to be subjected to data analysis.

Data Preparation

Among all the steps involved in data analysis, data preparation, although seemingly less problematic, in fact requires more resources and more time to be completed. Data are often collected from different data sources, each of which has data in it with a different representation and format. So, all of these data have to be prepared for the process of data analysis.

The preparation of the data is concerned with obtaining, cleaning, normalizing, and transforming data into an optimized dataset, that is, in a prepared format that's normally tabular and is suitable for the methods of analysis that have been scheduled during the design phase.

Many potential problems can arise, including invalid, ambiguous, or missing values, replicated fields, and out-of-range data.

Data Exploration/Visualization

Exploring the data involves essentially searching the data in a graphical or statistical presentation in order to find patterns, connections, and relationships. Data visualization is the best tool to highlight possible patterns.

In recent years, data visualization has been developed to such an extent that it has become a real discipline in itself. In fact, numerous technologies are utilized exclusively to display data, and many display types are applied to extract the best possible information from a dataset.

Data exploration consists of a preliminary examination of the data, which is important for understanding the type of information that has been collected and what it means. In combination with the information acquired during the definition problem, this categorization determines which method of data analysis is most suitable for arriving at a model definition.

Generally, this phase, in addition to a detailed study of charts through the visualization data, may consist of one or more of the following activities:

- Summarizing data

- Grouping data

- Exploring the relationship between the various attributes

- Identifying patterns and trends

Generally, data analysis requires summarizing statements regarding the data to be studied. *Summarization* is a process by which data are reduced to interpretation without sacrificing important information.

Clustering is a method of data analysis that is used to find groups united by common attributes (also called *grouping*).

Another important step of the analysis focuses on the *identification* of relationships, trends, and anomalies in the data. In order to find this kind of information, you often have to resort to the tools as well as perform another round of data analysis, this time on the data visualization itself.

Other methods of data mining, such as decision trees and association rules, automatically extract important facts or rules from the data. These approaches can be used in parallel with data visualization to uncover relationships between the data.

Predictive Modeling

Predictive modeling is a process used in data analysis to create or choose a suitable statistical model to predict the probability of a result.

After exploring the data, you have all the information needed to develop the mathematical model that encodes the relationship between the data. These models are useful for understanding the system under study, and in a specific way they are used for two main purposes. The first is to make predictions about the data values produced by the system; in this case, you will be dealing with *regression models* if the result is numeric or with *classification models* if the result is categorical. The second purpose is to classify new data products, and in this case, you will be using *classification models* if the results are identified by classes or *clustering models* if the results could be identified by segmentation. In fact, it is possible to divide the models according to the type of result they produce:

- *Classification models*: If the result obtained by the model type is categorical.

- *Regression models*: If the result obtained by the model type is numeric.

- *Clustering models*: If the result obtained by the model type is a segmentation.

Simple methods to generate these models include techniques such as linear regression, logistic regression, classification and regression trees, and k-nearest neighbors. But the methods of analysis are numerous, and each has specific characteristics that make it excellent for some types of data and analysis. Each of these methods will produce a specific model, and then their choice is relevant to the nature of the product model.

Some of these models will provide values corresponding to the real system and according to their structure. They will explain some characteristics of the system under study in a simple and clear way. Other models will continue to give good predictions, but their structure will be no more than a "black box" with limited ability to explain characteristics of the system.

Model Validation

Validation of the model, that is, the test phase, is an important phase that allows you to validate the model built on the basis of starting data. That is important because it allows you to assess the validity of the data produced by the model by comparing these data directly with the actual system. But this time, you are coming from the set of starting data on which the entire analysis has been established.

Generally, you refer to the data as the *training set* when you are using them to build the model, and as the *validation set* when you are using them to validate the model.

Thus, by comparing the data produced by the model with those produced by the system, you can evaluate the error, and using different test datasets, you can estimate the limits of validity of the generated model. In fact the correctly predicted values could be valid only within a certain range, or they could have different levels of matching depending on the range of values taken into account.

This process allows you not only to numerically evaluate the effectiveness of the model but also to compare it with any other existing models. There are several techniques in this regard; the most famous is the *cross-validation*. This technique is based on the division of the training set into different parts. Each of these parts, in turn, is used as the validation set and any other as the training set. In this iterative manner, you will have an increasingly perfected model.

Deployment

This is the final step of the analysis process, which aims to present the results, that is, the conclusions of the analysis. In the deployment process of the business environment, the analysis is translated into a benefit for the client who has commissioned it. In technical or scientific environments, it is translated into design solutions or scientific publications. That is, the deployment basically consists of putting into practice the results obtained from the data analysis.

There are several ways to deploy the results of data analysis or data mining. Normally, a data analyst's deployment consists of writing a report for management or for the customer who requested the analysis. This document conceptually describes the results obtained from the analysis of data. The report should be directed to the managers, who are then able to make decisions. Then, they will put into practice the conclusions of the analysis.

In the documentation supplied by the analyst, each of these four topics is discussed in detail:

- Analysis results

- Decision deployment

- Risk analysis

- Measuring the business impact

When the results of the project include the generation of predictive models, these models can be deployed as stand-alone applications or can be integrated into other software.

Quantitative and Qualitative Data Analysis

Data analysis is completely focused on data. Depending on the nature of the data, it is possible to make some distinctions.

When the analyzed data have a strictly numerical or categorical structure, then you are talking about *quantitative analysis*, but when you are dealing with values that are expressed through descriptions in natural language, then you are talking about *qualitative analysis*.

Precisely because of the different nature of the data processed by the two types of analyses, you can observe some differences between them.

Quantitative analysis has to do with data with a logical order or that can be categorized in some way. This leads to the formation of structures within the data. The order, categorization, and structures in turn provide more information and allow further processing of the data in a more mathematical way. This leads to the generation of models that provide *quantitative predictions,* thus allowing the data analyst to draw more objective conclusions.

Qualitative analysis instead has to do with data that generally do not have a structure, at least not one that is evident, and their nature is neither numeric nor categorical. For example, data under qualitative study could include written textual, visual, or audio data. This type of analysis must therefore be based on methodologies, often *ad hoc,* to extract information that will generally lead to models capable of providing *qualitative predictions.* That means the conclusions to which the data analyst can arrive may also include *subjective interpretations.* On the other hand, qualitative analysis can explore more complex systems and draw conclusions that are not possible using a strictly mathematical approach. Often this type of analysis involves the study of systems that are not easily measurable, such as social phenomena or complex structures.

Figure 1-2 shows the differences between the two types of analyses.

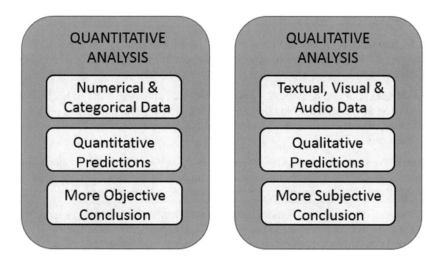

Figure 1-2. *Quantitative and qualitative analyses*

Open Data

In support of the growing demand for data, a huge number of data sources are now available on the Internet. These data sources freely provide information to anyone in need, and they are called *open data.*

Here is a list of some open data available online covering different topics. You can find a more complete list and details of the open data available online in Appendix B.

- **Kaggle** (www.kaggle.com/datasets) is a huge community of apprentices and expert data scientists who provide a vast amount of datasets and code that they use for their analyses. The extensive documentation and the introduction to every aspect of machine learning are also excellent. They also hold interesting competitions organized around the resolution of various problems.

- **DataHub** (datahub.io/search) is a community that makes a huge amount of datasets freely available, along with tools for their command-line management. The dataset topics cover various fields, ranging from the financial market, to population statistics, to the prices of cryptocurrencies.

- **Nasa Earth Observations** (https://neo.gsfc.nasa.gov/dataset_index.php/) provides a wide range of datasets that contain data collected from global climate and environmental observations.

- **World Health Organization** (www.who.int/data/collections) manages and maintains a wide range of data collections related to global health and well-being.

- **World Bank Open Data** (https://data.worldbank.org/) provides a listing of available World Bank datasets covering financial and banking data, development indicators, and information on the World Bank's lending projects from 1947 to the present.

- **Data.gov** (https://data.gov) is intended to collect and provide access to the U.S. government's Open Data, a broad range of government information collected at different levels (federal, state, local, and tribal).

- **European Union Open Data Portal** (https://data.europa.eu/en) collects and makes publicly available a wide range of datasets concerning the public sector of the European member states.

- **Healthdata.gov** (www.healthdata.gov/) provides data about health and health care for doctors and researchers so they can carry out clinical studies and solve problems regarding diseases, virus spread, and health practices, as well as improve the level of global health.

- **Google Trends Datastore** (https://googletrends.github.io/data/) collects and makes available the collected data divided by topic of the famous and very useful Google Trends, which is used to carry out analyses on its own account.

  Finally, recently Google has made available a search page dedicated to datasets, where you can search for a topic and obtain a series of datasets (or even data sources) that correspond as much as possible to what you are looking for. For example, in Figure 1-3, you can see how, when researching the price of houses, a series of datasets or data sources are suggested in real time.

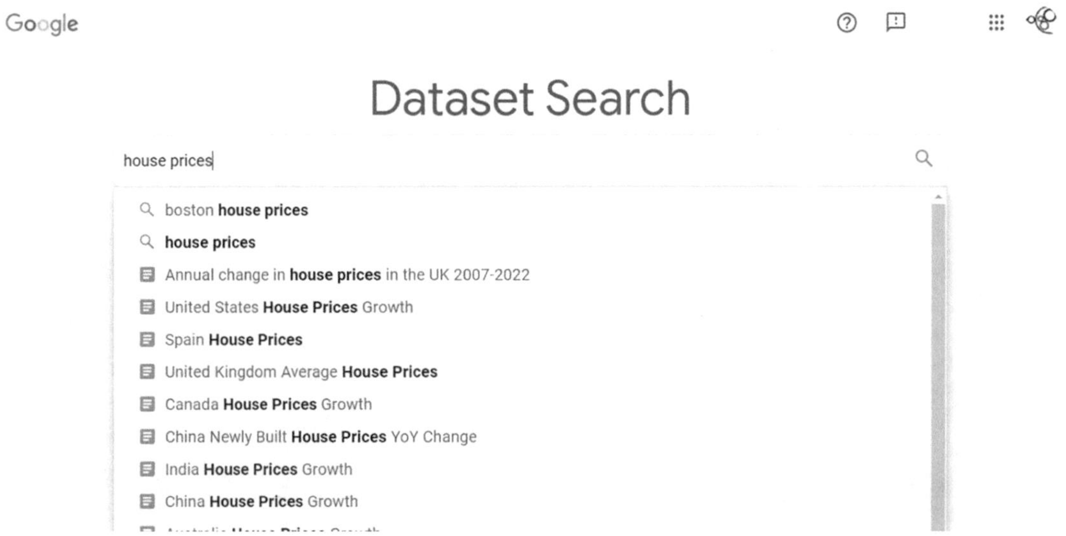

Google

Dataset Search

house prices

🔍 boston **house prices**

🔍 **house prices**

📄 Annual change in **house prices** in the UK 2007-2022

📄 United States **House Prices** Growth

📄 Spain **House Prices**

📄 United Kingdom Average **House Prices**

📄 Canada **House Prices** Growth

📄 China Newly Built **House Prices** YoY Change

📄 India **House Prices** Growth

📄 China **House Prices** Growth

Figure 1-3. *Example of a search for a dataset regarding the prices of houses on Google Dataset Search*

As an idea of open data sources available online, you can look at the *LOD cloud diagram* (`http://cas.lod-cloud.net`), which displays the connections of the data link among several open data sources currently available on the network (see Figure 1-4). The diagram contains a series of circular elements corresponding to the available data sources; their color corresponds to a specific topic of the data provided. The legend indicates the topic-color correspondence. When you click an element on the diagram, you see a page containing all the information about the selected data source and how to access it.

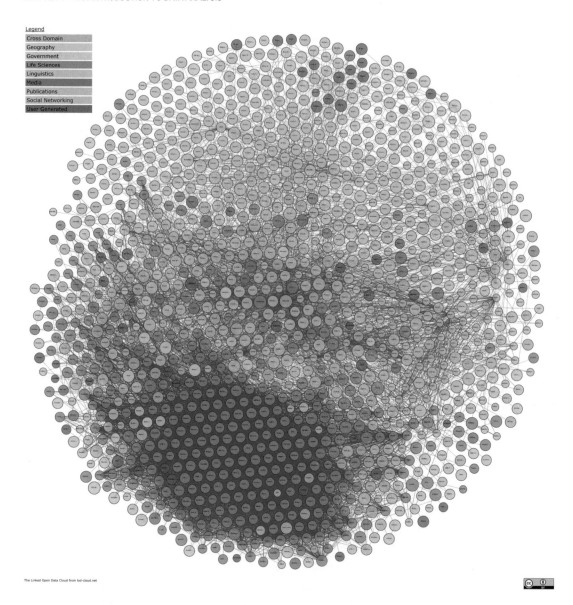

Figure 1-4. *Linked open data cloud diagram 2023, by Max Schmachtenberg, Christian Bizer, Anja Jentzsch, and Richard Cyganiak.* `http://cas.lod-cloud.net` *[CC-BY license]*

Python and Data Analysis

The main argument of this book is to develop all the concepts of data analysis by treating them in terms of Python. The Python programming language is widely used in scientific circles because of its large number of libraries that provide a complete set of tools for analysis and data manipulation.

Compared to other programming languages generally used for data analysis, such as R and MATLAB, Python not only provides a platform for processing data, but it also has features that make it unique compared to other languages and specialized applications. The development of an ever-increasing number of support libraries, the implementation of algorithms of more innovative methodologies, and the ability to interface with other programming languages (C and Fortran) all make Python unique among its kind.

Furthermore, Python is not only specialized for data analysis, but it also has many other applications, such as generic programming, scripting, interfacing to databases, and more recently web development, thanks to web frameworks like Django. So it is possible to develop data analysis projects that are compatible with the web server with the possibility to integrate them on the web.

For those who want to perform data analysis, Python, with all its packages, is considered the best choice for the foreseeable future.

Conclusions

In this chapter, you learned what data analysis is and, more specifically, the various processes that comprise it. Also, you have begun to see the role that data play in building a prediction model and how their careful selection is at the basis of a careful and accurate data analysis.

In the next chapter, you take this vision of Python and the tools it provides to perform data analysis.

■ ■ ■

## CHAPTER 2

Introduction to the Python World

The Python language, and the world around it, is made by interpreters, tools, editors, libraries, notebooks, and so on. This Python world has expanded greatly in recent years, enriching and taking forms that developers who approach it for the first time can sometimes find complicated and somewhat misleading. Thus, if you are approaching Python for the first time, you might feel lost among so many choices, especially about where to start.

This chapter gives you an overview of the entire Python world. You'll first gain an introduction to the Python language and its unique characteristics. You'll learn where to start, what an interpreter is, and how to begin writing your first lines of code in Python before being presented with some new and more advanced forms of interactive writing with respect to shells, such as IPython and the IPython Notebook.

Python—The Programming Language

The Python programming language was created by Guido Von Rossum in 1991 and started with a previous language called ABC. This language can be characterized by a series of adjectives:

- Interpreted

- Portable

- Object-oriented

- Interactive

- Interfaced

- Open source

- Easy to understand and use

Python is an *interpreted* programming language, that is, it's pseudo-compiled. Once you write the code, you need an *interpreter* to run it. An interpreter is a program that is installed on each machine; it interprets and runs the source code. Unlike with languages such as C, C++, and Java, there is no compile time with Python.

Python is a highly *portable* programming language. The decision to use an interpreter as an interface for reading and running code has a key advantage: portability. In fact, you can install an interpreter on any platform (Linux, Windows, and Mac) and the Python code will not change. Because of this, Python is often used with many small-form devices, such as the Raspberry Pi and other microcontrollers.

© Fabio Nelli 2023
F. Nelli, *Python Data Analytics*, https://doi.org/10.1007/978-1-4842-9532-8_2

Python is an *object-oriented* programming language. In fact, it allows you to specify classes of objects and implement their inheritance. But unlike C++ and Java, there are no constructors or destructors. Python also allows you to implement specific constructs in your code to manage exceptions. However, the structure of the language is so flexible that it allows you to program with alternative approaches with respect to the object-oriented one. For example, you can use functional or vectorial approaches.

Python is an *interactive* programming language. Thanks to the fact that Python uses an interpreter to be executed, this language can take on very different aspects depending on the context in which it is used. In fact, you can write long lines of code, similar to what you might do in languages like C++ or Java, and then launch the program, or you can enter the command line at once and execute a command, immediately getting the results. Then, depending on the results, you can decide what command to run next. This highly interactive way to execute code makes the Python computing environment similar to MATLAB. This feature of Python is one reason it's popular with the scientific community.

Python is a programming language that can be *interfaced*. In fact, this programming language can be interfaced with code written in other programming languages such as C/C++ and FORTRAN. Even this was a winning choice. In fact, thanks to this aspect, Python can compensate for what is perhaps its only weak point, the speed of execution. The nature of Python, as a highly dynamic programming language, can sometimes lead to execution of programs up to 100 times slower than the corresponding static programs compiled with other languages. The solution to this kind of performance problem is to interface Python to the compiled code of other languages by using it as if it were its own.

Python is an *open-source* programming language. CPython, which is the reference implementation of the Python language, is completely free and open source. Additionally every module or library in the network is open source and their code is available online. Every month, an extensive developer community includes improvements to make this language and all its libraries even richer and more efficient. CPython is managed by the nonprofit Python Software Foundation, which was created in 2001 and has given itself the task of promoting, protecting, and advancing the Python programming language.

Finally, Python is a simple language to use and learn. This aspect is perhaps the most important, because it is the most direct aspect that a developer, even a novice, faces. The high intuitiveness and ease of reading of Python code often leads to "sympathy" for this programming language, and consequently most newcomers to programming choose to use it. However, its simplicity does not mean narrowness, since Python is a language that is spreading in every field of computing. Furthermore, Python is doing all of this very simply, in comparison to existing programming languages such as C++, Java, and FORTRAN, which by their nature are very complex.

The Interpreter and the Execution Phases of the Code

Unlike programming languages such as Java or C, whose code must be compiled before being executed, Python is a language that allows direct execution of instructions. In fact, it is possible to execute code written in Python it two ways. You can execute entire programs (.py files) by running the python command followed by the file name, or you can open a session through a special command console, characterized by a >>> prompt (running the python command with no arguments). In this console, you can enter one instruction at a time, obtaining the result immediately by executing it directly.

In both cases, you have the immediate execution of the inserted code, without having to go through explicit compilation or other operations.

This direct execution operation can be schematized in four phases:

- Lexing or tokenization

- Parsing

- Compiling

- Interpreting

Lexing, or *tokenization*, is the initial phase in which the Python (human-readable) code is converted into a sequence of logical entities, the so-called lexical tokens (see Figure 2-1).

Parsing is the next stage in which the syntax and grammar of the lexical tokens are checked by a parser, which produces an abstract syntax tree (AST) as a result.

Compiling is the phase in which the compiler reads the AST and, based on this information, generates the Python bytecode (.pyc or .pyo files), which contains very basic execution instructions. Although this is a compilation phase, the generated bytecode is still platform-independent, which is very similar to what happens in the Java language.

The last phase is *interpreting,* in which the generated bytecode is executed by a *Python virtual machine (PVM).*

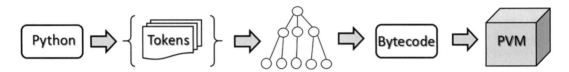

Figure 2-1. *The steps performed by the Python interpreter*

You can find good documentation on this process at www.ics.uci.edu/~pattis/ICS-31/lectures/tokens.pdf.

All these phases are performed by the interpreter, which in the case of Python is a fundamental component. When referring to the Python interpreter, this usually means the /urs/bin/python binary. In reality, there are currently several versions of this Python interpreter, each of which is profoundly different in its nature and specifications.

CPython

The standard Python interpreter is CPython, and it was written in C. This made it possible to use C-based libraries over Python. CPython is available on a variety of platforms, including ARM, iOS, and RISC. Despite this, CPython has been optimized on portability and other specifications, but not on speed.

Cython

The strongly intrinsic nature of C in the CPython interpreter has been taken further with the Cython project. This project is based on creating a compiler that translates Python code into C. This code is then executed within a Cython environment at runtime. This type of compilation system makes it possible to introduce C semantics into the Python code to make it even more efficient. This system has led to the merging of two worlds of programming language with the birth of *Cython,* which can be considered a new programming language. You can find documentation about it online. I advise you to visit cython.readthedocs.io/en/latest/.

Pyston

Pyston (www.pyston.org/) is a fork of the CPython interpreter that implements performance optimization. This project arises precisely from the need to obtain an interpreter that can replace CPython over time to remedy its poor performance in terms of execution speed. Recent results seem to confirm these predictions, reporting a 30 percent improvement in performance in the case of large, real-world applications. Unfortunately, due to the lack of compatible binary packages, Pyston packages have to be rebuilt during the download phase.

Jython

In parallel to Cython, there is a version built and compiled in Java, called *Jython*. It was created by Jim Hugunin in 1997 (`www.jython.org/`). Jython is an implementation of the Python programming language in Java; it is further characterized by using Java classes instead of Python modules to implement extensions and packages of Python.

IronPython

Even the .NET framework offers the possibility of being able to execute Python code inside it. For this purpose, you can use the IronPython interpreter (`https://ironpython.net/`). This interpreter allows .NET developers to develop Python programs on the Visual Studio platform, integrating perfectly with the other development tools of the .NET platform.

Initially built by Jim Hugunin in 2006 with the release of version 1.0, the project was later supported by a small team at Microsoft until version 2.7 in 2010. Since then, numerous other versions have been released up to the current 3.4, all ported forward by a group of volunteers on Microsoft's CodePlex repository.

PyPy

The PyPy interpreter is a JIT (just-in-time) compiler, and it converts the Python code directly to machine code at runtime. This choice was made to speed up the execution of Python. However, this choice has led to the use of a smaller subset of Python commands, defined as *RPython*. For more information on this, consult the official website at `www.pypy.org/`.

RustPython

As the name suggests, RustPython (`rustpython.github.io/`) is a Python interpreter written in Rust. This programming language is quite new but it is gaining popularity. RustPython is an interpreter like CPython but can also be used as a JIT compiler. It also allows you to run Python code embedded in Rust programs and compile the code into WebAssembly, so you can run Python code directly from web browsers.

Installing Python

In order to develop programs in Python, you have to install it on your operating system. Linux distributions and macOS X machines should have a preinstalled version of Python. If not, or if you want to replace that version with another, you can easily install it. The process for installing Python differs from operating system to operating system. However, it is a rather simple operation.

On Debian-Ubuntu Linux systems, the first thing to do is to check whether Python is already installed on your system and what version is currently in use.

Open a terminal (by pressing ALT+CTRL+T) and enter the following command:

```
python3 --version
```

If you get the version number as output, then Python is already present on the Ubuntu system. If you get an error message, Python hasn't been installed yet.

In this last case

```
sudo apt install python3
```

If, on the other hand, the current version is old, you can update it with the latest version of your Linux distribution by entering the following command:

```
sudo apt --only-upgrade install python3
```

Finally, if instead you want to install a specific version on your system, you have to explicitly indicate it in the following way:

```
sudo apt install python3.10
```

On Red Hat and CentOS Linux systems working with `rpm` packages, run this command instead:

```
yum install python3
```

If you are running Windows or macOS X, you can go to the official Python site (`www.python.org`) and download the version you prefer. The packages in this case are installed automatically.

However, today there are distributions that provide a number of tools that make the management and installation of Python, all libraries, and associated applications easier. I strongly recommend you choose one of the distributions available online.

Python Distributions

Due to the success of the Python programming language, many Python tools have been developed to meet various functionalities over the years. There are so many that it's virtually impossible to manage all of them manually.

In this regard, many Python distributions efficiently manage hundreds of Python packages. In fact, instead of individually downloading the interpreter, which includes only the standard libraries, and then needing to individually install all the additional libraries, it is much easier to install a Python distribution.

At the heart of these distributions are the *package managers,* which are nothing more than applications that automatically manage, install, upgrade, configure, and remove Python packages that are part of the distribution.

Their functionality is very useful, since the user simply makes a request regarding a particular package (which could be an installation for example). Then the package manager, usually via the Internet, performs the operation by analyzing the necessary version, alongside all dependencies with any other packages, and downloads them if they are not present.

Anaconda

Anaconda is a free distribution of Python packages distributed by Continuum Analytics (`www.anaconda.com`). This distribution supports Linux, Windows, and macOS X operating systems. Anaconda, in addition to providing the latest packages released in the Python world, comes bundled with most of the tools you need to set up a Python development environment.

Indeed, when you install the Anaconda distribution on your system, you can use many tools and applications described in this chapter, without worrying about having to install and manage them separately. The basic distribution includes *Spyder*, an IDE used to develop complex Python programs, *Jupyter Notebook*, a wonderful tool for working interactively with Python in a graphical and orderly way, and *Anaconda Navigator*, a graphical panel for managing packages and virtual environments.

The management of the entire Anaconda distribution is performed by an application called *conda*. This is the package manager and the environment manager of the Anaconda distribution and it handles all of the packages and their versions.

```
conda install <package name>
```

One of the most interesting aspects of this distribution is the ability to manage multiple development environments, each with its own version of Python. With Anaconda, you can work simultaneously and independently with different Python versions at the same time, by creating several virtual environments. You can create, for instance, an environment based on Python 3.11 even if the current Python version is still 3.10 in your system. To do this, you write the following command via the console:

```
conda create -n py311 python=3.11 anaconda
```

This will generate a new Anaconda virtual environment with all the packages related to the Python 3.11 version. This installation will not affect the Python version installed on your system and won't generate any conflicts. When you no longer need the new virtual environment, you can simply uninstall it, leaving the Python system installed on your operating system completely unchanged. Once it's installed, you can activate the new environment by entering the following command:

```
source activate py311
```

On Windows, use this command instead:

```
activate py311
C:\Users\Fabio>activate py311
  (py311) C:\Users\Fabio>
```

You can create as many versions of Python as you want; you need only to change the parameter passed with the python option in the conda create command. When you want to return to work with the original Python version, use the following command:

```
source deactivate
```

On Windows, use this command:

```
(py311) C:\Users\Fabio>deactivate
Deactivating environment "py311"...
C:\Users\Fabio>
```

Anaconda Navigator

Although at the base of the Anaconda distribution there is the conda command for the management of packages and virtual environments, working through the command console is not always practical and efficient. As you will see in the following chapters of the book, Anaconda provides a graphical tool called Anaconda Navigator, which allows you to manage the virtual environments and related packages in a graphical and very simplified way (see Figure 2-2).

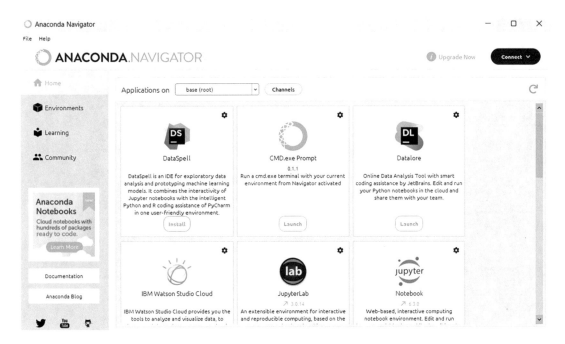

Figure 2-2. *Home panel of Anaconda Navigator*

Anaconda Navigator is mainly composed of four panels:

- Home

- Environments

- Learning

- Community

Each of them is selectable through the list of buttons clearly visible on the left.

The Home panel presents all the Python (and also R) development applications installed (or available) for a given virtual environment. By default, Anaconda Navigator will show the base operating system environment, referred as base(root) in the top-center drop-down menu (see Figure 2-2).

The second panel, called Environments, shows all the virtual environments created in the distribution (see Figure 2-3). From there, it is possible to select the virtual environment to activate by clicking it directly. It will display all the packages installed (or available) on that virtual environment, with the relative versions.

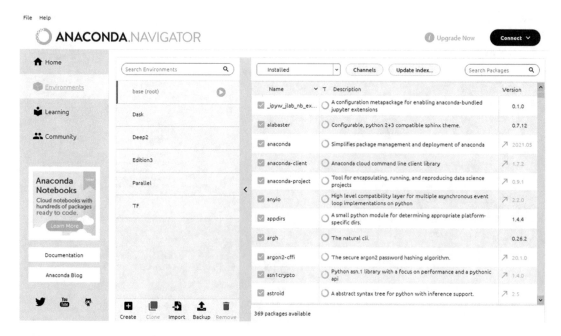

Figure 2-3. *Environments panel on Anaconda Navigator*

Also from the Environments panel it is possible to create new virtual environments, selecting the basic Python version. Similarly, the same virtual environments can be deleted, cloned, backed up, or imported using the menu shown in Figure 2-4.

Figure 2-4. *Button menu for managing virtual environments in Anaconda Navigator*

But that is not all. Anaconda Navigator is not only a useful application for managing Python applications, virtual environments, and packages. In the third panel, called Learning (see Figure 2-5), it provides links to the main sites of many useful Python libraries (including those covered in this book). By clicking one of these links, you can access a lot of documentation. This is always useful to have on hand if you program in Python on a daily basis.

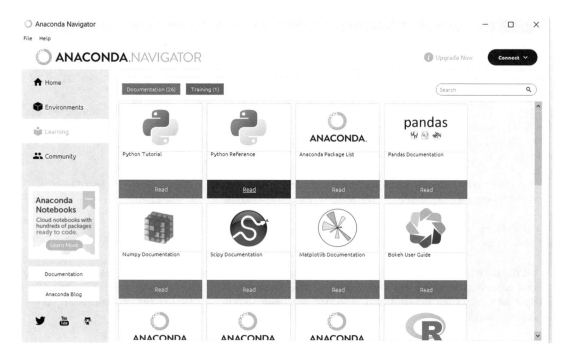

Figure 2-5. *Learning panel of Anaconda Navigator*

An identical panel to this is the next one, called Community. There are links here too, but this time to forums from the main Python development and Data Analytics communities.

The Anaconda platform, with its multiple applications and Anaconda Navigator, allows developers to take advantage of this simple and organized work environment and be well prepared for the development of Python code. It is no coincidence that this platform has become almost a standard for those belonging to the sector.

Using Python

Python is rich, but simple and very flexible. It allows you to expand your development activities in many areas of work (data analysis, scientific, graphic interfaces, etc.). Precisely for this reason, Python can be used in many different contexts, often according to the taste and ability of the developer. This section presents the various approaches to using Python in the course of the book. According to the various topics discussed in different chapters, these different approaches will be used specifically, as they are more suited to the task at hand.

Python Shell

The easiest way to approach the Python world is to open a session in the Python shell, which is a terminal running a command line. In fact, you can enter one command at a time and test its operation immediately. This mode makes clear the nature of the interpreter that underlies Python. In fact, the interpreter can read one command at a time, keeping the status of the variables specified in the previous lines, a behavior similar to that of MATLAB and other calculation software.

This approach is helpful when approaching Python the first time. You can test commands one at a time without having to write, edit, and run an entire program, which could be composed of many lines of code.

This mode is also good for testing and debugging Python code one line at a time, or simply to make calculations. To start a session on the terminal, simply type this on the command line:

```
C:\Users\nelli>python
Python 3.10 | packaged by Anaconda, Inc. | (main, Mar  1 2023, 18:18:21) [MSC v.1916 64 bit
(AMD64)] on win32
Type "help", "copyright", "credits" or "license" for more information.
>>>
```

The Python shell is now active and the interpreter is ready to receive commands in Python. Start by entering the simplest of commands, but a classic for getting started with programming.

```
>>> print("Hello World!")
Hello World!
```

If you have the Anaconda platform available on your system, you can open a Python shell related to a specific virtual environment you want to work on. In this case, from Anaconda Navigator, in the Home panel, activate the virtual environment from the drop-down menu and click the Launch button of the CMD.exe Prompt application, as shown in Figure 2-6.

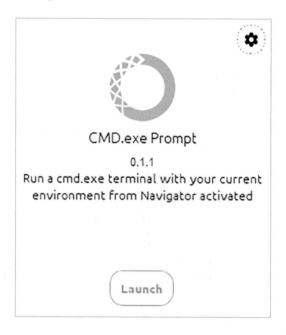

Figure 2-6. CMD.exe Prompt application in Anaconda Navigator

A command console will open with the name of the active virtual environment prefixed in brackets in the prompt. From there, you can run the python command to activate the Python shell.

```
(Edition3) C:\Users\nelli>python
Python 3.11.0 | packaged by Anaconda, Inc. | (main, Mar  1 2023, 18:18:21) [MSC v.1916 64
bit (AMD64)] on win32
Type "help", "copyright", "credits" or "license" for more information.
>>>
```

Run an Entire Program

The best way to become familiar with Python is to write an entire program and then run it from the terminal. First write a program using a simple text editor. For example, you can use the code shown in Listing 2-1 and save it as MyFirstProgram.py.

Listing 2-1. MyFirstProgram.py

```python
myname = input("What is your name?\n")
print("Hi %s, I'm glad to say: Hello world!" %myname)
```

Now you've written your first program in Python, and you can run it directly from the command line by calling the python command and then the name of the file containing the program code.

```
python MyFirstProgram.py
```

From the output, the program will ask for your name. Once you enter it, it will say hello.

```
What is your name?
Fabio Nelli
Hi Fabio Nelli, I'm glad to say: Hello world!
```

Implement the Code Using an IDE

A more comprehensive approach than the previous ones is to use an IDE (an *Integrated Development Environment*). These editors provide a work environment on which to develop your Python code. They are rich in tools that make developers' lives easier, especially when debugging. In the following sections, you see in detail which IDEs are currently available.

Interact with Python

The last approach to using Python, and in my opinion, perhaps the most innovative, is the interactive one. In fact, in addition to the three previous approaches, this approach provides you with the opportunity to interact directly with the Python code.

In this regard, the Python world has been greatly enriched with the introduction of *IPython*. IPython is a very powerful tool, designed specifically to meet the needs of interacting between the Python interpreter and the developer, which under this approach takes the role of analyst, engineer, or researcher. IPython and its features are explained in more detail in a later section.

Writing Python Code

In the previous section, you saw how to write a simple program in which the string "Hello World" was printed. Now in this section, you get a brief overview of the basics of the Python language.

This section is not intended to teach you to program in Python, or to illustrate syntax rules of the programming language, but just to give you a quick overview of some basic principles of Python necessary to continue with the topics covered in this book.

If you already know the Python language, you can safely skip this introductory section. Instead, if you are not familiar with programming and you find it difficult to understand the topics, I highly recommend that you visit online documentation, tutorials, and courses of various kinds.

Make Calculations

You have already seen that the print() function is useful for printing almost anything. Python, in addition to being a printing tool, is a great calculator. Start a session on the Python shell and begin to perform these mathematical operations:

```
>>> 1 + 2
3
>>> (1.045 * 3)/4
0.78375
>>> 4 ** 2
16
>>> ((4 + 5j) * (2 + 3j))
(-7+22j)
>>> 4 < (2*3)
True
```

Python can calculate many types of data, including complex numbers and conditions with Boolean values. As you can see from these calculations, the Python interpreter directly returns the result of the calculations without the need to use the print() function. The same thing applies to values contained in variables. It's enough to call the variable to see its contents.

```
>>> a = 12 * 3.4
>>> a
40.8
```

Import New Libraries and Functions

You saw that Python is characterized by the ability to extend its functionality by importing numerous packages and modules. To import a module in its entirety, you have to use the import command.

```
>>> import math
```

In this way, all the functions contained in the math package are available in your Python session so you can call them directly. Thus, you have extended the standard set of functions available when you start a Python session. These functions are called with the following expression.

```
library_name.function_name()
```

For example, you can now calculate the sine of the value contained in the variable a.

```
>>> math.sin(a)
```

As you can see, the function is called along with the name of the library. Sometimes you might find the following expression for declaring an import.

```
>>> from math import *
```

Even if this works properly, it is to be avoided for good practice. In fact, writing an import in this way involves the importation of all functions without necessarily defining the library to which they belong.

```
>>> sin(a)
0.040693257349864856
```

This form of import can lead to very large errors, especially if the imported libraries are numerous. In fact, it is not unlikely that different libraries have functions with the same name, and importing all of these would result in an override of all functions with the same name that were previously imported. Therefore, the behavior of the program could generate numerous errors or worse, abnormal behavior.

Actually, this way to import is generally used for only a limited number of functions, that is, functions that are strictly necessary for the functioning of the program, thus avoiding the importation of an entire library when it is completely unnecessary.

```
>>> from math import sin
```

Data Structure

You saw in the previous examples how to use simple variables containing a single value. Python provides a number of extremely useful data structures. These data structures can contain lots of data simultaneously and sometimes even data of different types. The various data structures provided are defined differently depending on how their data are structured internally.

- List
- Set
- Strings
- Tuples
- Dictionary
- Deque
- Heap

This is only a small part of all the data structures that can be made with Python. Among all these data structures, the most commonly used are *dictionaries* and *lists*.

The type *dictionary*, defined also as *dicts*, is a data structure in which each particular value is associated with a particular label, called a *key*. The data collected in a dictionary have no internal order but are only definitions of key/value pairs.

```
>>> dict = {'name':'William', 'age':25, 'city':'London'}
```

If you want to access a specific value within the dictionary, you have to indicate the name of the associated key.

```
>>> dict["name"]
'William'
```

If you want to iterate the pairs of values in a dictionary, you have to use the for-in construct. This is possible through the use of the items() function.

```
>>> for key, value in dict.items():
...     print(key,value)
...
name William
age 25
city London
```

The type *list* is a data structure that contains a number of objects in a precise order to form a sequence to which elements can be added and removed. Each item is marked with a number corresponding to the order of the sequence, called the *index*.

```
>>> list = [1,2,3,4]
>>> list
[1, 2, 3, 4]
```

If you want to access the individual elements, it is sufficient to specify the index in square brackets (the first item in the list has 0 as its index), while if you take out a portion of the list (or a sequence), it is sufficient to specify the range with the indices i and j corresponding to the extremes of the portion.

```
>>> list[2]
3
>>> list[1:3]
[2, 3]
```

If you are using negative indices instead, this means you are considering the last item in the list and gradually moving to the first.

```
>>> list[-1]
4
```

In order to do a scan of the elements of a list, you can use the for-in construct.

```
>>> items = [1,2,3,4,5]
>>> for item in items:
...         print(item + 1)
...
2
3
4
5
6
```

Functional Programming

The for-in loop shown in the previous example is very similar to loops found in other programming languages. However, if you want to be a "Python" developer, you have to avoid using explicit loops. Python offers alternative approaches, specifying programming techniques such as *functional programming* (expression-oriented programming).

The tools that Python provides to develop functional programming comprise a series of functions:

- map(function, list)
- filter(function, list)
- reduce(function, list)
- lambda
- list comprehension

The for loop that you just saw has a specific purpose, which is to apply an operation on each item and then somehow gather the result. This can be done by the map() function.

```
>>> items = [1,2,3,4,5]
>>> def inc(x): return x+1
...
>>> list(map(inc,items))
[2, 3, 4, 5, 6]
```

In the previous example, it first defines the function that performs the operation on every single element, and then it passes it as the first argument to map(). Python allows you to define the function directly within the first argument using lambda as a function. This greatly reduces the code and compacts the previous construct into a single line of code.

```
>>> list(map((lambda x: x+1),items))
[2, 3, 4, 5, 6]
```

Two other functions working in a similar way are filter() and reduce(). The filter() function extracts the elements of the list for which the function returns True. The reduce() function instead considers all the elements of the list to produce a single result. To use reduce(), you must import the functools module.

```
>>> list(filter((lambda x: x < 4), items))
[1, 2, 3]
>>> from functools import reduce
>>> reduce((lambda x,y: x/y), items)
0.008333333333333333
```

Both of these functions implement other types by using the for loop. They replace these cycles and their functionality, which can be alternatively expressed with simple functions. That is what constitutes *functional programming*.

The final concept of functional programming is *list comprehension*. This concept is used to build lists in a very natural and simple way, referring to them in a manner similar to how mathematicians describe datasets. The values in the sequence are defined through a particular function or operation.

```
>>> S = [x**2 for x in range(5)]
>>> S
[0, 1, 4, 9, 16]
```

Indentation

A peculiarity for those coming from other programming languages is the role that *indentation* plays. Whereas you used to manage the indentation for purely aesthetic reasons, making the code somewhat more readable, in Python indentation assumes an integral role in the implementation of the code, by dividing it into logical blocks. In fact, while in Java, C, and C++, each line of code is separated from the next by a semicolon (;), in Python you should not specify any symbol that separates them, included the braces to indicate a logical block.

These roles in Python are handled through indentation; that is, depending on the starting point of the code line, the interpreter determines whether it belongs to a logical block or not.

```
>>> a = 4
>>> if a > 3:
...     if a < 5:
...         print("I'm four")
... else:
...     print("I'm a little number")
...
I'm four
>>> if a > 3:
...     if a < 5:
...         print("I'm four")
...     else:
...         print("I'm a big number")
...
I'm four
```

In this example you can see that, depending on how the `else` command is indented, the conditions assume two different meanings (specified by me in the strings themselves).

IPython

IPython is a further development of Python that includes a number of tools:

- The IPython shell, which is a powerful interactive shell resulting in a greatly enhanced Python terminal.

- A QtConsole, which is a hybrid between a shell and a GUI, allowing you to display graphics inside the console instead of in separate windows.

- An IPython Notebook, called Jupyter Notebook, which is a web interface that allows you to mix text, executable code, graphics, and formulas in a single representation.

IPython Shell

This shell apparently resembles a Python session run from a command line, but actually, it provides many other features that make this shell much more powerful and versatile than the classic one. To launch this shell, just type ipython on the command line.

```
> ipython
Python 3.11.0 | packaged by Anaconda, Inc. | (main, Mar  1 2023, 18:18:21) [MSC v.1916 64
bit (AMD64)]
Type 'copyright', 'credits' or 'license' for more information
IPython 8.12.0 -- An enhanced Interactive Python. Type '?' for help.
In [1]:
```

As you can see, a particular prompt appears with the value In [1]. This means that it is the first line of input. Indeed, IPython offers a system of numbered prompts (indexed) with input and output caching.

```
In [1]: print("Hello World!")
Hello World!
In [2]: 3/2
Out[2]: 1.5
In [3]: 5.0/2
Out[3]: 2.5
In [4]:
```

The same thing applies to values in output that are indicated with the values Out[1], Out [2], and so on. IPython saves all inputs that you enter by storing them as variables. In fact, all the inputs entered were included as fields in a list called In.

```
In [4]: In
Out[4]: ['', 'print "Hello World!"', '3/2', '5.0/2', 'In']
```

The indices of the list elements are the values that appear in each prompt. Thus, to access a single line of input, you can simply specify that value.

```
In [5]: In[3]
Out[5]: '5.0/2'
```

For output, you can apply the same concept.

```
In [6]: Out
Out[6]:
{2: 1.5,
 3: 2.5,
 4: ['', 'print("Hello World!")', '3/2', '5.0/2', 'In', 'In[3]', 'Out'], 5: u'5.0/2'}
```

The Jupyter Project

IPython has grown enormously in recent times, and with the release of IPython 3.0, everything is moving toward a new project called Jupyter (`https://jupyter.org`)—see Figure 2-7.

Figure 2-7. *The Jupyter project logo*

IPython will continue to exist as a Python shell and as a kernel of Jupyter, but the Notebook and the other language-agnostic components belonging to the IPython project will move to form the new Jupyter project.

Jupyter QtConsole

In order to launch this application from the command line, you must enter the following command:

```
jupyter qtconsole
```

The application consists of a GUI that has all the functionality present in the IPython shell. See Figure 2-8.

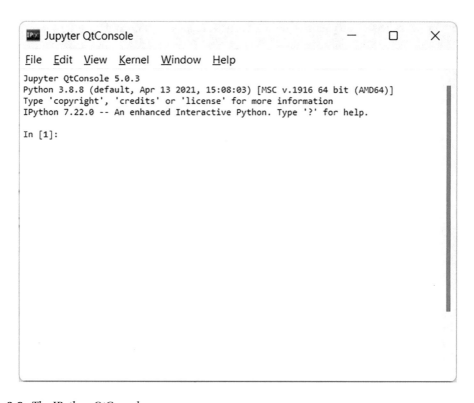

Figure 2-8. *The IPython QtConsole*

Jupyter Notebook

Jupyter Notebook is the latest evolution of this interactive environment (see Figure 2-9). In fact, with Jupyter Notebook, you can merge executable code, text, formulas, images, and animations into a single web document. This is useful for many purposes, such as presentations, tutorials, debugging, and so forth.

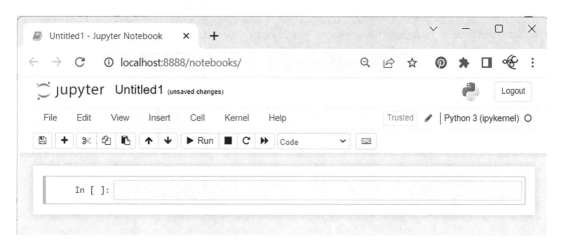

Figure 2-9. *The web page showing the Jupyter Notebook*

To start Jupyter Notebook on your web browser, run the following command from the console:

```
jupyter notebook
```

If instead you are working with Anaconda Navigator, in the Home panel, click the Launch button of the Jupyter Notebook application to start it (see Figure 2-10).

Figure 2-10. *Jupyter Notebook application in Anaconda Navigator*

Jupyter Lab

Another application that brings together the characteristics of all the applications seen so far is Jupyter Lab. It runs on browsers like Jupyter Notebook, but it's a real development environment, where you can manage files, data, and code in the form of files, sessions, notebooks, and so on.

To start Jupyter Lab on your web browser, run the following command from the console:

```
jupyter lab
```

This application is also present on Anaconda Navigator together with the others, and to start it from one of the virtual environments, simply click the Launch button of the corresponding icon shown in Figure 2-11.

Figure 2-11. *Jupyter Lab icon in Anaconda Navigator*

Starting the application will open the default browser (if it's not already open) and load the `https://localhost:8892/lab` page, which corresponds to Jupyter Lab, as shown in Figure 2-12.

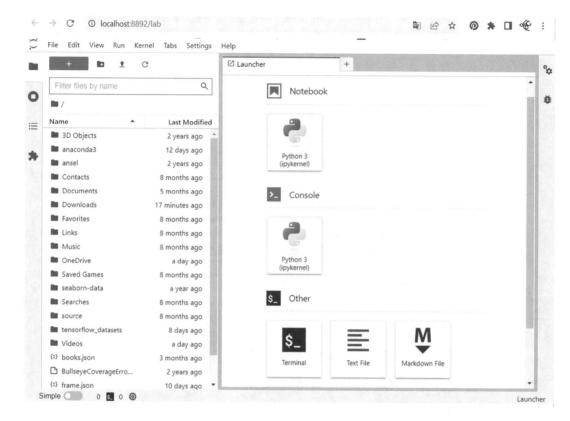

Figure 2-12. *Jupyter Lab application*

PyPI—The Python Package Index

The Python Package Index (PyPI) is a software repository that contains all the software needed for programming in Python—for example, all Python packages belonging to other Python libraries. The content repository is managed directly by the developers of individual packages that deal with updating the repository with the latest versions of their released libraries. For a list of the packages contained in the repository, go to the official page of PyPI at https://pypi.python.org/pypi.

As far as the administration of these packages, you can use the pip application, which is the package manager of PyPI.

By launching it from the command line, you can manage all the packages and individually decide if a package should be installed, upgraded, or removed. Pip will check if the package is already installed, or if it needs to be updated, to control dependencies, and to assess whether other packages are necessary. Furthermore, it manages the downloading and installation processes.

```
$ pip install <<package_name>>
$ pip search <<package_name>>
$ pip show <<package_name>>
$ pip unistall <<package_name>>
```

The IDEs for Python

Although most Python developers are used to implementing their code directly from the shell (Python or IPython), some IDEs (Interactive Development Environments) are also available. In fact, in addition to a text editor, these graphics editors also provide a series of tools that are very useful during the drafting of the code. For example, the auto-completion of code, viewing the documentation associated with the commands, debugging, and breakpoints are only some of the tools that this kind of application can provide.

Spyder

Spyder (Scientific Python Development Environment) is an IDE that has similar features to the IDE of MATLAB (see Figure 2-13). The text editor is enriched with syntax highlighting and code analysis tools. Also, you can integrate ready-to-use widgets in your graphic applications.

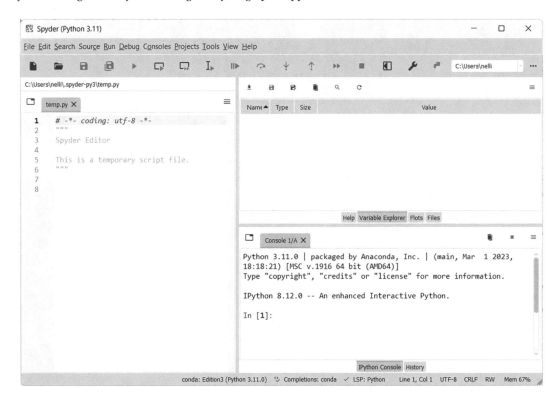

Figure 2-13. *The Spyder IDE*

Eclipse (pyDev)

Those of you who have developed in other programming languages certainly know Eclipse, a universal IDE developed entirely in Java (therefore requiring Java installation on your PC) that provides a development environment for many programming languages (see Figure 2-14). There is also an Eclipse version for developing in Python, thanks to the installation of an additional plugin called *pyDev*.

Figure 2-14. *The Eclipse IDE*

Sublime

This text editor is one of the preferred environments for Python programmers (see Figure 2-15). In fact, there are several plugins available for this application that make Python implementation easy and enjoyable.

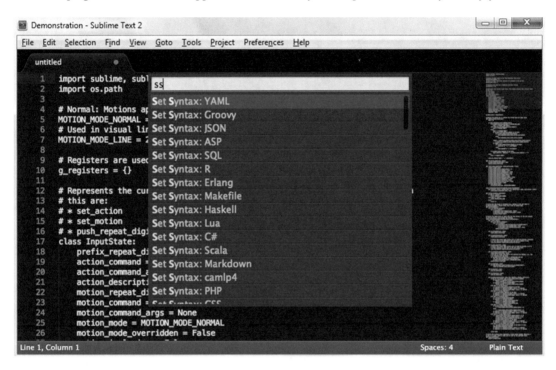

Figure 2-15. *The Sublime IDE*

Liclipse

Liclipse, similarly to Spyder, is a development environment specifically designed for the Python language (see Figure 2-16). It is very similar to the Eclipse IDE but it is fully adapted for a specific use in Python, without needing to install plugins like PyDev. So its installation and settings are much simpler than Eclipse.

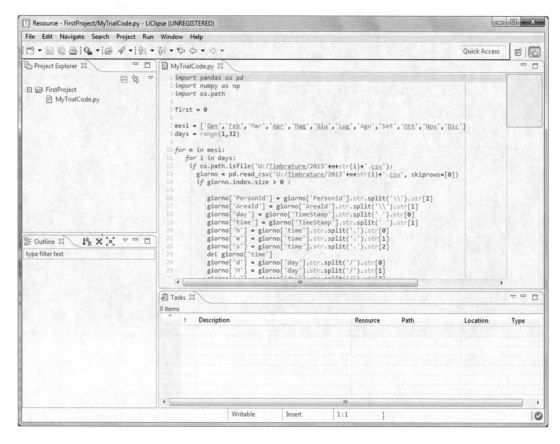

Figure 2-16. *The Liclipse IDE*

NinjaIDE

NinjaIDE (NinjaIDE is "Not Just Another IDE"), which characterized by a name that is a recursive acronym, is a specialized IDE for the Python language (see Figure 2-17). It's a very recent application on which the efforts of many developers are focused. Already very promising, it is likely that in the coming years, this IDE will be a source of many surprises.

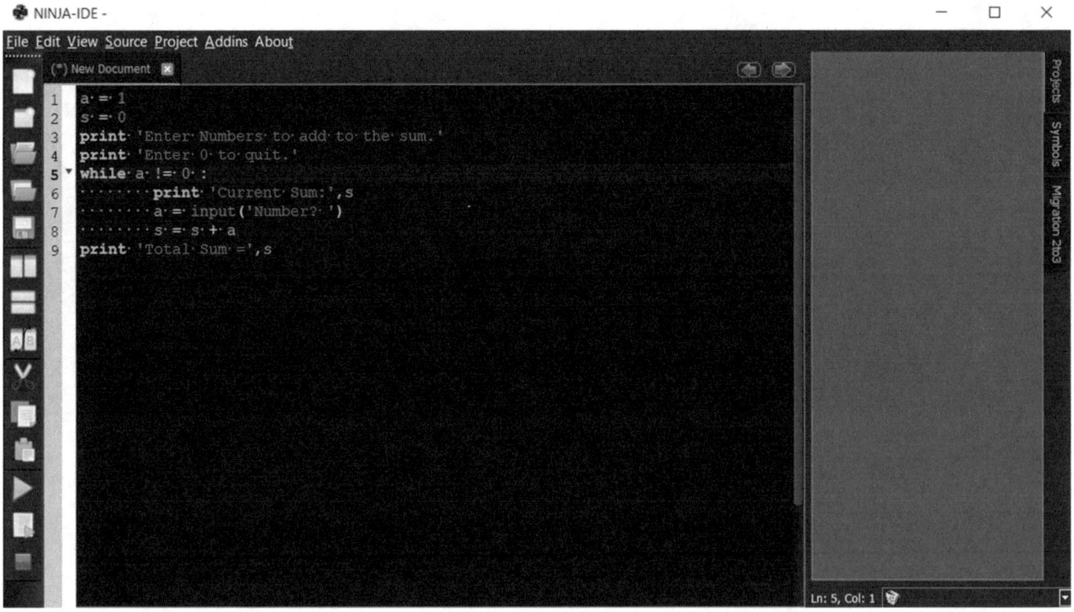

Figure 2-17. *The Ninja IDE*

Komodo IDE

Komodo is a very powerful IDE full of tools that make it a complete and professional development environment (see Figure 2-18). Paid software and written in C++, the Komodo development environment is adaptable to many programming languages, including Python.

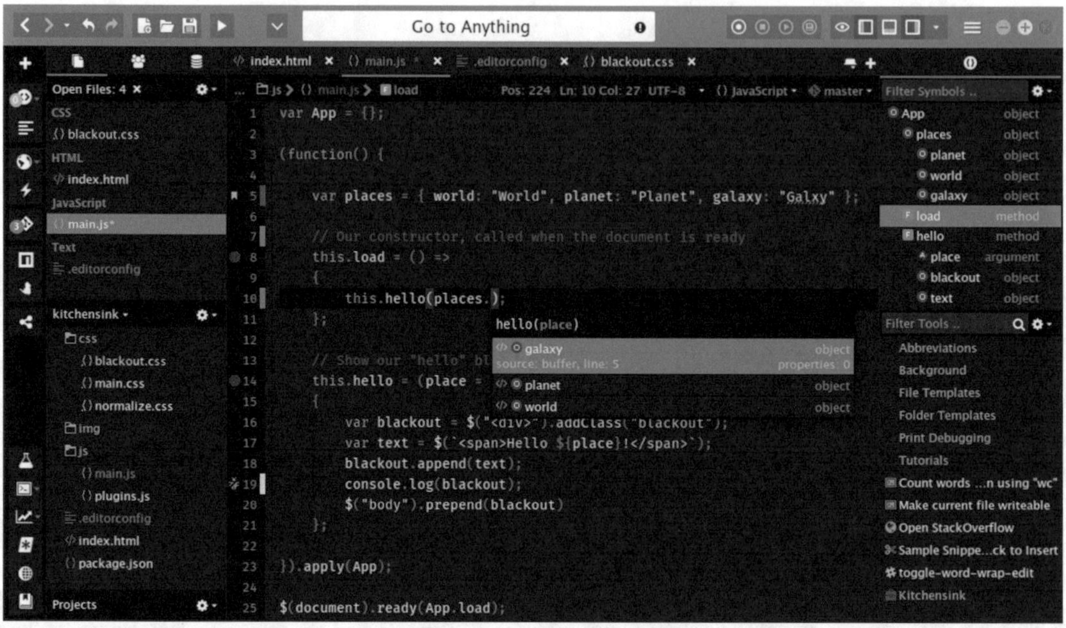

Figure 2-18. *The Komodo IDE*

SciPy

SciPy (pronounced "sigh pie") is a set of open-source Python libraries specialized for scientific computing. Many of these libraries are the protagonists of many chapters of the book, given that their knowledge is critical to data analysis. Together they constitute a set of tools for calculating and displaying data. It has little to envy from other specialized environments for calculation and data analysis (such as R or MATLAB). Among the libraries that are part of the SciPy group, there are three in particular that are discussed in the following chapters:

- NumPy
- matplotlib
- Pandas

NumPy

This library, whose name means *numerical Python,* constitutes the core of many other Python libraries that have originated from it. Indeed, NumPy is the foundation library for scientific computing in Python since it provides data structures and high-performing functions that the basic package of the Python cannot provide. In fact, as you will see later in the book, NumPy defines a specific data structure that is an *N*-dimensional array defined as *ndarray*.

Knowledge of this library is essential in terms of numerical calculations since its correct use can greatly influence the performance of your computations. Throughout the book, this library is almost omnipresent because of its unique characteristics, so an entire chapter is devoted to it (Chapter 3).

This package provides some features added to the standard Python:

- *Ndarray*: A multidimensional array much faster and more efficient than those provided by the basic package of Python.

- *Element-wise computation*: A set of functions for performing this type of calculation with arrays and mathematical operations between arrays.

- *Reading-writing datasets*: A set of tools for reading and writing data stored in the hard disk.

- *Integration with other languages such as C, C++, and FORTRAN*: A set of tools to integrate code developed with these programming languages.

Pandas

This package provides complex data structures and functions specifically designed to make the work on them easy, fast, and effective. This package is the core of data analysis in Python. Therefore, the study and application of this package is the main goal on which you will work throughout the book (especially in Chapters 4, 5, and 6). Knowledge of its every detail, especially when it is applied to data analysis, is a fundamental objective of this book.

The fundamental concept of this package is the *DataFrame*, a two-dimensional tabular data structure with row and column labels.

Pandas applies the high-performance properties of the NumPy library to the manipulation of data in spreadsheets or in relational databases (SQL databases). In fact, by using sophisticated indexing, it will be easy to carry out many operations on this kind of data structure, such as reshaping, slicing, aggregations, and the selection of subsets.

matplotlib

This package is the Python library that is currently the most popular for producing plots and other data visualizations in 2D. Because data analysis requires visualization tools, this library best suits this purpose. In Chapter 7, you learn about this rich library in detail so you will know how to represent the results of your analysis in the best way.

Conclusions

During the course of this chapter, all the fundamental aspects characterizing the Python world have been illustrated. The basic concepts of the Python programming language were introduced, with brief examples explaining its innovative aspects and how it stands out compared to other programming languages. In addition, different ways of using Python at various levels were presented. First you saw how to use a simple command-line interpreter, then a set of simple graphical user interfaces were shown until you got to complex development environments, known as IDEs, such as Spyder, Liclipse, and NinjaIDE.

Even the highly innovative project Jupyter (IPython) was presented, showing you how you can develop Python code interactively, in particular with the Jupyter Notebook.

Moreover, the modular nature of Python was highlighted with the ability to expand the basic set of standard functions provided by Python's external libraries. In this regard, the PyPI online repository was shown along with other Python distributions such as Anaconda and Enthought Canopy.

In the next chapter, you deal with the first library that is the basis of all numerical calculations in Python: NumPy. You learn about the ndarray, a data structure that is the basis of the more complex data structures used in data analysis in the following chapters.

■ ■ ■

## CHAPTER 3

The NumPy Library

NumPy is a basic package for scientific computing with Python and especially for data analysis. In fact, this library is the basis of a large amount of mathematical and scientific Python packages, and among them, as you will see later in the book, is the pandas library. This library, specialized for data analysis, is fully developed using the concepts introduced by NumPy. In fact, the built-in tools provided by the standard Python library could be too simple or inadequate for most of the calculations in data analysis.

Having knowledge of the NumPy library is important to being able to use all scientific Python packages, and particularly, to use and understand the pandas library. The pandas library is the main subject of the following chapters.

If you are already familiar with this library, you can proceed directly to the next chapter; otherwise you can view this chapter as a way to review the basic concepts or to regain familiarity with it by running the examples in this chapter.

NumPy: A Little History

At the dawn of the Python language, the developers needed to perform numerical calculations, especially when this language was being used by the scientific community.

The first attempt was Numeric, developed by Jim Hugunin in 1995, which was followed by an alternative package called *Numarray*. Both packages were specialized for the calculation of arrays, and each had strengths depending on in which case they were used. Thus, they were used differently depending on the circumstances. This ambiguity led then to the idea of unifying the two packages. Travis Oliphant started to develop the NumPy library for this purpose. Its first release (v 1.0) occurred in 2006.

From that moment on, NumPy proved to be the extension library of Python for scientific computing, and it is currently the most widely used package for the calculation of multidimensional arrays and large arrays. In addition, the package comes with a range of functions that allow you to perform operations on arrays in a highly efficient way and perform high-level mathematical calculations.

Currently, NumPy is open source and licensed under BSD. There are many contributors who have expanded the potential of this library. At present, NumPy has arrived at release 1.24. As you can see in Figure 3-1, this library is in continuous development, with approximately one release every six months.

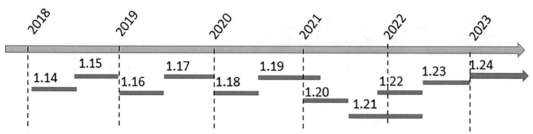

Figure 3-1. *NumPy releases in the last five years, with the new NumPy logo [CC BY-SA 4.0 Isabela Presedo-Floyd]*

The NumPy Installation

This library doesn't have any requirements except that you have a Python platform to run on, so installing it shouldn't cause any problems. Generally, this module is present as a basic package in most Python distributions; however, if not, you can install it later.

However, regardless of the platform you are using, as with all the other libraries that you will use throughout the book, it is also recommended for NumPy to use the platform of the Anaconda distribution. This allows you to cleanly manage the NumPy installation aspect, as well as easily create and manage the virtual environments on which to install it. In this way, it is possible to test and develop your code with different Python versions and NumPy releases without having to uninstall and reinstall everything each time.

If you have Anaconda, just write the following to install NumPy:

```
conda install numpy
```

If instead you want to work without the support of this distribution, use the command-line pip command to install the NumPy library (see https://pypi.org/project/numpy/):

```
pip install numpy
```

Once NumPy is installed on your distribution, to import the NumPy module in your Python session, write the following:

```
>>> import numpy as np
```

If, on the other hand, you are writing code, in order to access NumPy and its functions, you have to insert this instruction at the beginning of the Python code.

ndarray: The Heart of the Library

The NumPy library is based on one main object: *ndarray* (which stands for N-dimensional array). This object is a multidimensional, homogeneous array with a predetermined number of items: homogeneous because virtually all the items in it are of the same type and the same size. In fact, the data type is specified by another NumPy object called *dtype* (data-type); each ndarray is associated with only one type of dtype.

The number of the dimensions and items in an array is defined by its *shape*, a tuple of N-positive integers that specifies the size for each dimension. The dimensions are defined as *axes* and the number of axes as *rank*.

Moreover, another peculiarity of NumPy arrays is that their size is fixed, that is, once you define their size at the time of creation, it remains unchanged. This behavior is different from Python lists, which can grow or shrink in size.

The easiest way to define a new ndarray is to use the array() function, passing a Python list containing the elements to be included in it as an argument.

```
>>> a = np.array([1, 2, 3])
>>> a
array([1, 2, 3])
```

You can easily check that a newly created object is an ndarray by passing the new variable to the type() function.

```
>>> type(a)
<class 'numpy.ndarray'>
```

In order to know the associated dtype to the newly created ndarray, you have to use the dtype attribute.

■ **Note** The result of dtype, shape, and other attributes can vary among different operating systems and Python distributions.

```
>>> a.dtype
dtype('int32')
```

The just-created array has one axis, and then its rank is 1, while its shape should be (3,1). To obtain these values from the corresponding array, it is sufficient to use the ndim attribute for getting the axes, the size attribute to determine the array length, and the shape attribute to get its shape.

```
>>> a.ndim
1
>>> a.size
3
>>> a.shape
(3,)
```

What you have just seen is the simplest case of a one-dimensional array. But the use of arrays can be easily extended to several dimensions. For example, if you define a two-dimensional array 2x2:

```
>>> b = np.array([[1.3, 2.4],[0.3, 4.1]])
>>> b
array([[1.2, 2.4],
       [0.3, 3. ]])
>>> b.dtype
dtype('float64')
>>> b.ndim
2
>>> b.size
4
>>> b.shape
(2, 2)
```

This array has rank 2, since it has two axes, each of length 2.

Another important attribute is itemsize, which can be used with ndarray objects. It defines the size in bytes of each item in the array, and data is the buffer containing the actual elements of the array. This second attribute is still not generally used, because to access the data in the array you use the indexing mechanism, which you will see in the next sections.

```
>>> b.itemsize
8
>>> b.data
<memory at 0x000001A8AD526A80>
```

Create an Array

To create a new array, you can follow different paths. The most common path is the one you saw in the previous section through a list or sequence of lists as arguments to the array() function.

```
>>> c = np.array([[1, 2, 3],[4, 5, 6]])
>>> c
array([[1, 2, 3],
       [4, 5, 6]])
```

The array() function, in addition to lists, can accept tuples and sequences of tuples.

```
>>> d = np.array(((1, 2, 3),(4, 5, 6)))
>>> d
array([[1, 2, 3],
       [4, 5, 6]])
```

It can also accept sequences of tuples and interconnected lists.

```
>>> e = np.array([(1, 2, 3), [4, 5, 6], (7, 8, 9)])
>>> e
array([[1, 2, 3],
       [4, 5, 6],
       [7, 8, 9]])
```

48

Types of Data

So far you have seen only simple integer and float numeric values, but NumPy arrays are designed to contain a wide variety of data types (see Table 3-1). For example, you can use the data type string:

```
>>> g = np.array([['a', 'b'],['c', 'd']])
>>> g
array([['a', 'b'],
       ['c', 'd']], dtype='<U1')>>> g.dtype
dtype('<U1')
>>> g.dtype.name
'str32'
```

Table 3-1. *Data Types Supported by NumPy*

Data Type	Description
bool_	Boolean (true or false) stored as a byte
int_	Signed integer type (same as C long and Python int; normally either int64 or int32 depending on the platform)
intc	Signed integer type, identical to C int (normally int32 or int64)
intp	Integer used for indexing (same as C size_t; normally either int32 or int64)
int8	Alias for the signed integer type with 8 bits (–128 to 127)
int16	Alias for the signed integer type with 16 bits (–32768 to 32767)
int32	Alias for the signed integer type with 32 bits (–2147483648 to 2147483647)
int64	Alias for the signed integer type with 64 bits (–9223372036854775808 to 9223372036854775807)
uint8	Alias for the unsigned integer type with 8 bits (0 to 255)
uint16	Alias for the unsigned integer type with 16 bits (0 to 65535)
uint32	Alias for the unsigned integer type with 32 bits (0 to 4294967295)
uint64	Alias for the unsigned integer type with 64 bits (0 to 18446744073709551615)
float_	Shorthand for float64
float16	Half precision float: sign bit, 5-bit exponent, 10-bit mantissa
float32	Single precision float: sign bit, 8-bit exponent, 23-bit mantissa
float64	Double precision float: sign bit, 11-bit exponent, 52-bit mantissa
complex_	Shorthand for complex128
complex64	Complex number, represented by two 32-bit floats (real and imaginary components)
complex128	Complex number, represented by two 64-bit floats (real and imaginary components)

The dtype Option

The `array()` function does not accept a single argument. You have seen that each ndarray object is associated with a dtype object that uniquely defines the type of data that will occupy each item in the array. By default, the `array()` function can associate the most suitable type according to the values contained in the sequence of lists or tuples. Actually, you can explicitly define the dtype using the dtype option as an argument of the function.

For example, if you want to define an array with complex values, you can use the dtype option as follows:

```
>>> f = np.array([[1, 2, 3],[4, 5, 6]], dtype=complex)
>>> f
array([[ 1.+0.j,  2.+0.j,  3.+0.j],
       [ 4.+0.j,  5.+0.j,  6.+0.j]])
```

Intrinsic Creation of an Array

The NumPy library provides a set of functions that generate ndarrays with initial content, created with different values depending on the function. Throughout the chapter, and throughout the book, you'll discover that these features will be very useful. In fact, they allow a single line of code to generate large amounts of data.

The `zeros()` function, for example, creates a full array of zeros with dimensions defined by the shape of the argument. For example, to create a two-dimensional array 3x3, you can use:

```
>>> np.zeros((3, 3))
array([[ 0.,  0.,  0.],
       [ 0.,  0.,  0.],
       [ 0.,  0.,  0.]])
```

While the `ones()` function creates an array full of ones in a very similar way.

```
>>> np.ones((3, 3))
array([[ 1.,  1.,  1.],
       [ 1.,  1.,  1.],
       [ 1.,  1.,  1.]])
```

By default, the two functions created arrays with the `float64` data type. A feature that is particularly useful is `arange()`. This function generates NumPy arrays with numerical sequences that respond to particular rules depending on the passed arguments. For example, if you want to generate a sequence of values between 0 and 10, you will be passed only one argument to the function—the value with which you want to end the sequence.

```
>>> np.arange(0, 10)
array([0, 1, 2, 3, 4, 5, 6, 7, 8, 9])
```

If instead of starting from 0 you want to start from another value, you simply specify two arguments: the first is the starting value and the second is the final value.

```
>>> np.arange(4, 10)
array([4, 5, 6, 7, 8, 9])
```

It is also possible to generate a sequence of values with precise intervals between them. If the third argument of the arange() function is specified, this will represent the gap between one value and the next one in the sequence of values.

```
>>> np.arange(0, 12, 3)
array([0, 3, 6, 9])
```

In addition, this third argument can also be a float.

```
>>> np.arange(0, 6, 0.6)
array([ 0. ,  0.6,  1.2,  1.8,  2.4,  3. ,  3.6,  4.2,  4.8,  5.4])
```

So far you have only created one-dimensional arrays. To generate two-dimensional arrays, you can still continue to use the arange() function but combined with the reshape() function. This function divides a linear array in different parts in the manner specified by the shape argument.

```
>>> np.arange(0, 12).reshape(3, 4)
array([[ 0,  1,  2,  3],
       [ 4,  5,  6,  7],
       [ 8,  9, 10, 11]])
```

Another function very similar to arange() is linspace(). This function still takes as its first two arguments the initial and end values of the sequence, but the third argument, instead of specifying the distance between one element and the next, defines the number of elements into which you want the interval to be split.

```
>>> np.linspace(0,10,5)
array([ 0. ,  2.5,  5. ,  7.5, 10. ])
```

Finally, another method to obtain arrays already containing values is to fill them with random values. This is possible using the random() function of the numpy.random module. This function will generate an array with as many elements as specified in the argument.

```
>>> np.random.random(3)
array([ 0.78610272,  0.90630642,  0.80007102])
```

The numbers obtained will vary with every run. To create a multidimensional array, you simply pass the size of the array as an argument.

```
>>> np.random.random((3,3))
array([[ 0.07878569,  0.7176506 ,  0.05662501],
       [ 0.82919021,  0.80349121,  0.30254079],
       [ 0.93347404,  0.65868278,  0.37379618]])
```

Basic Operations

So far you have seen how to create a new NumPy array and how items are defined in it. Now it is the time to see how to apply various operations to these arrays.

Arithmetic Operators

The first operations that you will perform on arrays are the arithmetic operators. The most obvious are adding and multiplying an array by a scalar.

```
>>> a = np.arange(4)
>>> a
array([0, 1, 2, 3])
>>> a+4
array([4, 5, 6, 7])
>>> a*2
array([0, 2, 4, 6])
```

These operators can also be used between two arrays. In NumPy, these operations are *element-wise*, that is, the operators are applied only between corresponding elements. These objects occupy the same position, so that the end result is a new array containing the results in the same location of the operands (see Figure 3-2).

```
>>> b = np.arange(4,8)
>>> b
array([4, 5, 6, 7])
>>> a + b
array([ 4,  6,  8, 10])
>>> a - b
array([-4, -4, -4, -4])
>>> a * b
array([ 0,  5, 12, 21])
```

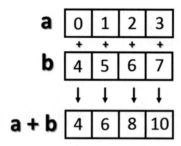

Figure 3-2. *Element-wise addition*

Moreover, these operators are also available for functions, provided that the value returned is a NumPy array. For example, you can multiply the array by the sine or the square root of the elements of array b.

```
>>> a * np.sin(b)
array([-0.        , -0.95892427, -0.558831  ,  1.9709598 ])
>>> a * np.sqrt(b)
array([ 0.        ,  2.23606798,  4.89897949,  7.93725393])
```

Moving on to the multidimensional case, even here the arithmetic operators continue to operate element-wise.

```
>>> A = np.arange(0, 9).reshape(3, 3)
>>> A
array([[0, 1, 2],
       [3, 4, 5],
       [6, 7, 8]])
>>> B = np.ones((3, 3))
>>> B
array([[ 1.,  1.,  1.],
       [ 1.,  1.,  1.],
       [ 1.,  1.,  1.]])
>>> A * B
array([[ 0.,  1.,  2.],
       [ 3.,  4.,  5.],
       [ 6.,  7.,  8.]])
```

The Matrix Product

The choice of operating element-wise is a peculiar aspect of the NumPy library. In fact, in many other tools for data analysis, the * operator is understood as a *matrix product* when it is applied to two matrices. Using NumPy, this kind of product is instead indicated by the dot() function. This operation is not element-wise.

```
>>> np.dot(A,B)
array([[  3.,   3.,   3.],
       [ 12.,  12.,  12.],
       [ 21.,  21.,  21.]])
```

The result at each position is the sum of the products of each element of the corresponding row of the first matrix with the corresponding element of the corresponding column of the second matrix. Figure 3-3 illustrates the process carried out during the matrix product (run for two elements).

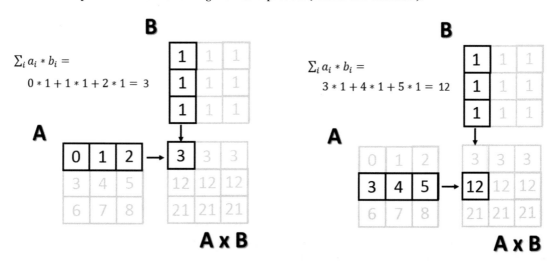

Figure 3-3. *Calculating matrix elements as a result of a matrix product*

An alternative way to write the matrix product is to use the dot() function as an object's function of one of the two matrices.

```
>>> A.dot(B)
array([[  3.,   3.,   3.],
       [ 12.,  12.,  12.],
       [ 21.,  21.,  21.]])
```

Note that because the matrix product is not a commutative operation, the order of the operands is important. Indeed, A * B is not equal to B * A.

```
>>> np.dot(B,A)
array([[  9.,  12.,  15.],
       [  9.,  12.,  15.],
       [  9.,  12.,  15.]])
```

Increment and Decrement Operators

Actually, there are no such operators in Python, because there are no operators called ++ or −−. To increase or decrease values, you have to use operators such as += and −=. These operators are not different from ones you saw earlier, except that instead of creating a new array with the results, they reassign the results to the same array.

```
>>> a = np.arange(4)
>>> a
array([0, 1, 2, 3])
>>> a += 1
>>> a
array([1, 2, 3, 4])
>>> a -= 1
>>> a
array([0, 1, 2, 3])
```

Therefore, using these operators is much more extensive than the simple incremental operators that increase the values by one unit, and they can be applied in many cases. For instance, you need them every time you want to change the values in an array without generating a new array.

```
>>> a += 4
>>> a
array([4, 5, 6, 7])
>>> a *= 2
>>> a
array([ 8, 10, 12, 14])
```

Universal Functions (ufunc)

A universal function, generally called *ufunc*, is a function operating on an array in an element-by-element fashion. This means that it acts individually on each single element of the input array to generate a corresponding result in a new output array. In the end, you obtain an array of the same size as the input.

There are many mathematical and trigonometric operations that meet this definition; for example, calculating the square root with sqrt(), the logarithm with log(), or the sin with sin().

```
>>> a = np.arange(1, 5)
>>> a
array([1, 2, 3, 4])
>>> np.sqrt(a)
array([ 1.        ,  1.41421356,  1.73205081,  2.        ])
>>> np.log(a)
array([ 0.        ,  0.69314718,  1.09861229,  1.38629436])
>>> np.sin(a)
array([ 0.84147098,  0.90929743,  0.14112001, -0.7568025 ])
```

Many common math functions are already implemented in the NumPy library.

Aggregate Functions

Aggregate functions perform an operation on a set of values, an array for example, and produce a single result. Therefore, the sum of all the elements in an array is an aggregate function. Many functions of this kind are implemented in the ndarray class and so can be invoked directly from the array on which you want to perform the calculation.

```
>>> a = np.array([3.3, 4.5, 1.2, 5.7, 0.3])
>>> a.sum()
15.0
>>> a.min()
0.3
>>> a.max()
5.7
>>> a.mean()
3.0
>>> a.std()
2.0079840636817816
```

Indexing, Slicing, and Iterating

In the previous sections, you saw how to create an array and how to perform operations on it. In this section, you see how to manipulate these objects. You learn how to select elements through indexes and slices, in order to obtain the values contained in them or to make assignments in order to change their values. Finally, you also see how you can make iterations within them.

Indexing

Array indexing always uses square brackets ([ ]) to index the elements of the array so that the elements can then be referred individually for various uses, such as extracting a value, selecting items, or even assigning a new value.

When you create a new array, an appropriate scale index is also automatically created (see Figure 3-4).

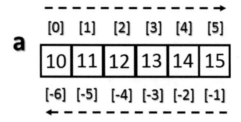

Figure 3-4. *Indexing a monodimensional ndarray*

In order to access a single element of an array, you can refer to its index.

```
>>> a = np.arange(10, 16)
>>> a
array([10, 11, 12, 13, 14, 15])
>>> a[4]
14
```

The NumPy arrays also accept negative indexes. These indexes have the same incremental sequence from 0 to –1, –2, and so on, but in practice they cause the final element to move gradually toward the initial element, which is the one with the more negative index value.

```
>>> a[-1]
15
>>> a[-6]
10
```

To select multiple items at once, you can pass an array of indexes in square brackets.

```
>>> a[[1, 3, 4]]
array([11, 13, 14])
```

Moving on to the two-dimensional case, namely the matrices, they are represented as rectangular arrays consisting of rows and columns, defined by two axes, where axis 0 is represented by the rows and axis 1 is represented by the columns. Thus, indexing in this case is represented by a pair of values: the first value is the index of the row and the second is the index of the column. Therefore, if you want to access the values or select elements in the matrix, you still use square brackets, but this time there are two values [row index, column index] (see Figure 3-5).

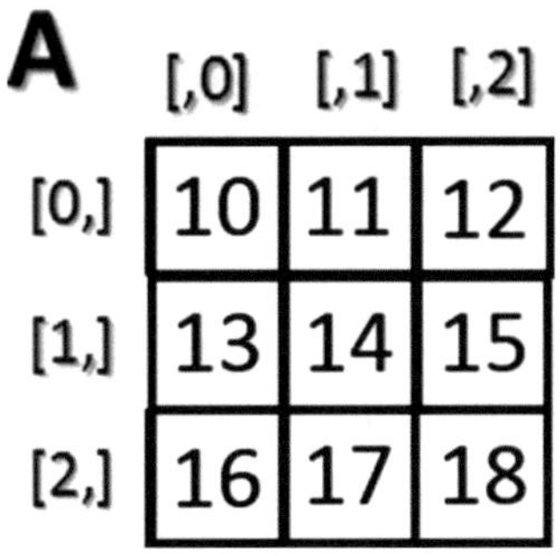

Figure 3-5. *Indexing a two-dimensional array*

```
>>> A = np.arange(10, 19).reshape((3, 3))
>>> A
array([[10, 11, 12],
       [13, 14, 15],
       [16, 17, 18]])
```

If you want to remove the element of the third column in the second row, you have to insert the pair [1, 2].

```
>>> A[1, 2]
15
```

Slicing

Slicing allows you to extract portions of an array to generate new arrays. When you use the Python lists to slice arrays, the resulting arrays are copies, but in NumPy, the arrays are views of the same underlying buffer.

Depending on the portion of the array that you want to extract (or view), you must use the slice syntax; that is, you use a sequence of numbers separated by colons (:) within square brackets.

If you want to extract a portion of the array, for example one that goes from the second to the sixth element, you have to insert the index of the starting element, that is 1, and the index of the final element, that is 5, separated by a colon (:).

```
>>> a = np.arange(10, 16)
>>> a
array([10, 11, 12, 13, 14, 15])
>>> a[1:5]
array([11, 12, 13, 14])
```

Now if you want to extract an item from the previous portion and skip a specific number of following items, then extract the next and skip again, you can use a third number that defines the gap in the sequence of the elements. For example, with a value of 2, the array will take the elements in an alternating fashion.

```
>>> a[1:5:2]
array([11, 13])
```

To better understand the slice syntax, you also should look at cases where you do not use explicit numerical values. If you omit the first number, NumPy implicitly interprets this number as 0 (i.e., the initial element of the array). If you omit the second number, this will be interpreted as the maximum index of the array; and if you omit the last number, this will be interpreted as 1. All the elements will be considered without intervals.

```
>>> a[::2]
array([10, 12, 14])
>>> a[:5:2]
array([10, 12, 14])
>>> a[:5:]
array([10, 11, 12, 13, 14])
```

In the case of a two-dimensional array, the slicing syntax still applies, but it is separately defined for the rows and columns. For example, if you want to extract only the first row:

```
>>> A = np.arange(10, 19).reshape((3, 3))
>>> A
array([[10, 11, 12],
       [13, 14, 15],
       [16, 17, 18]])
>>> A[0,:]
array([10, 11, 12])
```

As you can see in the second index, if you leave only the colon without defining a number, you will select all the columns. Instead, if you want to extract all the values of the first column, you have to write the inverse.

```
>>> A[:,0]
array([10, 13, 16])
```

Instead, if you want to extract a smaller matrix, you need to explicitly define all intervals with indexes that define them.

```
>>> A[0:2, 0:2]
array([[10, 11],
       [13, 14]])
```

If the indexes of the rows or columns to be extracted are not contiguous, you can specify an array of indexes.

```
>>> A[[0,2], 0:2]
array([[10, 11],
       [16, 17]])
```

Iterating an Array

In Python, iterating the items in an array is really very simple; you just need to use the `for` construct.

```
>>> for i in a:
...       print(i)
...
10
11
12
13
14
15
```

Of course, even here, moving to the two-dimensional case, you could think of applying the solution of two nested loops with the `for` construct. The first loop will scan the rows of the array, and the second loop will scan the columns. Actually, if you apply the `for` loop to a matrix, it will always perform a scan according to the first axis.

```
>>> for row in A:
...       print(row)
...
[10 11 12]
[13 14 15]
[16 17 18]
```

If you want to make an iteration element by element, you can use the following construct, using the `for` loop on `A.flat`.

```
>>> for item in A.flat:
...       print(item)
...
10
11
12
13
14
15
16
17
18
```

However, despite all this, NumPy offers an alternative and more elegant solution than the `for` loop. Generally, you need to apply an iteration to apply a function on the rows, on the columns, or on an individual item. If you want to launch an aggregate function that returns a value calculated for every single column or for every single row, there is an optimal way that leaves it to NumPy to manage the iteration: the `apply_along_axis()` function.

This function takes three arguments: the aggregate function, the axis on which to apply the iteration, and the array. If the axis option equals 0, then the iteration evaluates the elements column by column, whereas if axis equals 1 then the iteration evaluates the elements row by row. For example, you can calculate the average values first by column and then by row.

```
>>> np.apply_along_axis(np.mean, axis=0, arr=A)
array([ 13.,   14.,   15.])
>>> np.apply_along_axis(np.mean, axis=1, arr=A)
array([ 11.,   14.,   17.])
```

The previous case uses a function already defined in the NumPy library, but nothing prevents you from defining your own functions. You also used an aggregate function. However, nothing forbids you from using an ufunc. In this case, iterating by column and by row produces the same result. In fact, using a ufunc performs one iteration element-by-element.

```
>>> def foo(x):
...      return x/2
...
>>> np.apply_along_axis(foo, axis=1, arr=A)
array([[5.,   5.5, 6. ],
       [6.5, 7.,   7.5],
       [8.,   8.5, 9. ]])
>>> np.apply_along_axis(foo, axis=0, arr=A)
array([[5.,   5.5, 6.],
       [6.5, 7.,   7.5],
       [8.,   8.5, 9.]])
```

As you can see, the ufunc function halves the value of each element of the input array, regardless of whether the iteration is performed by row or by column.

Conditions and Boolean Arrays

So far you have used indexing and slicing to select or extract a subset of an array. These methods use numerical indexes. An alternative way to selectively extract the elements in an array is to use the conditions and Boolean operators.

Suppose you wanted to select all the values that are less than 0.5 in a 4x4 matrix containing random numbers between 0 and 1.

```
>>> A = np.random.random((4, 4))
>>> A
array([[ 0.03536295,  0.0035115 ,  0.54742404,  0.68960999],
       [ 0.21264709,  0.17121982,  0.81090212,  0.43408927],
       [ 0.77116263,  0.04523647,  0.84632378,  0.54450749],
       [ 0.86964585,  0.6470581 ,  0.42582897,  0.22286282]])
```

Once a matrix of random numbers is defined, if you apply an operator condition, you will receive as a return value a Boolean array containing true values in the positions in which the condition is satisfied. In this example, that is all the positions in which the values are less than 0.5.

```
>>> A < 0.5
array([[ True,  True, False, False],
       [ True,  True, False,  True],
       [False,  True, False, False],
       [False, False,  True,  True]], dtype=bool)
```

Actually, the Boolean arrays are used implicitly for making selections of parts of arrays. In fact, by inserting the previous condition directly inside the square brackets, you can extract all elements smaller than 0.5, so as to obtain a new array.

```
>>> A[A < 0.5]
array([ 0.03536295,  0.0035115 ,  0.21264709,  0.17121982,  0.43408927,
        0.04523647,  0.42582897,  0.22286282])
```

Shape Manipulation

You already saw, when creating a two-dimensional array, that it is possible to convert a one-dimensional array into a matrix, thanks to the reshape() function.

```
>>> a = np.random.random(12)
>>> a
array([ 0.77841574,  0.39654203,  0.38188665,  0.26704305,  0.27519705,
        0.78115866,  0.96019214,  0.59328414,  0.52008642,  0.10862692,
        0.41894881,  0.73581471])
>>> A = a.reshape(3, 4)
>>> A
array([[ 0.77841574,  0.39654203,  0.38188665,  0.26704305],
       [ 0.27519705,  0.78115866,  0.96019214,  0.59328414],
       [ 0.52008642,  0.10862692,  0.41894881,  0.73581471]])
```

The reshape() function returns a new array and can therefore create new objects. However, if you want to modify the object by modifying the shape, you have to assign a tuple containing the new dimensions directly to its shape attribute.

```
>>> a.shape = (3, 4)
>>> a
array([[ 0.77841574,  0.39654203,  0.38188665,  0.26704305],
       [ 0.27519705,  0.78115866,  0.96019214,  0.59328414],
       [ 0.52008642,  0.10862692,  0.41894881,  0.73581471]])
```

As you can see, this time it is the starting array that changes shape and no object is returned. The inverse operation is also possible; that is, you can convert a two-dimensional array into a one-dimensional array. You do this by using the ravel() function.

```
>>> a = a.ravel()
>>> a
array([ 0.77841574,  0.39654203,  0.38188665,  0.26704305,  0.27519705,
        0.78115866,  0.96019214,  0.59328414,  0.52008642,  0.10862692,
        0.41894881,  0.73581471])
```

Or you can even act directly on the shape attribute of the array itself.

```
>>>  a.shape = (A.size)
>>> a
array([ 0.77841574,  0.39654203,  0.38188665,  0.26704305,  0.27519705,
        0.78115866,  0.96019214,  0.59328414,  0.52008642,  0.10862692,
        0.41894881,  0.73581471])
```

Another important operation is transposing a matrix, which is inverting the columns with the rows. NumPy provides this feature with the transpose() function.

```
>>> A.transpose()
array([[ 0.77841574,  0.27519705,  0.52008642],
       [ 0.39654203,  0.78115866,  0.10862692],
       [ 0.38188665,  0.96019214,  0.41894881],
       [ 0.26704305,  0.59328414,  0.73581471]])
```

Array Manipulation

Often you need to create an array using already created arrays. In this section, you see how to create new arrays by joining or splitting arrays that are already defined.

Joining Arrays

You can merge multiple arrays to form a new one that contains all of the arrays. NumPy uses the concept of *stacking*, providing a number of functions in this regard. For example, you can perform vertical stacking with the vstack() function, which combines the second array as new rows of the first array. In this case, the array grows in the vertical direction. By contrast, the hstack() function performs horizontal stacking; that is, the second array is added to the columns of the first array.

```
>>> A = np.ones((3, 3))
>>> B = np.zeros((3, 3))
>>> np.vstack((A, B))
array([[ 1.,   1.,   1.],
       [ 1.,   1.,   1.],
       [ 1.,   1.,   1.],
       [ 0.,   0.,   0.],
       [ 0.,   0.,   0.],
       [ 0.,   0.,   0.]])
>>> np.hstack((A,B))
array([[ 1.,   1.,   1.,   0.,   0.,   0.],
       [ 1.,   1.,   1.,   0.,   0.,   0.],
       [ 1.,   1.,   1.,   0.,   0.,   0.]])
```

Two other functions performing stacking between multiple arrays are column_stack() and row_stack(). These functions operate differently than the two previous functions. Generally these functions are used with one-dimensional arrays, which are stacked as columns or rows in order to form a new two-dimensional array.

```
>>> a = np.array([0, 1, 2])
>>> b = np.array([3, 4, 5])
>>> c = np.array([6, 7, 8])
>>> np.column_stack((a, b, c))
array([[0, 3, 6],
       [1, 4, 7],
       [2, 5, 8]])
>>> np.row_stack((a, b, c))
array([[0, 1, 2],
       [3, 4, 5],
       [6, 7, 8]])
```

Splitting Arrays

In the previous section, you saw how to assemble multiple arrays through stacking. Now you see how to divide an array into several parts. In NumPy, you use splitting to do this. Here too, you have a set of functions that work both horizontally with the hsplit() function and vertically with the vsplit() function.

```
>>> A = np.arange(16).reshape((4, 4))
>>> A
array([[ 0,  1,  2,  3],
       [ 4,  5,  6,  7],
       [ 8,  9, 10, 11],
       [12, 13, 14, 15]])
```

Thus, if you want to split the array horizontally, meaning the width of the array is divided into two parts, the 4x4 matrix A will be split into two 2x4 matrices.

```
>>> [B,C] = np.hsplit(A, 2)
>>> B
array([[ 0,  1],
       [ 4,  5],
       [ 8,  9],
       [12, 13]])
>>> C
array([[ 2,  3],
       [ 6,  7],
       [10, 11],
       [14, 15]])
```

Instead, if you want to split the array vertically, meaning the height of the array is divided into two parts, the 4x4 matrix A will be split into two 4x2 matrices.

```
>>> [B,C] = np.vsplit(A, 2)
>>> B
array([[0, 1, 2, 3],
       [4, 5, 6, 7]])
>>> C
array([[ 8,  9, 10, 11],
       [12, 13, 14, 15]])
```

A more complex command is the split() function, which allows you to split the array into nonsymmetrical parts. Passing the array as an argument, you also have to specify the indexes of the parts to be divided. If you use the axis = 1 option, then the indexes will be columns; if instead the option is axis = 0, then they will be row indexes.

For example, if you want to divide the matrix into three parts, the first of which will include the first column, the second will include the second and the third column, and the third will include the last column, you must specify three indexes in the following way.

```
>>> [A1,A2,A3] = np.split(A,[1,3],axis=1)
>>> A1
array([[ 0],
       [ 4],
       [ 8],
       [12]])
>>> A2
array([[ 1,  2],
       [ 5,  6],
       [ 9, 10],
       [13, 14]])
>>> A3
array([[ 3],
       [ 7],
       [11],
       [15]])
```

You can do the same thing by row.

```
>>> [A1,A2,A3] = np.split(A,[1,3],axis=0)
>>> A1
array([[0, 1, 2, 3]])
>>> A2
array([[ 4,  5,  6,  7],
       [ 8,  9, 10, 11]])
>>> A3
array([[12, 13, 14, 15]])
```

This feature also includes the functionalities of the vsplit() and hsplit() functions.

General Concepts

This section describes the general concepts underlying the NumPy library. The difference between copies and views is when they return values. The mechanism of broadcasting, which occurs implicitly in many NumPy functions, is also covered in this section.

Copies or Views of Objects

As you may have noticed with NumPy, especially when you are manipulating an array, you can return a copy or a view of the array. None of the NumPy assignments produces copies of arrays, nor any element contained in them.

```
>>> a = np.array([1, 2, 3, 4])
>>> b = a
>>> b
array([1, 2, 3, 4])
>>> a[2] = 0
>>> b
array([1, 2, 0, 4])
```

If you assign one array a to another array b, you are not copying it; array b is just another way to call array a. In fact, by changing the value of the third element of a, you change the third value of b too. When you slice an array, the object returned is a view of the original array.

```
>>> c = a[0:2]
>>> c
array([1, 2])
>>> a[0] = 0
>>> c
array([0, 2])
```

As you can see, even when slicing, you are actually pointing to the same object. If you want to generate a complete and distinct array, use the copy() function.

```
>>> a = np.array([1, 2, 3, 4])
>>> c = a.copy()
>>> c
array([1, 2, 3, 4])
>>> a[0] = 0
>>> c
array([1, 2, 3, 4])
```

In this case, even when you change the items in array a, array c remains unchanged.

Vectorization

Vectorization, along with *broadcasting*, is the basis of the internal implementation of NumPy. Vectorization is the absence of an explicit loop during the development of the code. These loops actually cannot be omitted, but are implemented internally and then are replaced by other constructs in the code. The application of vectorization leads to more concise and readable code, and you can say that it will appear more "Pythonic" in its appearance. In fact, thanks to the vectorization, many operations take on a more mathematical expression. For example, NumPy allows you to express the multiplication of two arrays as shown:

```
a * b
```

Or even two matrices:

```
A * B
```

In other languages, such operations would be expressed with many nested loops and the `for` construct. For example, the first operation would be expressed in the following way:

```
for (i = 0; i < rows; i++){
  c[i] = a[i]*b[i];
}
```

While the product of matrices would be expressed as follows:

```
for( i=0; i < rows; i++){
   for(j=0; j < columns; j++){
      c[i][j] = a[i][j]*b[i][j];
   }
}
```

You can see that using NumPy makes the code more readable and more mathematical.

Broadcasting

Broadcasting allows an operator or a function to act on two or more arrays even if these arrays do not have the same shape. That said, not all the dimensions can be subjected to broadcasting; they must meet certain rules.

You saw that using NumPy, you can classify multidimensional arrays through a shape that is a tuple representing the length of the elements of each dimension.

Two arrays can be subjected to broadcasting when all their dimensions are compatible, i.e., the length of each dimension must be equal or one of them must be equal to 1. If neither of these conditions is met, you get an exception that states that the two arrays are not compatible.

```
>>> A = np.arange(16).reshape(4, 4)
>>> b = np.arange(4)
>>> A
array([[ 0,  1,  2,  3],
       [ 4,  5,  6,  7],
       [ 8,  9, 10, 11],
       [12, 13, 14, 15]])
>>> b
array([0, 1, 2, 3])
```

In this case, you obtain two arrays:

```
4 x 4
4
```

There are two rules of broadcasting. First you must add a 1 to each missing dimension. If the compatibility rules are now satisfied, you can apply broadcasting and move to the second rule. For example:

```
4 x 4
4 x 1
```

The rule of compatibility is met. Then you can move to the second rule of broadcasting. This rule explains how to extend the size of the smallest array so that it's the size of the biggest array, so that the element-wise function or operator is applicable.

The second rule assumes that the missing elements (size, length 1) are filled with replicas of the values contained in extended sizes (see Figure 3-6).

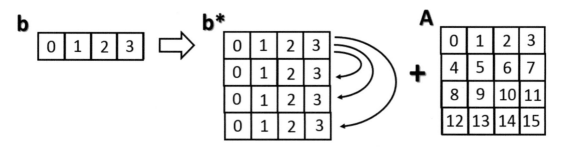

Figure 3-6. *Applying the second broadcasting rule*

Now that the two arrays have the same dimensions, the values inside may be added together.

```
>>> A + b
array([[ 0,  2,  4,  6],
       [ 4,  6,  8, 10],
       [ 8, 10, 12, 14],
       [12, 14, 16, 18]])
```

This is a simple case in which one of the two arrays is smaller than the other. There may be more complex cases in which the two arrays have different shapes and each is smaller than the other only in certain dimensions.

```
>>> m = np.arange(6).reshape(3, 1, 2)
>>> n = np.arange(6).reshape(3, 2, 1)
>>> m
array([[[0, 1]],
       [[2, 3]],
       [[4, 5]]])
>>> n
array([[[0],
        [1]],
       [[2],
        [3]],
       [[4],
        [5]]])
```

Even in this case, by analyzing the shapes of the two arrays, you can see that they are compatible and therefore the rules of broadcasting can be applied.

```
3 x 1 x 2
3 x 2 x 1
```

In this case, both arrays undergo the extension of dimensions (broadcasting).

```
m* = [[[0,1],            n* = [[[0,0],
        [0,1]],                  [1,1]],
       [[2,3],                  [[2,2],
        [2,3]],                  [3,3]],
       [[4,5],                  [[4,4],
        [4,5]]]                  [5,5]]]
```

Then you can apply, for example, the addition operator between the two arrays, operating element-wise.

```
>>> m + n
array([[[ 0,  1],
        [ 1,  2]],
       [[ 4,  5],
        [ 5,  6]],
       [[ 8,  9],
        [ 9, 10]]])
```

Structured Arrays

So far in the various examples in the previous sections, you saw monodimensional and two-dimensional arrays. NumPy allows you to create arrays that are much more complex not only in size, but in the structure, called *structured arrays*. This type of array contains *structs* or records instead of individual items.

For example, you can create a simple array of structs as items. Thanks to the dtype option, you can specify a list of comma-separated specifiers to indicate the elements that will constitute the struct, along with data type and order.

```
bytes                   b1
int                     i1, i2, i4, i8
unsigned ints           u1, u2, u4, u8
floats                  f2, f4, f8
complex                 c8, c16
fixed length strings    a<n>
```

For example, if you want to specify a struct consisting of an integer, a character string of length 6, and a Boolean value, you specify the three types of data in the dtype option with the right order using the corresponding specifiers.

■ **Note**   The result of dtype and other format attributes can vary among different operating systems and Python distributions.

```
>>> structured = np.array([(1, 'First', 0.5, 1+2j),(2, 'Second', 1.3, 2-2j), (3, 'Third',
0.8, 1+3j)],dtype=('i2, a6, f4, c8'))
>>> structured
array([(1, b'First', 0.5, 1+2.j),
```

```
        (2, b'Second', 1.3, 2.-2.j),
        (3, b'Third', 0.8, 1.+3.j)],
     dtype=[('f0', '<i2'), ('f1', 'S6'), ('f2', '<f4'), ('f3', '<c8')])
```

You can also use the data type explicitly by specifying int8, uint8, float16, complex64, and so forth.

```
>>> structured = np.array([(1, 'First', 0.5, 1+2j),(2, 'Second', 1.3,2-2j), (3, 'Third',
0.8, 1+3j)],dtype=('
int16, a6, float32, complex64'))
>>> structured
array([(1, b'First', 0.5, 1.+2.j),
        (2, b'Second', 1.3, 2.-2.j),
        (3, b'Third', 0.8, 1.+3.j)],
     dtype=[('f0', '<i2'), ('f1', 'S6'), ('f2', '<f4'), ('f3', '<c8')])
```

Both cases have the same result. Inside the array, you see a dtype sequence containing the name of each item of the struct with the corresponding data type.

Writing the appropriate reference index, you obtain the corresponding row, which contains the struct.

```
>>> structured[1]
(2, b'Second', 1.3, 2.-2.j)
```

The names that are assigned automatically to each item of the struct can be considered the names of the columns of the array. Using them as a structured index, you can refer to all the elements of the same type, or of the same column.

```
>>> structured['f1']
array([b'First', b'Second', b'Third'], dtype='|S6')
```

As you have just seen, the names are assigned automatically with an f (which stands for field) and a progressive integer that indicates the position in the sequence. In fact, it would be more useful to specify the names with something more meaningful. This is possible and you can do it at the time of array declaration:

```
>>> structured = np.array([(1,'First',0.5,1+2j),(2,'Second',1.3,2-2j),(3,'Third',0.8,1+3j)],
dtype=[(
'id','i2'),('position','a6'),('value','f4'),('complex','c8')])
>>> structured
array([(1, b'First', 0.5, 1.+2.j),
       (2, b'Second', 1.3, 2.-2.j),
       (3, b'Third', 0.8, 1.+3.j)],
     dtype=[('id', '<i2'), ('position', 'S6'), ('value', '<f4'), ('complex', '<c8')])
```

Or you can do it at a later time, redefining the tuples of names assigned to the dtype attribute of the structured array.

```
>>> structured.dtype.names = ('id','order','value','complex')
```

Now you can use meaningful names for the various field types:

```
>>> structured['order']
array([b'First', b'Second', b'Third'],  dtype='|S6')
```

Reading and Writing Array Data on Files

A very important aspect of NumPy that has not been discussed yet is the process of reading data contained in a file. This procedure is very useful, especially when you have to deal with large amounts of data collected in arrays. This is a very common data analysis operation, since the size of the dataset to be analyzed is almost always huge, and therefore it is not advisable or even possible to manage it manually.

NumPy provides a set of functions that allow data analysts to save the results of their calculations in a text or binary file. Similarly, NumPy allows you to read and convert written data in a file into an array.

Loading and Saving Data in Binary Files

NumPy provides a pair of functions, called save() and load(), that enable you to save and then later retrieve data stored in binary format.

Once you have an array to save, for example, one that contains the results of your data analysis processing, you simply call the save() function and specify as arguments the name of the file and the array. The file will automatically be given the .npy extension.

```
>>> data = np.random.random((3,3))
>>> data
array([[0.47941017, 0.43759768, 0.76636206],
       [0.51928993, 0.06358527, 0.72109914],
       [0.64501488, 0.94113659, 0.42052306]])
>>> np.save('saved_data',data)
```

When you need to recover the data stored in a .npy file, you use the load() function by specifying the file name as the argument, this time adding the .npy extension.

```
>>> loaded_data = np.load('saved_data.npy')
>>> loaded_data
array([[0.47941017, 0.43759768, 0.76636206],
       [0.51928993, 0.06358527, 0.72109914],
       [0.64501488, 0.94113659, 0.42052306]])
```

Reading Files with Tabular Data

Many times, the data that you want to read or save are in textural format (TXT or CSV, for example). You might save the data in this format, instead of binary, because the files can then be accessed outside independently if you are working with NumPy or with any other application. Take for example the case of a set of data in the CSV (Comma-Separated Values) format, in which data are collected in a tabular format and the values are separated by commas (see Listing 3-1).

Listing 3-1. ch3_data.csv

```
id,value1,value2,value3
1,123,1.4,23
2,110,0.5,18
3,164,2.1,19
```

To be able to read your data in a text file and insert values into an array, NumPy provides a function called genfromtxt(). Normally, this function takes three arguments—the name of the file containing the data, the character that separates the values from each other (in this case, a comma), and whether the data contain column headers.

```
>>> data = np.genfromtxt('ch3_data.csv', delimiter=',', names=True)
>>> data
array([(1.0, 123.0, 1.4, 23.0), (2.0, 110.0, 0.5, 18.0),
       (3.0, 164.0, 2.1, 19.0)],
      dtype=[('id', '<f8'), ('value1', '<f8'), ('value2', '<f8'), ('value3', '<f8')])
```

As you can see from the result, you get a structured array in which the column headings have become the field names.

This function implicitly performs two loops: the first loop reads a line at a time, and the second loop separates and converts the values contained in it, inserting the consecutive elements created specifically. One positive aspect of this feature is that if some data are missing, the function can handle them.

Take for example the previous file (see Listing 3-2) with some items removed. Save it as data2.csv.

Listing 3-2. ch3_data2.csv

```
id,value1,value2,value3
1,123,1.4,23
2,110,,18
3,,2.1,19
```

Launching these commands, you can see how the genfromtxt() function replaces the blanks in the file with nan values.

```
>>> data2 = np.genfromtxt('ch3_data2.csv', delimiter=',', names=True)
>>> data2
array([(1.0, 123.0, 1.4, 23.0), (2.0, 110.0, nan, 18.0),
       (3.0, nan, 2.1, 19.0)],
      dtype=[('id', '<f8'), ('value1', '<f8'), ('value2', '<f8'), ('value3', '<f8')])
```

At the bottom of the array, you can find the column headings contained in the file. These headers can be considered labels that act as indexes to extract the values by column.

```
>>> data2['id']
array([ 1.,  2.,  3.])
```

Instead, by using the numerical indexes in the classic way, you extract data corresponding to the rows.

```
>>> data2[0]
(1.0, 123.0, 1.4, 23.0)
```

Conclusions

In this chapter, you learned about all the main aspects of the NumPy library and became familiar with a range of features that form the basis of many other aspects you'll face in the course of the book. In fact, many of these concepts are from other scientific and computing libraries that are more specialized, but that have been structured and developed on the basis of this library.

You saw how, thanks to ndarray, you can extend the functionalities of Python, making it a suitable language for scientific computing and data analysis.

Knowledge of NumPy is therefore crucial for anyone who wants to take on the world of data analysis.

The next chapter introduces a new library, called pandas, which is structured on NumPy and so encompasses all the basic concepts illustrated in this chapter. However, pandas extends these concepts so they are more suitable to data analysis.

CHAPTER 4

■ ■ ■

The pandas Library—An Introduction

This chapter gets into the heart of this book: the *pandas library*. This fantastic Python library is a perfect tool for anyone who wants to perform data analysis using Python as a programming language.

First you'll learn about the fundamental aspects of this library and how to install it on your system, and then you'll become familiar with the two data structures, called *series* and *dataframes*. During the course of the chapter, you'll work with a basic set of functions provided by the pandas library, in order to perform the most common data processing tasks. Getting familiar with these operations is a key goal of the rest of the book. This is why it is very important to repeat this chapter until you feel comfortable with its content.

Furthermore, with a series of examples, you'll learn some particularly new concepts introduced in the pandas library: indexing data structures. You'll learn how to get the most of this feature for data manipulation in this chapter and in the next chapters.

Finally, you'll see how to extend the concept of indexing to multiple levels at the same time, through the process called hierarchical indexing.

pandas: The Python Data Analysis Library

pandas is an open-source Python library for highly specialized data analysis. It is currently the reference point that all professionals using the Python language need to study for the statistical purposes of analysis and decision making.

This library was designed and developed primarily by Wes McKinney starting in 2008. In 2012, Sien Chang, one of his colleagues, was added to the development. Together they set up one of the most used libraries in the Python community.

pandas arises from the need to have a specific library to analyze data that provides, in the simplest possible way, all the instruments for data processing, data extraction, and data manipulation.

This Python package is designed on the basis of the NumPy library. This choice was critical to the success and the rapid spread of pandas. In fact, this choice not only makes this library compatible with most other modules, but also takes advantage of the high quality of the NumPy module.

Another fundamental choice was to design ad hoc data structures for data analysis. In fact, instead of using existing data structures built into Python or provided by other libraries, two new data structures were developed.

These data structures are designed to work with relational data or labeled data, thus allowing you to manage data with features similar to those designed for SQL relational databases and Excel spreadsheets.

F. Nelli, *Python Data Analytics*, https://doi.org/10.1007/978-1-4842-9532-8_4

Throughout the book in fact, you will see a series of basic operations for data analysis, which are normally used on database tables and spreadsheets. pandas in fact provides an extended set of functions and methods that allow you to perform these operations efficiently.

So pandas' main purpose is to provide all the building blocks for anyone approaching the data analysis world.

Installation of pandas

The easiest and most general way to install the pandas library is to use a prepackaged solution, that is, installing it through an Anaconda distribution. In fact, over the years this distribution has developed more and more around the data analysis environment, becoming the reference platform for those who work in this area. In addition to pandas, in fact, there are many other libraries available that specialize in data analysis, machine learning, and data visualization. It also provides useful development and analysis tools, as well as Jupyter Notebook.

Installation from Anaconda

For those who choose to use the Anaconda distribution, managing the installation is very simple. The simplest way is the graphical one, activating Anaconda Navigator and then selecting from the Environments panel the virtual environment on which you want to install the library, as shown in Figure 4-1. This will activate the Python virtual environment on which to install pandas and then run the examples in the book.

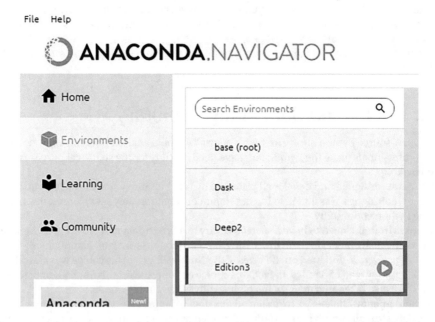

Figure 4-1. *Selection and activation of the Python virtual environment with Anaconda Navigator*

Once the desired virtual environment is activated, go to the right side of Anaconda Navigator and select All from the top drop-down menu. This will display the list of all available packages (installed and not) with their version corresponding to the chosen Python version. Search for pandas (see Step 1 in Figure 4-2). Almost instantly, all the pandas-related packages should appear. Select the one corresponding to the pandas library (as shown in Step 2 of Figure 4-2). At this point, start the installation of the package by clicking the Apply button at the bottom right (as shown in Point 3 of Figure 4-2).

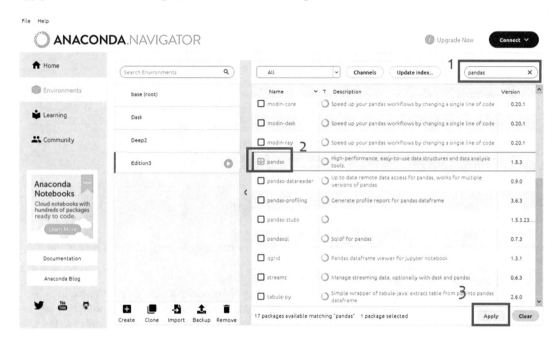

Figure 4-2. *Search and select the pandas package and then start the installation with Anaconda Navigator*

After a few seconds a window will appear with the list of packages to install and their versions (pandas and dependencies), as shown in Figure 4-3. Click the Apply button to confirm the installation. A scroll bar at the bottom will show the progress of the installation.

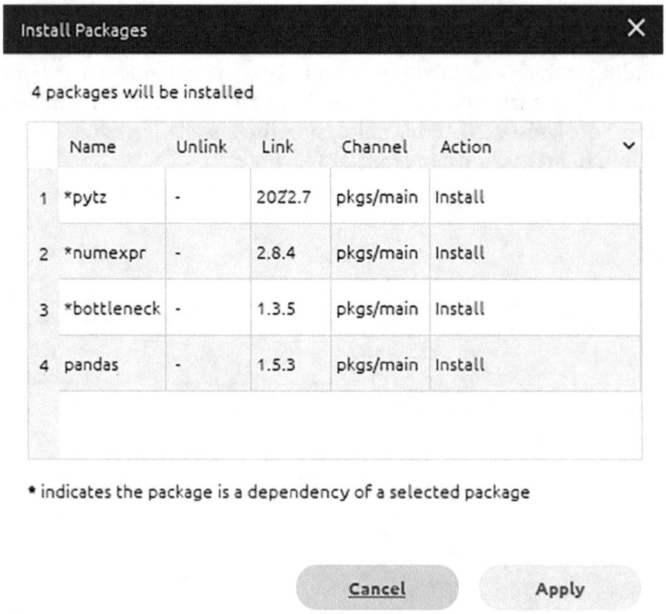

Figure 4-3. *List of packages to install and their versions shown when installing a package in Anaconda Navigator*

If you prefer, even within the Anaconda distribution, there is a console from which to check and install packages. Still from Anaconda Navigator, in the Home panel, select the CMD.exe Prompt to open a command console (as shown in Figure 4-4). Another window will open with the console related to the virtual environment you activated, from which you can enter all the commands manually.

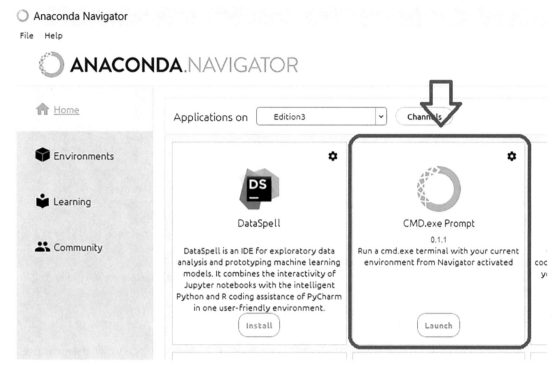

Figure 4-4. *Launching the Python virtual environment command console from Anaconda Navigator*

In my case, because I'm currently on a Windows system and I'm working on a Python virtual environment that I called Edition3, I get the following prompt.

```
(Edition3) C:\Users\nelli>
```

First you have to see if the pandas module is installed and, if so, which version. To do this, type the following command from the terminal:

```
conda list pandas
```

Because I have the module installed on my PC (Windows), I get the following result:

```
# packages in environment at C:\Users\nelli\anaconda3\envs\Edition3:
#
# Name                    Version                   Build  Channel
pandas                    1.5.3                            py311heda8569_0
```

If you do not have pandas installed, you need to install it. Enter the following command to do so:

```
conda install pandas
```

Anaconda will immediately check all dependencies, managing the installation of other modules, without you having to worry too much.

```
## Package Plan ##

  environment location: C:\Users\nelli\anaconda3\envs\Edition3

  added / updated specs:
    - pandas

The following NEW packages will be INSTALLED:

  bottleneck          pkgs/main/win-64::bottleneck-1.3.5-py311h5bb9823_0 None
  numexpr             pkgs/main/win-64::numexpr-2.8.4-py311hffd1eac_0 None
  pandas              pkgs/main/win-64::pandas-1.5.3-py311heda8569_0 None
  pytz                pkgs/main/win-64::pytz-2022.7-py311haa95532_0 NoneProceed ([y]/n)?
```

Enter y to continue with the installation.

If you want to upgrade your package to a newer version, the command to do so is very simple and intuitive:

```
conda update pandas
```

The system will check the version of pandas and the version of all the modules on which it depends and then suggest any updates. It will then ask if you want to proceed with the update.

Installation from PyPI

If you are not using the Anaconda platform, the easiest way to install the pandas library on your Python environment is via PyPI using the `pip` command. From the console, enter the following command:

```
pip install pandas
```

Getting Started with pandas

As you saw during installation, there are several approaches on how to work with pandas. You can choose to open a Jupyter notebook, work with the QtConsole (IPython GUI), or more simply open a session on a simple Python console and enter the instructions one at a time. There is no absolute best way to proceed; all of these methods have strengths and weaknesses depending on the case. The most important thing is to work with the code interactively, by entering a command one by one. This way, you have the opportunity to become familiar with the individual functions and data structures that are explained in this chapter.

Furthermore, the data and functions defined in the various examples remain valid throughout the chapter, which means you don't have to define them each time. You are invited, at the end of each example, to repeat the various commands, modify them if appropriate, and control how the values in the data structures vary during operation. This approach is great for getting familiar with the different topics covered in this chapter, leaving you the opportunity to interact freely with what you are reading.

■ **Note**   This chapter assumes that you have some familiarity with Python and NumPy in general. If you have any difficulty, read Chapters 2 and 3 of this book.

First, open a session on the Python shell and then import the pandas library. The general practice for importing the pandas module is as follows:

```
>>> import pandas as pd
>>> import numpy as np
```

Thus, in this chapter and throughout the book, every time you see pd and np, you'll make reference to an object or method referring to these two libraries, even though you will often be tempted to import the pandas module in this way:

```
>>> from pandas import *
```

Thus, you no longer have to reference a function, object, or method with pd; this approach is not considered good practice by the Python community in general. If you are working on Jupyter, import the two libraries into the first cell of the notebook and run it, as shown in Figure 4-5.

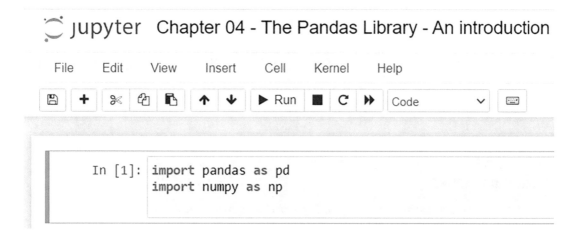

Figure 4-5. *Importing the NumPy and pandas libraries into a Jupyter Notebook*

From now on, any line of code inserted in the examples in the book will correspond to a cell in the notebook. Just as you click ENTER on the Python console to immediately see the result of the entered command, in the same way you write the command into a single cell of the Notebook and execute it.

Introduction to pandas Data Structures

The heart of pandas is the two primary data structures on which all transactions, which are generally made during the analysis of data, are centralized:

- Series
- Dataframes

The *series*, as you will see, constitutes the data structure designed to accommodate a sequence of one-dimensional data, while the *dataframe*, a more complex data structure, is designed to contain cases with several dimensions.

Although these data structures are not the universal solution to all problems, they do provide a valid and robust tool for most applications. In fact, they remain very simple to understand and use. In addition, many cases of more complex data structures can still be traced to these simple two cases.

However, their peculiarities are based on a particular feature—integration in their structure of index objects and labels. You will see that this feature causes these data structures to be easily manipulated.

The Series

The *series* is the object of the pandas library designed to represent one-dimensional data structures, similar to an array but with some additional features. Its internal structure is simple (see Figure 4-6) and is composed of two arrays associated with each other. The main array holds the data (data of any NumPy type) to which each element is associated with a label, contained within the other array, called the *index*.

Series	
index	value
0	12
1	-4
2	7
3	9

Figure 4-6. *The structure of the series object*

Declaring a Series

To create the series specified in Figure 4-1, you simply call the Series() constructor and pass as an argument an array containing the values to be included in it.

```
>>> s = pd.Series([12,-4,7,9])
>>> s
0    12
1    -4
2     7
3     9
dtype: int64
```

As you can see from the output of the series, on the left there are the values in the index, which is a series of labels, and on the right are the corresponding values.

If you do not specify any index during the definition of the series, by default, pandas will assign numerical values increasing from 0 as labels. In this case, the labels correspond to the indexes (position in the array) of the elements in the series object.

Often, however, it is preferable to create a series using meaningful labels in order to distinguish and identify each item regardless of the order in which they were inserted into the series.

In this case it will be necessary, during the constructor call, to include the index option and assign an array of strings containing the labels.

```
>>> s = pd.Series([12,-4,7,9], index=['a','b','c','d'])
>>> s
a    12
b    -4
c     7
d     9
dtype: int64
```

If you want to individually see the two arrays that make up this data structure, you can call the two attributes of the series as follows: index and values.

```
>>> s.values
array([12, -4,  7,  9], dtype=int64)
>>> s.index
Index(['a', 'b', 'c', 'd'], dtype='object')
)
```

Selecting the Internal Elements

You can select individual elements as ordinary NumPy arrays, specifying the key.

```
>>> s[2]
7
```

Or you can specify the label corresponding to the position of the index.

```
>>> s['b']
-4
```

In the same way you select multiple items in a NumPy array, you can specify the following:

```
>>> s[0:2]
a    12
b    -4
dtype: int64
```

In this case, you can use the corresponding labels, but specify the list of labels in an array.

```
>>> s[['b','c']]
b    -4
c     7
dtype: int64
```

Assigning Values to the Elements

Now that you understand how to select individual elements, you also know how to assign new values to them. In fact, you can select the value by index or by label.

```
>>> s[1] = 0
>>> s
a    12
b     0
c     7
d     9
dtype: int64
>>> s['b'] = 1
>>> s
a    12
b     1
c     7
d     9
dtype: int64
```

Defining a Series from NumPy Arrays and Other Series

You can define a new series starting with NumPy arrays or with an existing series.

```
>>> arr = np.array([1,2,3,4])
>>> s3 = pd.Series(arr)
>>> s3
0    1
1    2
2    3
3    4
dtype: int64
>>> s4 = pd.Series(s)
>>> s4
a    12
b     4
c     7
d     9
dtype: int64
```

Always keep in mind that the values contained in the NumPy array or in the original series are not copied, but are passed by reference. That is, the object is inserted dynamically within the new series object. If it changes, for example its internal element varies in value, those changes will also be present in the new series object.

```
>>> s3
0    1
1    2
2    3
3    4
dtype: int64
```

```
>>> arr[2] = -2
>>> s3
0    1
1    2
2   -2
3    4
dtype: int64
```

As you can see in this example, by changing the third element of the arr array, the code also modified the corresponding element in the s3 series.

Filtering Values

Thanks to the choice of the NumPy library as the base of the pandas library and, as a result, for its data structures, many operations that are applicable to NumPy arrays are extended to the series. One of these is filtering values contained in the data structure through conditions.

For example, if you need to know which elements in the series are greater than 8, you write the following:

```
>>> s[s > 8]
a    12
d     9
dtype: int64
```

Operations and Mathematical Functions

Other operations such as operators (+, -, *, and /) and mathematical functions that are applicable to NumPy array can be extended to series.

You can simply write the arithmetic expression for the operators.

```
>>> s / 2
a     6.0
b    -2.0
c     3.5
d     4.5
dtype: float64
```

However, with the NumPy mathematical functions, you must specify the function referenced with np and the instance of the series passed as an argument.

```
>>> np.log(s)
a    2.484907
b    0.000000
c    1.945910
d    2.197225
dtype: float64
```

Evaluating Vales

There are often duplicate values in a series. Then you may need to have more information about the samples, including existence of any duplicates and whether a certain value is present in the series.

In this regard, you can declare a series in which there are many duplicate values.

```
>>> serd = pd.Series([1,0,2,1,2,3], index=['white','white','blue','green','green','yellow'])
>>> serd
white     1
white     0
blue      2
green     1
green     2
yellow    3
dtype: int64
```

To know all the values contained in the series, excluding duplicates, you can use the unique() function. The return value is an array containing the unique values in the series, although not necessarily in order.

```
>>> serd.unique()
array([1, 0, 2, 3], dtype=int64)
```

A function that's similar to unique() is value_counts(), which not only returns unique values but also calculates the occurrences within a series.

```
>>> serd.value_counts()
2    2
1    2
3    1
0    1
dtype: int64
```

Finally, isin() evaluates the membership, that is, the given list of values. This function tells you if the values are contained in the data structure. Boolean values that are returned can be very useful when filtering data in a series or in a column of a dataframe.

```
>>> serd.isin([0,3])
white     False
white      True
blue      False
green     False
green     False
yellow     True
dtype: bool
>>> serd[serd.isin([0,3])]
white     0
yellow    3
dtype: int64
```

NaN Values

The previous case tried to run the logarithm of a negative number and received NaN as a result. This specific value NaN (Not a Number) is used in pandas data structures to indicate the presence of an empty field or something that's not definable numerically.

Generally, these NaN values are a problem and must be managed in some way, especially during data analysis. These data are often generated when extracting data from a questionable source or when the source is missing data. Furthermore, as you have just seen, the NaN values can also be generated in special cases, such as calculations of logarithms of negative values, or exceptions during execution of some calculation or function. In later chapters, you see how to apply different strategies to address the problem of NaN values.

Despite their problematic nature, however, pandas allows you to explicitly define NaNs and add them to a data structure, such as a series. Within the array containing the values, you enter np.NaN wherever you want to define a missing value.

```
>>> s2 = pd.Series([5,-3,np.NaN,14])
>>> s2
0     5.0
1    -3.0
2    NaN
3    14.0
dtype: float64
```

The isnull() and notnull() functions are very useful for identifying the indexes without a value.

```
>>> s2.isnull()
0    False
1    False
2     True
3    False
dtype: bool
>>> s2.notnull()
0     True
1     True
2    False
3     True
dtype: bool
```

In fact, these functions return two series with Boolean values that contain the True and False values, depending on whether the item is a NaN value or less. The isnull() function returns True for NaN values in the series; inversely, the notnull() function returns True if they are not NaN. These functions are often placed inside filters to make a condition.

```
>>> s2[s2.notnull()]
0     5.0
1    -3.0
3    14.0
dtype: float64
>>> s2[s2.isnull()]
2    NaN
dtype: float64
```

Series as Dictionaries

An alternative way to think of a series is to think of it as an object dict (dictionary). This similarity is also exploited during the definition of an object series. In fact, you can create a series from a previously defined dict.

```
>>> mydict = {'red': 2000, 'blue': 1000, 'yellow': 500,
 'orange': 1000}
>>> myseries = pd.Series(mydict)
>>> myseries
red        2000
blue       1000
yellow      500
orange     1000
dtype: int64
```

As you can see from this example, the array of the index is filled with the keys, while the data are filled with the corresponding values. You can also define the array indexes separately. In this case, controlling correspondence between the keys of the dict and labels array of indexes will run. If there is a mismatch, pandas will add the NaN value.

```
>>> colors = ['red','yellow','orange','blue','green']
>>> myseries = pd.Series(mydict, index=colors)
>>> myseries
red        2000.0
yellow      500.0
orange     1000.0
blue       1000.0
green        NaN
dtype: float64
```

Operations Between Series

You have seen how to perform arithmetic operations between series and scalar values. The same thing is possible by performing operations between two series, but in this case even the labels come into play.

In fact, one of the great potentials of this type of data structures is that series can align data addressed differently between them by identifying their corresponding labels.

In the following example, you add two series having only some elements in common with the label.

```
>>> mydict2 = {'red':400,'yellow':1000,'black':700}
>>> myseries2 = pd.Series(mydict2)
>>> myseries + myseries2
black       NaN
blue        NaN
green       NaN
orange      NaN
red        2400.0
yellow     1500.0
dtype: float64
```

You get a new object series in which only the items with the same label are added. All other labels present in one of the two series are still added to the result but have a NaN value.

The Dataframe

The *dataframe* is a tabular data structure very similar to a spreadsheet. This data structure is designed to extend series to multiple dimensions. In fact, the dataframe consists of an ordered collection of columns (see Figure 4-7), each of which can contain a value of a different type (numeric, string, Boolean, etc.).

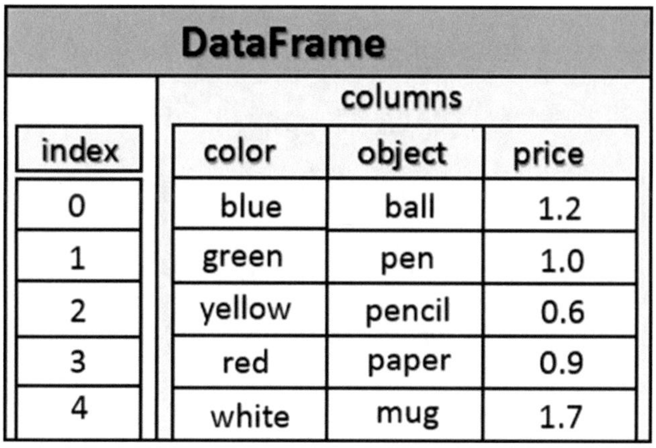

Figure 4-7. *The dataframe structure*

Unlike series, which have an index array containing labels associated with each element, the dataframe has two index arrays. The first index array, associated with the lines, has very similar functions to the index array in series. In fact, each label is associated with all the values in the row. The second array contains a series of labels, each associated with a particular column.

A dataframe may also be understood as a dict of series, where the keys are the column names and the values are the series that form the columns of the dataframe. Furthermore, all elements in each series are mapped according to an array of labels, called the *index*.

Defining a Dataframe

The most common way to create a new dataframe is to pass a dict object to the DataFrame() constructor. This dict object contains a key for each column that you want to define, with an array of values for each of them.

```
>>> data = {'color' : ['blue','green','yellow','red','white'],
                  'object' : ['ball','pen','pencil','paper','mug'],
                  'price' : [1.2,1.0,0.6,0.9,1.7]}
>>> frame = pd.DataFrame(data)
>>> frame
    color  object price
0    blue    ball   1.2
1   green     pen   1.0
```

```
2  yellow  pencil   0.6
3     red   paper   0.9
4   white     mug   1.7
```

If you are working with Jupyter and run this command, you will not get the classic output identical to the one you get with a Python console. Instead, you get a graphical representation of the dataframe, as shown in Figure 4-8.

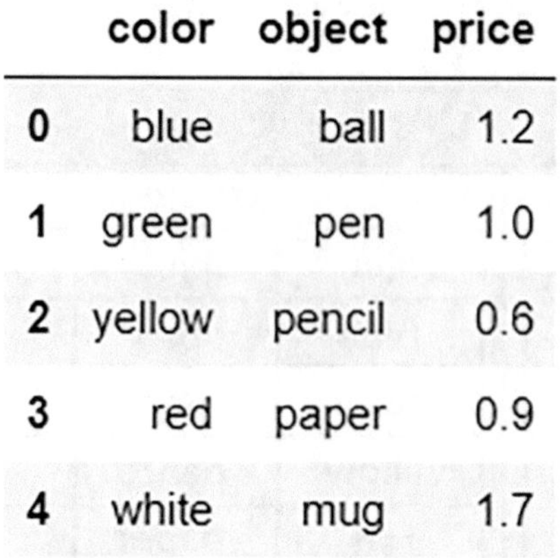

	color	object	price
0	blue	ball	1.2
1	green	pen	1.0
2	yellow	pencil	0.6
3	red	paper	0.9
4	white	mug	1.7

Figure 4-8. *Graphical representation of the dataframe as a result on a Jupyter Notebook*

If the dict object from which you want to create a dataframe contains more data than you are interested in, you can make a selection. In the constructor of the dataframe, you can specify a sequence of columns using the columns option. The columns will be created in the order of the sequence regardless of how they are contained in the dict object.

```
>>> frame2 = pd.DataFrame(data, columns=['object','price'])
>>> frame2
   object price
0    ball   1.2
1     pen   1.0
2  pencil   0.6
3   paper   0.9
4     mug   1.7
```

Even for dataframe objects, if the labels are not explicitly specified in the index array, pandas automatically assigns a numeric sequence starting from 0. Instead, if you want to assign labels to the indexes of a dataframe, you have to use the index option and assign it an array containing the labels.

```
>>> frame2 = pd.DataFrame(data, index=['one','two','three','four','five'])
>>> frame2
        color  object  price
one      blue    ball    1.2
two     green     pen    1.0
three  yellow  pencil    0.6
four      red   paper    0.9
five    white     mug    1.7
```

Now that I have introduced the two new options called index and columns, it is easy to imagine an alternative way to define a dataframe. Instead of using a dict object, you can define three arguments in the constructor, in the following order—a data matrix, an array containing the labels assigned to the index option, and an array containing the names of the columns assigned to the columns option.

In many examples, as you will see from now on in this book, to create a matrix of values quickly and easily, you can use np.arange(16).reshape((4,4)), which generates a 4x4 matrix of numbers increasing from 0 to 15.

```
>>> frame3 = pd.DataFrame(np.arange(16).reshape((4,4)),
...                    index=['red','blue','yellow','white'],
...                    columns=['ball','pen','pencil','paper'])
>>> frame3
        ball  pen  pencil  paper
red        0    1       2      3
blue       4    5       6      7
yellow     8    9      10     11
white     12   13      14     15
```

Selecting Elements

If you want to know the name of all the columns of a dataframe, you can specify the columns attribute on the instance of the dataframe object.

```
>>> frame.columns
Index(['colors', 'object', 'price'], dtype='object')
```

Similarly, to get the list of indexes, you should specify the index attribute.

```
>>> frame.index
RangeIndex(start=0, stop=5, step=1)
```

You can also get the entire set of data contained within the data structure using the values attribute.

```
>>> frame.values
array([['blue', 'ball', 1.2],
       ['green', 'pen', 1.0],
       ['yellow', 'pencil', 0.6],
       ['red', 'paper', 0.9],
       ['white', 'mug', 1.7]], dtype=object)
```

Or, if you are interested in selecting only the contents of a column, you can write the name of the column.

```
>>> frame['price']
0    1.2
1    1.0
2    0.6
3    0.9
4    1.7
Name: price, dtype: float64
```

As you can see, the return value is a series object. Another way to do this is to use the column name as an attribute of the instance of the dataframe.

```
>>> frame.price
0    1.2
1    1.0
2    0.6
3    0.9
4    1.7
Name: price, dtype: float64
```

For rows within a dataframe, it is possible to use the loc attribute with the index value of the row that you want to extract.

```
>>> frame.loc[2]
color      yellow
object     pencil
price         0.6
Name: 2, dtype: object
```

The object returned is again a series in which the names of the columns have become the label of the array index, and the values have become the data of series.

To select multiple rows, you specify an array with the sequence of rows to insert:

```
>>> frame.loc[[2,4]]
    color   object  price
2   yellow  pencil    0.6
4   white      mug    1.7
```

If you need to extract a portion of a dataframe, selecting the lines that you want to extract, you can use the reference numbers of the indexes. In fact, you can consider a row as a portion of a dataframe that has the index of the row as the source (in the next 0) value and the line above the one you want as a second value (in the next one).

```
>>> frame[0:1]
  color object  price
0  blue   ball    1.2
```

As you can see, the return value is an object dataframe containing a single row. If you want more than one line, you must extend the selection range.

```
>>> frame[1:3]
    color   object  price
1   green      pen    1.0
2   yellow  pencil    0.6
```

Finally, if what you want to achieve is a single value within a dataframe, you first use the name of the column and then the index or the label of the row.

```
>>> frame['object'][3]
'paper'
```

Assigning Values

Once you understand how to access the various elements that make up a dataframe, you follow the same logic to add or change the values in it.

For example, you have already seen that within the dataframe structure, an array of indexes is specified by the index attribute, and the row containing the name of the columns is specified with the columns attribute. Well, you can also assign a label, using the name attribute, to these two substructures to identify them.

```
>>> frame.index.name = 'id'
>>> frame.columns.name = 'item'
>>> frame
item    color  object  price
id
0        blue    ball    1.2
1       green     pen    1.0
2      yellow  pencil    0.6
3         red   paper    0.9
4       white     mug    1.7
```

One of the best features of the data structures of pandas is their high flexibility. In fact, you can always intervene at any level to change the internal data structure. For example, a very common operation is to add a new column.

You can do this by simply assigning a value to the instance of the dataframe and specifying a new column name.

```
>>> frame['new'] = 12
>>> frame
   colors  object price   new
0    blue    ball   1.2    12
1   green     pen   1.0    12
2  yellow  pencil   0.6    12
3     red   paper   0.9    12
4   white     mug   1.7    12
```

As you can see from this result, there is a new column called new with the value within 12 replicated for each of its elements.

If, however, you want to update the contents of a column, you have to use an array.

```
>>> frame['new'] = [3.0,1.3,2.2,0.8,1.1]
>>> frame
   color  object price  new
0    blue    ball   1.2  3.0
1   green     pen   1.0  1.3
2  yellow  pencil   0.6  2.2
3     red   paper   0.9  0.8
4   white     mug   1.7  1.1
```

You can follow a similar approach if you want to update an entire column, for example, by using the np. arange() function to update the values of a column with a predetermined sequence.

The columns of a dataframe can also be created by assigning a series to one of them, for example by specifying a series containing an increasing series of values through the use of np.arange().

```
>>> ser = pd.Series(np.arange(5))
>>> ser
0    0
1    1
2    2
3    3
4    4
dtype: int32
>>> frame['new'] = ser
>>> frame
    color  object  price  new
0    blue    ball    1.2    0
1   green     pen    1.0    1
2  yellow  pencil    0.6    2
3     red   paper    0.9    3
4   white     mug    1.7    4
```

Finally, to change a single value, you simply select the item and give it the new value. The operation seems very simple and intuitive. To access the element, I could think of inserting the column and then the row indexes and thus obtaining the current value.

And in fact if I write the following command:

```
>>> frame['price'][2]
0.6
```

I actually get the value of the corresponding element of the dataframe. But if I go to make an assignment on this element, in order to modify its value, I get a warning message.

```
>>> frame['price'][2] = 3.3
SettingWithCopyWarning:
A value is trying to be set on a copy of a slice from a DataFrame
```

If check inside the dataframe, however, I see that the element has changed its value.

```
>>> frame
    color  object  price  new
0    blue    ball    1.2    0
1   green     pen    1.0    1
2  yellow  pencil    3.3    2
3     red   paper    0.9    3
4   white     mug    1.7    4
```

In reality, the message warns you that this nomenclature could lead to assignment errors in the passage between internal slices that generate copies or views. In this simple case it doesn't happen, but in more complex cases where you do more complex index assignments (with index lists and conditions), it could happen. So the most correct and cleanest way to write the previous command is to define the indexes of the dataframe section to select/assign through the loc() function

```
>>> frame.loc[ 2, 'price'] = 3.3
```

Membership of a Value

You have already seen the `isin()` function applied to the series to determine the membership of a set of values. Well, this feature is also applicable to dataframe objects.

```
>>> frame.isin([1.0,'pen'])
   color object  price  new
0  False  False  False  False
1  False   True   True   True
2  False  False  False  False
3  False  False  False  False
4  False  False  False  False
```

You get a dataframe containing Boolean values, where `True` indicates values that meet the membership. If you pass the value returned as a condition, you'll get a new dataframe containing only the values that satisfy the condition.

```
>>> frame[frame.isin([1.0,'pen'])]
  color object  price  new
0  NaN    NaN    NaN  NaN
1  NaN    pen    1.0  1.0
2  NaN    NaN    NaN  NaN
3  NaN    NaN    NaN  NaN
4  NaN    NaN    NaN  NaN
```

Deleting a Column

If you want to delete an entire column and all its contents, use the `del` command.

```
>>> del frame['new']
>>> frame
   colors  object price
0    blue    ball   1.2
1   green     pen   1.0
2  yellow  pencil   3.3
3     red   paper   0.9
4   white     mug   1.7
```

Filtering

Even with a dataframe, you can apply filtering through the application of certain conditions. For example, say you want to get all elements that have a column value below a certain limit, for example, where the prices are less than 1.2. You simply need to insert this condition into the index of the dataframe.

```
>>> frame[frame['price'] < 1.2]
>>> frame
   colors  object price
1   green     pen   1.0
3     red   paper   0.9
```

You will get a dataframe containing only elements with prices less than 1.2, keeping their original position. You have thus carried out a filtering operation on the elements of the dataframe.

Dataframe from a Nested dict

A very common data structure used in Python is a nested dict, as follows:

```
nestdict = { 'red': { 2012: 22, 2013: 33 },
                 'white': { 2011: 13, 2012: 22, 2013: 16},
                 'blue': {2011: 17, 2012: 27, 2013: 18}}
```

This data structure, when it is passed directly as an argument to the DataFrame() constructor, will be interpreted by pandas to treat external keys as column names and internal keys as labels for the indexes.

During the interpretation of the nested structure, it is possible that not all fields will find a successful match. pandas compensates for this inconsistency by adding the NaN value to the missing values.

```
>>> nestdict = {'red':{2012: 22, 2013: 33},
               'white':{2011: 13, 2012: 22, 2013: 16},
               'blue': {2011: 17, 2012: 27, 2013: 18}}
>>> frame2 = pd.DataFrame(nestdict)
>>> frame2
       red  white  blue
2012  22.0     22    27
2013  33.0     16    18
2011   NaN     13    17
```

Transposition of a Dataframe

An operation that you might need when you're dealing with tabular data structures is transposition (that is, columns become rows and rows become columns). pandas allows you to do this in a very simple way. You can get the transposition of the dataframe by adding the T attribute to its application.

```
>>> frame2.T
       2012  2013  2011
red    22.0  33.0   NaN
white  22.0  16.0  13.0
blue   27.0  18.0  17.0
```

The Index Objects

Now that you know what the series and the dataframe are and how they are structured, you can likely perceive the peculiarities of these data structures. Indeed, the majority of their excellent characteristics are due to the presence of an Index object that's integrated in these data structures.

The Index objects are responsible for the labels on the axes and other metadata as the name of the axes. You have already seen how an array containing labels is converted into an Index object and that you need to specify the index option in the constructor.

```
>>> ser = pd.Series([5,0,3,8,4], index=['red','blue','yellow','white','green'])
>>> ser.index
Index(['red', 'blue', 'yellow', 'white', 'green'], dtype='object')
```

Unlike all the other elements in the pandas data structures (series and dataframes), the Index objects are immutable. Once declared, they cannot be changed. This ensures their secure sharing between the various data structures.

Each Index object has a number of methods and properties that are useful when you need to know the values they contain.

Methods on Index

There are specific methods that enable you to get information about indexes from a data structure. For example, idmin() and idmax() are two functions that return, respectively, the index with the lowest value and the index with the highest value.

```
>>> ser.idxmin()
'blue'
>>> ser.idxmax()
'white'
```

Index with Duplicate Labels

So far, you have seen all cases in which indexes within a single data structure have a unique label. Although many functions require this condition to run, this condition is not mandatory on the data structures of pandas.

This example defines, by way of an example, a series with some duplicate labels.

```
>>> serd = pd.Series(range(6), index=['white','white','blue','green','green','yellow'])
>>> serd
white     0
white     1
blue      2
green     3
green     4
yellow    5
dtype: int64
```

Regarding the selection of elements in a data structure, if there are more values with the same label, you get a series in place of a single element.

```
>>> serd['white']
white     0
white     1
dtype: int64
```

The same logic applies to the dataframe, with duplicate indexes that will return the dataframe.

With small data structures, it is easy to identify duplicate indexes, but if the structure becomes gradually larger, this starts to become difficult. In this respect, pandas provides you with the is_unique attribute belonging to the Index objects. This attribute will tell you if there are indexes with duplicate labels inside the structure data (both series and dataframe).

```
>>> serd.index.is_unique
False
>>> frame.index.is_unique
True
```

Other Functionalities on Indexes

Compared to data structures commonly used with Python, you saw that pandas, as well as taking advantage of the high-performance quality offered by NumPy arrays, has chosen to integrate indexes in them.

This choice has proven somewhat successful. In fact, despite the enormous flexibility given by the dynamic structures that already exist, using the internal reference to the structure, such as that offered by the labels, allows developers who must perform operations to carry them out in a simpler and more direct way.

This section analyzes in detail a number of basic features that take advantage of this mechanism.

- Reindexing
- Dropping
- Alignment

Reindexing

It was previously stated that once it's declared in a data structure, the Index object cannot be changed. This is true, but by executing a reindexing, you can also overcome this problem.

In fact it is possible to obtain a new data structure from an existing one where indexing rules can be defined again.

```
>>> ser = pd.Series([2,5,7,4], index=['one','two','three','four'])
>>> ser
one      2
two      5
three    7
four     4
dtype: int64
```

In order to reindex this series, pandas provides you with the reindex() function. This function creates a new series object with the values of the previous series rearranged according to the new sequence of labels.

During reindexing, it is possible to change the order of the sequence of indexes, delete some of them, and add new ones. In the case of a new label, pandas adds NaN as the corresponding value.

```
>>> ser.reindex(['three','four','five','one'])
three    7.0
four     4.0
five     NaN
one      2.0
dtype: float64
```

As you can see from the value returned, the order of the labels has been completely rearranged. The value corresponding to the label two has been dropped and a new label called five is present in the series.

However, to measure the reindexing process, defining the list of the labels can be awkward, especially with a large dataframe. You can use a method that allows you to fill in or interpolate values automatically.

To better understand this mode of automatic reindexing, define the following series.

```
>>> ser3 = pd.Series([1,5,6,3],index=[0,3,5,6])
>>> ser3
0    1
3    5
5    6
6    3
dtype: int64
```

As you can see in this example, the index column is not a perfect sequence of numbers; in fact there are some missing values (1, 2, and 4). A common need would be to perform interpolation in order to obtain the complete sequence of numbers. To achieve this, you use reindexing with the method option set to ffill. Moreover, you need to set a range of values for indexes. In this case, to specify a set of values between 0 and 5, you can use range(6) as an argument.

```
>>> ser3.reindex(range(6),method='ffill')
0    1
1    1
2    1
3    5
4    5
5    6
dtype: int64
```

As you can see from the result, the indexes that were not present in the original series were added. By interpolation, those with the lowest index in the original series have been assigned as values. In fact, the indexes 1 and 2 have the value 1, which belongs to index 0.

If you want this index value to be assigned during the interpolation, you have to use the bfill method.

```
>>> ser3.reindex(range(6),method='bfill')
0    1
1    5
2    5
3    5
4    6
5    6
dtype: int64
```

In this case, the value assigned to the indexes 1 and 2 is the value 5, which belongs to index 3.

Extending the concepts of reindexing with series to the dataframe, you can have a rearrangement not only for indexes (rows), but also with regard to the columns, or even both. As previously mentioned, adding a new column or index is possible, but since there are missing values in the original data structure, pandas adds NaN values to them.

```
>>> frame.reindex(range(5), method='ffill',columns=['colors','price','new','object'])
   colors price  new   object
0    blue   1.2  blue    ball
```

```
1   green    1.0  green   pen
2   yellow   3.3  yellow  pencil
3      red   0.9  red     paper
4   white    1.7  white   mug
```

Dropping

Another operation that is connected to Index objects is dropping. Deleting a row or a column becomes simple, due to the labels used to indicate the indexes and column names.

Also in this case, pandas provides a specific function for this operation, called drop(). This method will return a new object without the items that you want to delete.

For example, take the case where you want to remove a single item from a series. To do this, define a generic series of four elements with four distinct labels.

```
>>> ser = pd.Series(np.arange(4.), index=['red','blue','yellow','white'])
>>> ser
red       0.0
blue      1.0
yellow    2.0
white     3.0
dtype: float64
```

Now say, for example, that you want to delete the item corresponding to the label yellow. Simply specify the label as an argument of the function drop() to delete it.

```
>>> ser.drop('yellow')
red      0.0
blue     1.0
white    3.0
dtype: float64
```

To remove more items, just pass an array with the corresponding labels.

```
>>> ser.drop(['blue','white'])
red       0.0
yellow    2.0
dtype: float64
```

Regarding the dataframe instead, the values can be deleted by referring to the labels of both axes. Declare the following frame by way of example.

```
>>> frame = pd.DataFrame(np.arange(16).reshape((4,4)),
...                  index=['red','blue','yellow','white'],
...                  columns=['ball','pen','pencil','paper'])
>>> frame
        ball  pen  pencil  paper
red        0    1       2      3
blue       4    5       6      7
yellow     8    9      10     11
white     12   13      14     15
```

To delete rows, you just pass the indexes of the rows.

```
>>> frame.drop(['blue','yellow'])
       ball  pen  pencil  paper
red       0    1       2      3
white    12   13      14     15
```

To delete columns, you always need to specify the indexes of the columns, but you must specify the axis from which to delete the elements, and this can be done using the axis option. So to refer to the column names, you should specify axis = 1.

```
>>> frame.drop(['pen','pencil'],axis=1)
        ball  paper
red        0      3
blue       4      7
yellow     8     11
white     12     15
```

Arithmetic and Data Alignment

Perhaps the most powerful feature involving the indexes in a data structure is that pandas can align indexes coming from two different data structures. This is especially true when you are performing an arithmetic operation on them. In fact, during these operations, not only can the indexes between the two structures be in a different order, but they also can be present in only one of the two structures.

As you can see from the examples that follow, pandas proves to be very powerful in aligning indexes during these operations. For example, you can start considering two series in which they are defined, respectively, two arrays of labels not perfectly matching each other.

```
>>> s1 = pd.Series([3,2,5,1],['white','yellow','green','blue'])
>>> s2 = pd.Series([1,4,7,2,1],['white','yellow','black','blue','brown'])
```

Now among the various arithmetic operations, consider the simple sum. As you can see from the two series just declared, some labels are present in both, while other labels are present only in one of the two. When the labels are present in both operators, their values will be added, while in the opposite case, they will also be shown in the result (new series), but with the value NaN.

```
>>> s1 + s2
black    NaN
blue     3.0
brown    NaN
green    NaN
white    4.0
yellow   6.0
dtype: float64
```

In the case of the dataframe, although it may appear more complex, the alignment follows the same principle, but is carried out both for the rows and for the columns.

```
>>> frame1 = pd.DataFrame(np.arange(16).reshape((4,4)),
...                  index=['red','blue','yellow','white'],
...                  columns=['ball','pen','pencil','paper'])
```

```
>>> frame2 = pd.DataFrame(np.arange(12).reshape((4,3)),
...                        index=['blue','green','white','yellow'],
...                        columns=['mug','pen','ball'])
>>> frame1
        ball  pen  pencil  paper
red       0    1      2      3
blue      4    5      6      7
yellow    8    9     10     11
white    12   13     14     15
>>> frame2
        mug  pen  ball
blue      0    1    2
green     3    4    5
white     6    7    8
yellow    9   10   11
>>> frame1 + frame2
        ball   mug  paper  pen  pencil
blue    6.0   NaN    NaN  6.0    NaN
green   NaN   NaN    NaN  NaN    NaN
red     NaN   NaN    NaN  NaN    NaN
white   20.0  NaN    NaN  20.0   NaN
yellow  19.0  NaN    NaN  19.0   NaN
```

Operations Between Data Structures

Now that you are familiar with the data structures such as series and dataframe and you have seen how various elementary operations can be performed on them, it's time to go to operations involving two or more of these structures.

For example, in the previous section, you saw how the arithmetic operators apply between two of these objects. This section deepens the topic of operations that can be performed between two data structures.

Flexible Arithmetic Methods

You've just seen how to use mathematical operators directly on the pandas data structures. The same operations can also be performed using appropriate methods, called *flexible arithmetic methods*.

- add()
- sub()
- div()
- mul()

In order to call these functions, you need to use a different specification than what you're used to dealing with when using mathematical operators. For example, instead of writing a sum between two dataframes, such as frame1 + frame2, you have to use the following format:

```
>>> frame1.add(frame2)
        ball   mug  paper  pen  pencil
blue    6.0   NaN    NaN  6.0    NaN
green   NaN   NaN    NaN  NaN    NaN
```

```
red      NaN NaN   NaN  NaN   NaN
white   20.0 NaN   NaN 20.0   NaN
yellow  19.0 NaN   NaN 19.0   NaN
```

As you can see, the results are the same as what you'd get using the addition operator +. You can also note that if the indexes and column names differ greatly from one series to another, you'll find yourself with a new dataframe full of NaN values. You'll see later in this chapter how to handle this kind of data.

Operations Between Dataframes and Series

Coming back to the arithmetic operators, pandas allows you to make transactions between different structures, such as between a dataframe and a series. For example, you can define these two structures in the following way.

```
>>> frame = pd.DataFrame(np.arange(16).reshape((4,4)),
...                       index=['red','blue','yellow','white'],
...                       columns=['ball','pen','pencil','paper'])
>>> frame
        ball pen pencil paper
red        0   1      2     3
blue       4   5      6     7
yellow     8   9     10    11
white     12  13     14    15
>>> ser = pd.Series(np.arange(4), index=['ball','pen','pencil','paper'])
>>> ser
ball      0
pen       1
pencil    2
paper     3
dtype: int64
```

The two newly defined data structures have been created specifically so that the indexes of series match the names of the columns of the dataframe. This way, you can apply a direct operation.

```
>>> frame - ser
        ball pen pencil paper
red        0   0      0     0
blue       4   4      4     4
yellow     8   8      8     8
white     12  12     12    12
```

As you can see, the elements of the series are subtracted from the values of the dataframe corresponding to the same index on the column. The value is subtracted for all values of the column, regardless of their index.

If an index is not present in one of the two data structures, the result will be a new column with that index and all its elements will be NaN.

```
>>> ser['mug'] = 9
>>> ser
ball      0
pen       1
```

```
pencil   2
paper    3
mug      9
dtype: int64
>>> frame - ser
        ball   mug   paper   pen   pencil
red        0   NaN       0     0        0
blue       4   NaN       4     4        4
yellow     8   NaN       8     8        8
white     12   NaN      12    12       12
```

Function Application and Mapping

This section covers the pandas library functions.

Functions by Element

The pandas library is built on the foundations of NumPy and then extends many of its features by adapting them to new data structures as series and dataframes. Among these are the *universal functions*, called *ufunc*. This class of functions operates by element in the data structure.

```
>>> frame = pd.DataFrame(np.arange(16).reshape((4,4)),
...                     index=['red','blue','yellow','white'],
...                     columns=['ball','pen','pencil','paper'])
>>> frame
        ball   pen   pencil   paper
red        0     1        2       3
blue       4     5        6       7
yellow     8     9       10      11
white     12    13       14      15
```

For example, you can calculate the square root of each value in the dataframe using the NumPy np.sqrt().

```
>>> np.sqrt(frame)
            ball        pen     pencil      paper
red     0.000000   1.000000   1.414214   1.732051
blue    2.000000   2.236068   2.449490   2.645751
yellow  2.828427   3.000000   3.162278   3.316625
white   3.464102   3.605551   3.741657   3.872983
```

Functions by Row or Column

The application of the functions is not limited to the ufunc functions, but also includes those defined by the user. The important point is that they operate on a one-dimensional array, giving a single number as a result. For example, you can define a lambda function that calculates the range covered by the elements in an array.

```
>>> f = lambda x: x.max() - x.min()
```

It is possible to define the function this way as well:

```
>>> def f(x):
...     return x.max() - x.min()
...
```

Using the apply() function, you can apply the function just defined on the dataframe.

```
>>> frame.apply(f)
ball      12
pen       12
pencil    12
paper     12
dtype: int64
```

The result this time is one value for the column, but if you prefer to apply the function by row instead of by column, you have to set the axis option to 1.

```
>>> frame.apply(f, axis=1)
red       3
blue      3
yellow    3
white     3
dtype: int64
```

It is not mandatory that the apply() method return a scalar value. It can also return a series. A useful case is to extend the application to many functions simultaneously. In this case, you have two or more values for each feature applied. This can be done by defining a function in the following manner:

```
>>> def f(x):
...     return pd.Series([x.min(), x.max()], index=['min','max'])
...
```

Then, you apply the function as before. But in this case, as an object returned, you get a dataframe instead of a series, in which there will be as many rows as the values returned by the function.

```
>>> frame.apply(f)
      ball  pen  pencil  paper
min      0    1       2      3
max     12   13      14     15
```

Statistics Functions

Most of the statistical functions for arrays are still valid for dataframe, so using the apply() function is no longer necessary. For example, functions such as sum() and mean() can calculate the sum and the average, respectively, of the elements contained within a dataframe.

```
>>> frame.sum()
ball      24
pen       28
pencil    32
```

```
paper      36
dtype: int64
>>> frame.mean()
ball       6.0
pen        7.0
pencil     8.0
paper      9.0
dtype: float64
```

There is also a function called describe() that allows you to obtain summary statistics at once.

```
>>> frame.describe()
            ball        pen     pencil      paper
count   4.000000   4.000000   4.000000   4.000000
mean    6.000000   7.000000   8.000000   9.000000
std     5.163978   5.163978   5.163978   5.163978
min     0.000000   1.000000   2.000000   3.000000
25%     3.000000   4.000000   5.000000   6.000000
50%     6.000000   7.000000   8.000000   9.000000
75%     9.000000  10.000000  11.000000  12.000000
max    12.000000  13.000000  14.000000  15.000000
```

Sorting and Ranking

Another fundamental operation that uses indexing is sorting. Sorting the data is often a necessity and it is very important to be able to do it easily. pandas provides the sort_index() function, which returns a new object that's identical to the start, but in which the elements are ordered.

Let's start by seeing how you can sort items in a series. The operation is quite trivial since the list of indexes to be ordered is only one.

```
>>> ser = pd.Series([5,0,3,8,4],
...      index=['red','blue','yellow','white','green'])
>>> ser
red        5
blue       0
yellow     3
white      8
green      4
dtype: int64
>>> ser.sort_index()
blue       0
green      4
red        5
white      8
yellow     3
dtype: int64
```

As you can see, the items were sorted in ascending alphabetical order based on their labels (from A to Z). This is the default behavior, but you can set the opposite order by setting the ascending option to False.

```
>>> ser.sort_index(ascending=False)
yellow    3
white     8
red       5
green     4
blue      0
dtype: int64
```

With the dataframe, the sorting can be performed independently on each of its two axes. So if you want to order by row following the indexes, you just continue to use the sort_index() function without arguments as you've seen before. Or if you prefer to order by columns, you need to set the axis options to 1.

```
>>> frame = pd.DataFrame(np.arange(16).reshape((4,4)),
...                      index=['red','blue','yellow','white'],
...                      columns=['ball','pen','pencil','paper'])
>>> frame
        ball  pen  pencil  paper
red        0    1       2      3
blue       4    5       6      7
yellow     8    9      10     11
white     12   13      14     15
>>> frame.sort_index()
        ball  pen  pencil  paper
blue       4    5       6      7
red        0    1       2      3
white     12   13      14     15
yellow     8    9      10     11
>>> frame.sort_index(axis=1)
        ball  paper  pen  pencil
red        0      3    1       2
blue       4      7    5       6
yellow     8     11    9      10
white     12     15   13      14
```

So far, you have learned how to sort the values according to the indexes. But very often you may need to sort the values contained in the data structure. In this case, you have to differentiate depending on whether you have to sort the values of a series or a dataframe.

If you want to order the series, you need to use the sort_values() function.

```
>>> ser.sort_values()
blue      0
yellow    3
green     4
red       5
white     8
dtype: int64
```

If you need to order the values in a dataframe, use the sort_values() function seen previously but with the by option. Then you have to specify the name of the column on which to sort.

```
>>> frame.sort_values(by='pen')
        ball  pen  pencil  paper
red        0    1       2      3
blue       4    5       6      7
yellow     8    9      10     11
white     12   13      14     15
```

If the sorting criteria will be based on two or more columns, you can assign an array containing the names of the columns to the by option.

```
>>> frame.sort_values(by=['pen','pencil'])
        ball  pen  pencil  paper
red        0    1       2      3
blue       4    5       6      7
yellow     8    9      10     11
white     12   13      14     15
```

The ranking is an operation closely related to sorting. It mainly consists of assigning a rank (that is, a value that starts at 0 and then increases gradually) to each element of the series. The rank will be assigned starting from the lowest value to the highest.

```
>>> ser.rank()
red       4.0
blue      1.0
yellow    2.0
white     5.0
green     3.0
dtype: float64
```

The rank can also be assigned in the order in which the data are already in the data structure (without a sorting operation). In this case, you just add the method option with the first value assigned.

```
>>> ser.rank(method='first')
red       4.0
blue      1.0
yellow    2.0
white     5.0
green     3.0
dtype: float64
```

By default, even the ranking follows an ascending sort. To reverse this criteria, set the ascending option to False.

```
>>> ser.rank(ascending=False)
red       2.0
blue      5.0
yellow    4.0
```

```
white     1.0
green     3.0
dtype: float64
```

Correlation and Covariance

Two important statistical calculations are correlation and covariance, expressed in pandas by the corr() and cov() functions. These kinds of calculations normally involve two series.

```
>>> seq2 = pd.Series([3,4,3,4,5,4,3,2],['2006','2007','2008',
'2009','2010','2011','2012','2013'])
>>> seq = pd.Series([1,2,3,4,4,3,2,1],['2006','2007','2008',
'2009','2010','2011','2012','2013'])
>>> seq.corr(seq2)
0.7745966692414835
>>> seq.cov(seq2)
0.8571428571428571
```

Covariance and correlation can also be applied to a single dataframe. In this case, they return their corresponding matrices in the form of two new dataframe objects.

```
>>> frame2 = pd.DataFrame([[1,4,3,6],[4,5,6,1],[3,3,1,5],[4,1,6,4]],
...                       index=['red','blue','yellow','white'],
...                       columns=['ball','pen','pencil','paper'])
>>> frame2
        ball  pen  pencil  paper
red        1    4       3      6
blue       4    5       6      1
yellow     3    3       1      5
white      4    1       6      4
>>> frame2.corr()
            ball       pen    pencil     paper
ball    1.000000 -0.276026  0.577350 -0.763763
pen    -0.276026  1.000000 -0.079682 -0.361403
pencil  0.577350 -0.079682  1.000000 -0.692935
paper  -0.763763 -0.361403 -0.692935  1.000000
>>> frame2.cov()
            ball       pen    pencil     paper
ball    2.000000 -0.666667  2.000000 -2.333333
pen    -0.666667  2.916667 -0.333333 -1.333333
pencil  2.000000 -0.333333  6.000000 -3.666667
paper  -2.333333 -1.333333 -3.666667  4.666667
```

Using the corrwith() method, you can calculate the pairwise correlations between the columns or rows of a dataframe with a series or another DataFrame().

```
>>> ser = pd.Series([0,1,2,3,9],
...                 index=['red','blue','yellow','white','green'])
>>> ser
red        0
```

```
blue      1
yellow    2
white     3
green     9
dtype: int64
>>> frame2.corrwith(ser)
ball      0.730297
pen      -0.831522
pencil    0.210819
paper    -0.119523
dtype: float64
>>> frame2.corrwith(frame)
ball      0.730297
pen      -0.831522
pencil    0.210819
paper    -0.119523
dtype: float64
```

"Not a Number" Data

In the previous sections, you saw how easily missing data can be formed. They are recognizable in the data structures by the NaN (Not a Number) value. So, having values that are not defined in a data structure is quite common in data analysis.

However, pandas is designed to better manage this eventuality. In fact, in this section, you learn how to treat these values so that many issues can be obviated. For example, in the pandas library, calculating descriptive statistics excludes NaN values implicitly.

Assigning a NaN Value

If you need to specifically assign a NaN value to an element in a data structure, you can use the np.NaN (or np.nan) value of the NumPy library.

```
>>> ser = pd.Series([0,1,2,np.NaN,9],
...                 index=['red','blue','yellow','white','green'])
>>> ser
red       0.0
blue      1.0
yellow    2.0
white     NaN
green     9.0
dtype: float64
>>> ser['white'] = None
>>> ser
red       0.0
blue      1.0
yellow    2.0
white     NaN
green     9.0
dtype: float64
```

108

Filtering Out NaN Values

There are various ways to eliminate the NaN values during data analysis. Eliminating them by hand, element by element, can be very tedious and risky, and you're never sure that you eliminated all the NaN values. This is where the dropna() function comes to your aid.

```
>>> ser.dropna()
red      0.0
blue     1.0
yellow   2.0
green    9.0
dtype: float64
```

You can also directly perform the filtering function by placing notnull() in the selection condition.

```
>>> ser[ser.notnull()]
red      0.0
blue     1.0
yellow   2.0
green    9.0
dtype: float64
```

If you're dealing with a dataframe, it gets a little more complex. If you use the dropna() function on this type of object, and there is only one NaN value on a column or row, it will eliminate it.

```
>>> frame3 = pd.DataFrame([[6,np.nan,6],[np.nan,np.nan,np.nan],[2,np.nan,5]],
...                       index = ['blue','green','red'],
...                       columns = ['ball','mug','pen'])
>>> frame3
       ball  mug  pen
blue   6.0   NaN  6.0
green  NaN   NaN  NaN
red    2.0   NaN  5.0
>>> frame3.dropna()
Empty DataFrame
Columns: [ball, mug, pen]
Index: []
```

Therefore, to avoid having entire rows and columns disappear completely, you should specify the how option, assigning a value of all to it. This tells the dropna() function to delete only the rows or columns in which *all* elements are NaN.

```
>>> frame3.dropna(how='all')
       ball  mug  pen
blue   6.0   NaN  6.0
red    2.0   NaN  5.0
```

Filling in NaN Occurrences

Rather than filter NaN values within data structures, with the risk of discarding them along with values that could be relevant in the context of data analysis, you can replace them with other numbers. For most purposes, the fillna() function is a great choice. This method takes one argument, the value with which to replace any NaN. It can be the same for all cases.

```
>>> frame3.fillna(0)
        ball  mug  pen
blue     6.0  0.0  6.0
green    0.0  0.0  0.0
red      2.0  0.0  5.0
```

Or you can replace NaN with different values depending on the column, specifying one by one the indexes and the associated values.

```
>>> frame3.fillna({'ball':1,'mug':0,'pen':99})
        ball  mug   pen
blue     6.0  0.0   6.0
green    1.0  0.0  99.0
red      2.0  0.0   5.0
```

Hierarchical Indexing and Leveling

Hierarchical indexing is a very important feature of pandas, as it allows you to have multiple levels of indexes on a single axis. It gives you a way to work with data in multiple dimensions while continuing to work in a two-dimensional structure.

Let's start with a simple example, creating a series containing two arrays of indexes, that is, creating a structure with two levels.

```
>>> mser = pd.Series(np.random.rand(8),
...          index=[['white','white','white','blue','blue','red','red',
             'red'],
...                 ['up','down','right','up','down','up','down','left']])
>>> mser
white  up      0.461689
       down    0.643121
       right   0.956163
blue   up      0.728021
       down    0.813079
red    up      0.536433
       down    0.606161
       left    0.996686
dtype: float64
>>> mser.index
Pd.MultiIndex(levels=[['blue', 'red', 'white'], ['down',
'left', 'right', 'up']],
...          labels=[[2, 2, 2, 0, 0, 1, 1, 1],
             [3, 0, 2, 3, 0, 3, 0, 1]])
```

Through the specification of hierarchical indexing, selecting subsets of values, is in a certain way, simplified.

In fact, you can select the values for a given value of the first index, and you do it in the classic way:

```
>>> mser['white']
up       0.461689
down     0.643121
right    0.956163
dtype: float64
```

Or you can select values for a given value of the second index, in the following manner:

```
>>> mser[:,'up']
white    0.461689
blue     0.728021
red      0.536433
dtype: float64
```

Intuitively, if you want to select a specific value, you specify both indexes.

```
>>> mser['white','up']
0.46168915430531676
```

Hierarchical indexing plays a critical role in reshaping data and group-based operations such as a pivot-table. For example, the data could be rearranged and used in a dataframe with a special function called unstack(). This function converts the series with a hierarchical index to a simple dataframe, where the second set of indexes is converted into a new set of columns.

```
>>> mser.unstack()
           down      left     right        up
blue    0.813079       NaN       NaN  0.728021
red     0.606161  0.996686       NaN  0.536433
white   0.643121       NaN  0.956163  0.461689
```

If you want to perform the reverse operation, which is to convert a dataframe to a series, use the stack() function.

```
>>> frame
        ball  pen  pencil  paper
red        0    1       2      3
blue       4    5       6      7
yellow     8    9      10     11
white     12   13      14     15
>>> frame.stack()
red     ball      0
        pen       1
        pencil    2
        paper     3
blue    ball      4
        pen       5
        pencil    6
        paper     7
```

```
yellow  ball      8
        pen       9
        pencil   10
        paper    11
white   ball     12
        pen      13
        pencil   14
        paper    15
dtype: int32
```

With dataframes, it is possible to define a hierarchical index both for the rows and for the columns. At the time the dataframe is declared, you have to define an array of arrays for the index and columns options.

```
>>> mframe = pd.DataFrame(np.random.randn(16).reshape(4,4),
...       index=[['white','white','red','red'], ['up','down','up','down']],
...       columns=[['pen','pen','paper','paper'],[1,2,1,2]])
>>> mframe
                  pen                 paper
               1        2          1          2
white up    -1.964055  1.312100 -0.914750 -0.941930
      down  -1.886825  1.700858 -1.060846 -0.197669
red   up    -1.561761  1.225509 -0.244772  0.345843
      down   2.668155  0.528971 -1.633708  0.921735
```

Reordering and Sorting Levels

Occasionally, you might need to rearrange the order of the levels on an axis or sort for values at a specific level.

The swaplevel() function accepts as arguments the names assigned to the two levels that you want to interchange and returns a new object with the two levels interchanged between them, while leaving the data unmodified.

```
>>> mframe.columns.names = ['objects','id']
>>> mframe.index.names = ['colors','status']
>>> mframe
objects             pen                paper
id                1        2          1          2
colors status
white  up      -1.964055  1.312100 -0.914750 -0.941930
       down    -1.886825  1.700858 -1.060846 -0.197669
red    up      -1.561761  1.225509 -0.244772  0.345843
       down     2.668155  0.528971 -1.633708  0.921735
>>> mframe.swaplevel('colors','status')
objects             pen                paper
id                1        2          1          2
status colors
up     white   -1.964055  1.312100 -0.914750 -0.941930
down   white   -1.886825  1.700858 -1.060846 -0.197669
up     red     -1.561761  1.225509 -0.244772  0.345843
down   red      2.668155  0.528971 -1.633708  0.921735
```

Instead, the sort_index() function orders the data considering only those of a certain level by specifying it as parameter

```
>>> mframe.sort_index(level='colors')
objects              pen                  paper
id                    1       2            1        2
colors status
red     down     2.668155  0.528971 -1.633708  0.921735
        up      -1.561761  1.225509 -0.244772  0.345843
white   down    -1.886825  1.700858 -1.060846 -0.197669
        up      -1.964055  1.312100 -0.914750 -0.941930
```

Summary Statistics with groupby Instead of with Level

Many descriptive statistics and summary statistics performed on a dataframe or on a series have still a level option, with which you can determine at what level the descriptive and summary statistics should be determined.

Until now, if you wanted to create a row-level statistic, you simply had to specify the level option by passing it the name of the level.

```
>>> mframe.sum(level='colors')
objects           pen                 paper
id                 1         2         1         2
colors
red          1.106394  1.754480 -1.878480  1.267578
white       -3.850881  3.012959 -1.975596 -1.139599
```

Unfortunately, if you run this command, you get a correct result, but the operation is deprecated and signals a warning message.

Future Warning: Using the level keyword in dataframe and series aggregations is deprecated and will be removed in a future version.

If, on the other hand, you want to work in line with new and future pandas versions, you need to change your approach. Instead of applying the selection level, you group the part on which you have to apply the sum operation in the following way:

```
>>>mframe.groupby('colors').sum()
objects           pen                 paper
id                 1         2         1         2
colors
red          1.106394  1.754480 -1.878480  1.267578
white       -3.850881  3.012959 -1.975596 -1.139599
```

The result is the same but no warning messages are obtained.

You must do the same thing when you want to make a statistic at a certain level of the columns, for example id. Instead of specifying the following command, which uses the level option:

```
>>> mframe.sum(level='id', axis=1)
```

You define a group on the second axis (axis=1) and on the index id. Again, you do this instead of specifying it in the level option, which has been the practice up to now. If you run the command, you get the same result, but without warning messages.

```
>>> mframe.groupby('id', axis=1).sum()
id                      1         2
colors status
white  up       -2.878806  0.370170
       down     -2.947672  1.503189
red    up       -1.806532  1.571352
       down      1.034447  1.450706
```

Conclusions

This chapter introduced the pandas library. You learned how to install it and saw a general overview of its characteristics.

You learned about the two basic data structures, called the series and dataframes, along with their operation and their main characteristics. Especially, you discovered the importance of indexing within these structures and how best to perform operations on them. Finally, you looked at the possibility of extending the complexity of these structures by creating hierarchies of indexes, thus distributing the data contained in them into different sublevels.

In the next chapter, you learn how to capture data from external sources such as files, and inversely, how to write the analysis results on them.

CHAPTER 5

■ ■ ■

pandas: Reading and Writing Data

In the previous chapter, you became familiar with the pandas library and with the basic functionalities that it provides for data analysis. You saw that dataframes and series are the heart of this library. These are the material on which to perform all data manipulations, calculations, and analyses.

In this chapter, you'll see all of the tools provided by pandas for reading data stored in many types of media (such as files and databases). In parallel, you'll also see how to write data structures directly on these formats, without worrying too much about the technologies used.

This chapter focuses on a series of I/O API functions that pandas provides to read and write data directly as dataframe objects. It starts by looking at text files, then moves gradually to more complex binary formats.

At the end of the chapter, you'll also learn how to interface with all common databases, both SQL and NoSQL, including examples that show you how to store data in a dataframe. At the same time, you'll learn how to read data contained in a database and retrieve them as a dataframe.

I/O API Tools

pandas is a library specialized for data analysis, so you expect that it is mainly focused on calculation and data processing. The processes of writing and reading data from/to external files can be considered part of data processing. In fact, you will see how, even at this stage, you can perform some operations in order to prepare the incoming data for manipulation.

Thus, this step is very important for data analysis and therefore a specific tool for this purpose must be present in the library pandas—a set of functions called I/O API. These functions are divided into two main categories: *readers* and *writers*.

Readers	Writers
read_csv	to_csv
read_excel	to_excel
read_hdf	to_hdf
read_sql	to_sql
read_json	to_json
read_html	to_html
read_stata	to_stata

(continued)

© Fabio Nelli 2023
F. Nelli, *Python Data Analytics*, https://doi.org/10.1007/978-1-4842-9532-8_5

Readers	Writers
read_clipboard	to_clipboard
read_pickle	to_pickle
read_msgpack	to_msgpack (experimental)
read_gbq	to_gbq (experimental)

CSV and Textual Files

Everyone has become accustomed over the years to writing and reading files in text form. In particular, data are generally reported in tabular form. If the values in a row are separated by commas, you have the CSV (comma-separated values) format, which is perhaps the best-known and most popular format.

Other forms of tabular data can be separated by spaces or tabs and are typically contained in text files of various types (generally with the .txt extension).

This type of file is the most common source of data and is easier to transcribe and interpret. In this regard, pandas provides a set of functions specific for this type of file.

- read_csv
- read_table
- to_csv

Reading Data in CSV or Text Files

From experience, the most common operation of a person approaching data analysis is to read the data contained in a CSV file, or at least in a text file.

But before you start dealing with files, you need to import the following libraries.

```
>>> import numpy as np
>>> import pandas as pd
```

In order to see how pandas handles this kind of data, start by creating a small CSV file in the working directory, as shown in Listing 5-1, and save it as ch05_01.csv.

Listing 5-1. ch05_01.csv

```
white,red,blue,green,animal
1,5,2,3,cat
2,7,8,5,dog
3,3,6,7,horse
2,2,8,3,duck
4,4,2,1,mouse
```

Because this file is comma-delimited, you can use the read_csv() function to read its content and convert it to a dataframe object.

```
>>> csvframe = pd.read_csv('ch05_01.csv')
>>> csvframe
   white  red  blue  green animal
0      1    5     2      3    cat
1      2    7     8      5    dog
2      3    3     6      7  horse
3      2    2     8      3   duck
4      4    4     2      1  mouse
```

This chapter continues to use the Python shell, but in the same way you can choose to use a Jupyter Notebook, inserting the code in the cells and executing them one after the other in an interactive way. In the latter case, you get a graphical display of the dataframes obtained as a result, like the one shown in Figure 5-1.

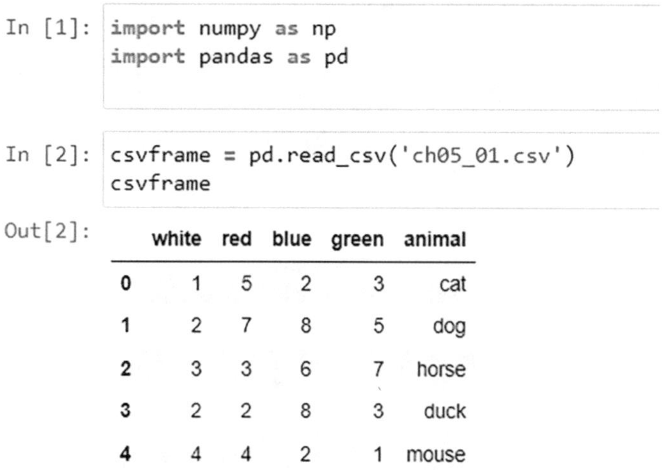

```
In [1]: import numpy as np
        import pandas as pd

In [2]: csvframe = pd.read_csv('ch05_01.csv')
        csvframe

Out[2]:
        white  red  blue  green  animal
     0      1    5     2      3     cat
     1      2    7     8      5     dog
     2      3    3     6      7   horse
     3      2    2     8      3    duck
     4      4    4     2      1   mouse
```

Figure 5-1. *Book code executed on a Jupyter Notebook*

As you can see, reading the data in a CSV file is rather trivial. CSV files are tabulated data in which the values on the same column are separated by commas. Because CSV files are considered text files, you can also use the read_table() function, but specify the delimiter.

```
>>> pd.read_table('ch05_01.csv',sep=',')
   white  red  blue  green animal
0      1    5     2      3    cat
1      2    7     8      5    dog
2      3    3     6      7  horse
3      2    2     8      3   duck
4      4    4     2      1  mouse
```

In this example, you can see that in the CSV file, headers that identify all the columns are in the first row. But this is not a general case; it often happens that the tabulated data begin directly in the first line (see Listing 5-2).

Listing 5-2. ch05_02.csv

```
1,5,2,3,cat
2,7,8,5,dog
3,3,6,7,horse
2,2,8,3,duck
4,4,2,1,mouse
>>> pd.read_csv('ch05_02.csv')
   1  5  2  3    cat
0  2  7  8  5    dog
1  3  3  6  7  horse
2  2  2  8  3   duck
3  4  4  2  1  mouse
```

In this case, you can make sure that it is pandas that assigns the default names to the columns by setting the header option to None.

```
>>> pd.read_csv('ch05_02.csv', header=None)
   0  1  2  3      4
0  1  5  2  3    cat
1  2  7  8  5    dog
2  3  3  6  7  horse
3  2  2  8  3   duck
4  4  4  2  1  mouse
```

In addition, you can specify the names directly by assigning a list of labels to the names option.

```
>>> pd.read_csv('ch05_02.csv', names=['white','red','blue','green','animal'])
   white  red  blue  green animal
0      1    5     2      3    cat
1      2    7     8      5    dog
2      3    3     6      7  horse
3      2    2     8      3   duck
4      4    4     2      1  mouse
```

In more complex cases, in which you want to create a dataframe with a hierarchical structure by reading a CSV file, you can extend the functionality of the read_csv() function by adding the index_col option, assigning all the columns to be converted into indexes.

To better understand this possibility, create a new CSV file with two columns to be used as indexes of the hierarchy. Then, save the file in the working directory as ch05_03.csv (see Listing 5-3).

Listing 5-3. ch05_03.csv

```
color,status,item1,item2,item3
black,up,3,4,6
black,down,2,6,7
white,up,5,5,5
white,down,3,3,2
white,left,1,2,1
red,up,2,2,2
red,down,1,1,4
```

Now you can read the CSV file as a dataframe by writing the following code.

```
>>> pd.read_csv('ch05_03.csv', index_col=['color','status'])
              item1  item2  item3
color status
black up         3      4      6
      down       2      6      7
white up         5      5      5
      down       3      3      2
      left       1      2      1
red   up         2      2      2
      down       1      1      4
```

Using Regexp to Parse TXT Files

In some cases, the files on which to parse the data will not show separators well defined as commas or semicolons. In these cases, the regular expressions come to your aid. In fact, you can specify a regexp within the read_table() function using the sep option.

To better understand regexp and see how you can apply it as criteria for value separation, let's start with a simple case. For example, suppose that your TXT file has values that are separated by spaces or tabs in an unpredictable order. In this case, you have to use the regexp, because that's the only way to take into account both separator types. You can do that using the wildcard /s*. /s stands for the space or tab character (if you want to indicate a tab, you use /t), while the asterisk indicates that there may be multiple characters (see Table 5-1 for other common wildcards). That is, the values may be separated by more spaces or more tabs.

Table 5-1. *Metacharacters*

.	Single character, except newline
\d	Digit
\D	Non-digit character
\s	Whitespace character
\S	Non-whitespace character
\n	New line character
\t	Tab character
\uxxxx	Unicode character specified by the hexadecimal number xxxx

Take for example an extreme case in which you have the values separated by tabs or spaces in random order (see Listing 5-4).

Listing 5-4. ch05_04.txt

```
white red blue green
    1   5   2     3
    2   7   8     5
    3   3   6     7
```

Now you can read the TXT file as a dataframe by writing the following code:

```
>>> pd.read_table('ch05_04.txt',sep='\s+', engine='python')
   white   red  blue   green
0      1     5     2       3
1      2     7     8       5
2      3     3     6       7
```

As you can see, the result is a perfect dataframe in which the values are perfectly ordered.

Now you will see an example that may seem strange or unusual, but it is not as rare as it may seem. This example can be very helpful in understanding the high potential of a regexp. In fact, you might typically think of separators as special characters like commas, spaces, tabs, and so on, but in reality you can consider separator characters like alphanumeric characters, or for example, integers such as 0.

In this example, you need to extract the numeric part from a TXT file, in which there is a sequence of characters with numerical values and the literal characters are completely fused.

Remember to set the header option to None whenever the column headings are not present in the TXT file (see Listing 5-5).

Listing 5-5. ch05_05.txt

```
000END123AAA122
001END124BBB321
002END125CCC333
```

Now you can read the TXT file as a dataframe by writing the following code:

```
>>> pd.read_table('ch05_05.txt', sep='\D+', header=None, engine='python')
   0    1    2
0  0  123  122
1  1  124  321
2  2  125  333
```

Another fairly common event is to exclude lines from parsing. In fact, you do not always want to include headers or unnecessary comments contained in a file (see Listing 5-6). With the skiprows option, you can exclude all the lines you want, just assigning an array containing the line numbers to not consider in parsing.

Pay attention when you are using this option. If you want to exclude the first five lines, you have to write skiprows = 5, but if you want to rule out the fifth line, you have to write skiprows = [5].

Listing 5-6. ch05_06.txt

```
########### LOG FILE ############
This file has been generated by automatic system
white,red,blue,green,animal
12-Feb-2015: Counting of animals inside the house
1,5,2,3,cat
2,7,8,5,dog
13-Feb-2015: Counting of animals outside the house
3,3,6,7,horse
2,2,8,3,duck
4,4,2,1,mouse
```

Now you can read this TXT file as a dataframe by writing the following code:

```
>>> pd.read_table('ch05_06.txt',sep=',',skiprows=[0,1,3,6])
   white   red  blue  green animal
0      1     5     2      3    cat
1      2     7     8      5    dog
2      3     3     6      7  horse
3      2     2     8      3   duck
4      4     4     2      1  mouse
```

Reading TXT Files Into Parts

When large files are processed, or when you are only interested in portions of these files, you often need to read the file into portions (chunks). This is both to apply any iterations and because you are not interested in parsing the entire file.

If, for example, you wanted to read only a portion of the file, you can explicitly specify the number of lines on which to parse. Thanks to the nrows and skiprows options, you can select the starting line n (n = SkipRows) and the lines to be read after it (nrows = i).

```
>>> pd.read_csv('ch05_02.csv',skiprows=[2],nrows=3,header=None)
   0  1  2  3     4
0  1  5  2  3   cat
1  2  7  8  5   dog
2  2  2  8  3  duck
```

Another interesting and fairly common operation is to split into portions that part of the text on which you want to parse. Then, for each portion a specific operation may be carried out, in order to obtain an iteration, portion by portion.

For example, you want to add the values in a column every three rows and then insert these sums in a series. This example is trivial and impractical but is very simple to understand. Once you have learned the underlying mechanism, you will be able to apply it in more complex cases.

```
>>> out = pd.Series(dtype='float64')
>>> i = 0
>>> pieces = pd.read_csv('ch05_01.csv',chunksize=3)
>>> for piece in pieces:
...     out.at[i] = piece['white'].sum()...    i = i + 1
...
>>> out
0    6
1    6
dtype: int64
```

Writing Data in CSV

In addition to reading the data contained in a file, it's also common to write a data file produced by a calculation, or in general the data contained in a data structure.

For example, you might want to write the data contained in a dataframe to a CSV file. To do this writing process, you use the to_csv() function, which accepts as an argument the name of the file you generate (see Listing 5-7).

```
>>> frame = pd.DataFrame(np.arange(16).reshape((4,4)),
              index = ['red', 'blue', 'yellow', 'white'],
              columns = ['ball', 'pen', 'pencil', 'paper'])
>>> frame.to_csv('ch05_07.csv')
```

If you open the new file called ch05_07.csv generated by the pandas library, you will see data, as in Listing 5-7.

Listing 5-7. ch05_07.csv

```
,ball,pen,pencil,paper
0,1,2,3
4,5,6,7
8,9,10,11
12,13,14,15
```

As you can see from the previous example, when you write a dataframe to a file, indexes and columns are marked on the file by default. This default behavior can be changed by setting the two options index and header to False (see Listing 5-8).

```
>>> frame.to_csv('ch05_07b.csv', index=False, header=False)
```

Listing 5-8. ch05_07b.csv

```
1,2,3
5,6,7
9,10,11
13,14,15
```

One point to remember when writing files is that NaN values present in a data structure are shown as empty fields in the file (see Listing 5-9).

```
>>> frame2 = pd.DataFrame([[6,np.nan,np.nan,6,np.nan],
...             [np.nan,np.nan,np.nan,np.nan,np.nan],
...             [np.nan,np.nan,np.nan,np.nan,np.nan],
...             [20,np.nan,np.nan,20.0,np.nan],
...             [19,np.nan,np.nan,19.0,np.nan]
...             ],
...                 index=['blue','green','red','white','yellow'],
...                 columns=['ball','mug','paper','pen','pencil'])
>>> frame2
        ball  mug  paper  pen  pencil
blue     6.0  NaN    NaN  6.0     NaN
green    NaN  NaN    NaN  NaN     NaN
red      NaN  NaN    NaN  NaN     NaN
white   20.0  NaN    NaN 20.0     NaN
yellow  19.0  NaN    NaN 19.0     NaN
>>> frame2.to_csv('ch05_08.csv')
```

Listing 5-9. ch05_08.csv

```
,ball,mug,paper,pen,pencil
blue,6.0,,,6.0,
green,,,,,
red,,,,,
white,20.0,,,20.0,
yellow,19.0,,,19.0,
```

However, you can replace this empty field with a value to your liking using the na_rep option in the to_csv() function. Common values include NULL, 0, or the same NaN (see Listing 5-10).

```
>>> frame3.to_csv('ch05_09.csv', na_rep ='NaN')
```

Listing 5-10. ch05_09.csv

```
,ball,mug,paper,pen,pencil
blue,6.0,NaN,NaN,6.0,NaN
green,NaN,NaN,NaN,NaN,NaN
red,NaN,NaN,NaN,NaN,NaN
white,20.0,NaN,NaN,20.0,NaN
yellow,19.0,NaN,NaN,19.0,NaN
```

■ **Note** In the cases specified, dataframe has always been the subject of discussion, because these are the data structures that are written to the file. But all these functions and options are also valid with regard to the series.

Reading and Writing HTML Files

pandas provides the corresponding pair of I/O API functions for the HTML format.

- read_html()
- to_html()

These two functions can be very useful. You will appreciate the ability to convert complex data structures such as dataframes directly into HTML tables without having to hack a long listing in HTML, especially if you're dealing with the web.

The inverse operation can be very useful too, because now the major source of data is just the web world. In fact, a lot of data on the Internet does not always have a "ready to use" form that is packaged in a TXT or CSV file. Very often, however, the data are reported as part of the text of web pages. So having a function for reading can prove to be really useful.

This activity is so widespread that it is currently identified as *web scraping*. This process is becoming a fundamental part of the set of processes and will be integrated in the first part of data analysis: data mining and data preparation.

■ **Note**   Many websites have now adopted the HTML5 format, to avoid any issues of missing modules and error messages. I strongly recommend you install the `html5lib` module. Anaconda specified:

```
conda install html5lib
```

Writing Data in HTML

Now you learn how to convert a dataframe into an HTML table. The internal structure of the dataframe is automatically converted into nested tags <TH>, <TR>, and <TD>, retaining any internal hierarchies. You do not need to know HTML to use this kind of function.

Because the data structures of the dataframe can be quite complex and large, it's great to have a function like this when you need to develop web pages.

To better understand this potential, here's an example. You can start by defining a simple dataframe. Thanks to the `to_html()` function, you can directly convert the dataframe into an HTML table.

```
>>> frame = pd.DataFrame(np.arange(4).reshape(2,2))
```

Because the I/O API functions are defined in the pandas data structures, you can call the `to_html()` function directly on the instance of the dataframe.

```
>>> print(frame.to_html())
<table border="1" class="dataframe">
  <thead>
    <tr style="text-align: right;">
      <th></th>
      <th>0</th>
      <th>1</th>
    </tr>
  </thead>
  <tbody>
    <tr>
      <th>0</th>
      <td> 0</td>
      <td> 1</td>
    </tr>
    <tr>
      <th>1</th>
      <td> 2</td>
      <td> 3</td>
    </tr>
  </tbody>
</table>
```

As you can see, the whole structure formed by the HTML tags needed to create an HTML table was generated correctly in order to respect the internal structure of the dataframe.

In the next example, you see how the table appears automatically generated within an HTML file. In this regard, the dataframe is a bit more complex than the previous one, where there are labels of the indexes and column names.

```
>>> frame = pd.DataFrame( np.random.random((4,4)),
...                      index = ['white','black','red','blue'],
...                      columns = ['up','down','right','left'])
>>> frame
            up        down      right      left
white  0.292434  0.457176  0.905139  0.737622
black  0.794233  0.949371  0.540191  0.367835
red    0.204529  0.981573  0.118329  0.761552
blue   0.628790  0.585922  0.039153  0.461598
```

Now you focus on writing an HTML page through the generation of a string. This is a simple and trivial example, but it is very useful to understand and to test the functionality of pandas directly on the web browser.

First of all, you create a string that contains the code of the HTML page.

```
>>> s = ['<HTML>']
>>> s.append('<HEAD><TITLE>My DataFrame</TITLE></HEAD>')
>>> s.append('<BODY>')
>>> s.append(frame.to_html())
>>> s.append('</BODY></HTML>')
>>> html = ''.join(s)
```

Now that all the listing of the HTML page is contained within the html variable, you can write directly on the file that will be called myFrame.html:

```
>>> html_file = open('myFrame.html','w')
>>> html_file.write(html)
>>> html_file.close()
```

A new HTML file will be in your working directory, called myFrame.html. Double-click it to open it directly from the browser. An HTML table will appear in the upper left, as shown in Figure 5-2.

	up	down	right	left
white	0.292434	0.457176	0.905139	0.737622
black	0.794233	0.949371	0.540191	0.367835
red	0.204529	0.981573	0.118329	0.761552
blue	0.628790	0.585922	0.039153	0.461598

Figure 5-2. The dataframe is shown as an HTML table in the web page

125

Reading Data from an HTML File

As you just saw, pandas can easily generate HTML tables starting from the dataframe. The opposite process is also possible; the function read_html() will parse an HTML page, looking for an HTML table. If found, it will convert that table into an object dataframe ready to be used in your data analysis.

More precisely, the read_html() function returns a list of dataframes even if there is only one table. The source that will be parsed can be different types. For example, you may have to read an HTML file in any directory. You can parse the HTML file you created in the previous example:

```
>>> web_frames = pd.read_html('myFrame.html')
>>> web_frames[0]
  Unnamed: 0        up      down     right      left
0      white  0.292434  0.457176  0.905139  0.737622
1      black  0.794233  0.949371  0.540191  0.367835
2        red  0.204529  0.981573  0.118329  0.761552
3       blue  0.628790  0.585922  0.039153  0.461598
```

As you can see, all of the tags that have nothing to do with HTML table are not considered. Furthermore, web_frames is a list of dataframes, although in your case, the dataframe that you are extracting is only one. However, you can select the item in the list that you want to use, calling it in the classic way. In this case, the item is unique and therefore the index will be 0.

However, the mode most commonly used regarding the read_html() function is that of a direct parsing of an URL on the web. In this way, the web pages in the network are directly parsed with the extraction of the tables in them.

For example, you can call a web page where there is an HTML table that shows a ranking list with some names and scores.

```
>>> ranking = pd.read_html('https://www.meccanismocomplesso.org/en/meccanismo-complesso-
sito-2/classifica-punteggio/')
>>> ranking[0]
     Unnamed: 0       Member  Points  Levels
0             1    BrunoOrsini    2075     NaN
1             2      Berserker     700     NaN
2             3    albertosallu    275     NaN
3             4            Jon     180     NaN
4             5           Mr.Y     180     NaN
..          ...            ...     ...     ...
110         111   Gigi Bertana       5     NaN
111         112        p.barut       5     NaN
112         113    Indri4Africa       5     NaN
113         114       ghirograf       5     NaN
114         115    Marco Corbet       5     NaN

[115 rows x 4 columns]
```

The same operation can be run on any web page that has one or more tables.

Reading Data from XML

In the list of I/O API functions, there is no specific tool regarding the XML (Extensible Markup Language) format. In fact, although it is not listed, this format is very important, because many structured data are available in XML format. This presents no problem, since Python has many other libraries (besides pandas) that manage the reading and writing of data in XML format.

One of these libraries is the lxml library, which stands out for its excellent performance during the parsing of very large files. In this section, you learn how to use this module to parse XML files and how to integrate it with pandas to get the dataframe containing the requested data. For more information about this library, I highly recommend visiting the official website of lxml at http://lxml.de/index.html.

Take for example the XML file shown in Listing 5-11. Write it down and save it with the name books.xml directly in your working directory.

Listing 5-11. books.xml

```
<?xml version="1.0"?>
<Catalog>
    <Book id="ISBN9872122367564">
        <Author>Ross, Mark</Author>
        <Title>XML Cookbook</Title>
        <Genre>Computer</Genre>
        <Price>23.56</Price>
        <PublishDate>2014-22-01</PublishDate>
    </Book>
    <Book id="ISBN9872122367564">
        <Author>Bracket, Barbara</Author>
        <Title>XML for Dummies</Title>
        <Genre>Computer</Genre>
        <Price>35.95</Price>
        <PublishDate>2014-12-16</PublishDate>
    </Book>
</Catalog>
```

In this example, you take the data structure described in the XML file and convert it directly into a dataframe. The first thing to do is use the submodule objectify of the lxml library, importing it in the following way.

```
>>> from lxml import objectify
```

Now you can parse the XML file with just the parse() function.

```
>>> xml = objectify.parse('books.xml')
>>> xml
<lxml.etree._ElementTree object at 0x0000000009734E08>
```

You get an object tree, which is an internal data structure of the lxml module.

Look in more detail at this type of object. To navigate in this tree structure, so as to select element by element, you must first define the root. You can do this with the getroot() function.

```
>>> root = xml.getroot()
```

Now that the root of the structure has been defined, you can access the various nodes of the tree, each corresponding to the tag contained in the original XML file. The items will have the same name as the corresponding tags. To select them, simply write the various separate tags with points, reflecting in a certain way the hierarchy of nodes in the tree.

```
>>> root.Book.Author
'Ross, Mark'
>>> root.Book.PublishDate
'2014-22-01'
```

In this way, you access nodes individually, but you can access various elements at the same time using getchildren(). With this function, you'll get all the child nodes of the reference element.

```
>>> root.getchildren()
[<Element Book at 0x9c66688>, <Element Book at 0x9c66e08>]
```

With the tag attribute, you get the name of the tag corresponding to the child node.

```
>>> [child.tag for child in root.Book.getchildren()]
['Author', 'Title', 'Genre', 'Price', 'PublishDate']
```

While with the text attribute, you get the value contained between the corresponding tags.

```
>>> [child.text for child in root.Book.getchildren()]
['Ross, Mark', 'XML Cookbook', 'Computer', '23.56', '2014-22-01']
```

However, regardless of the ability to move through the lxml.etree tree structure, what you need is to convert it into a dataframe. Define the following function, which has the task of analyzing the contents of an etree to fill a dataframe line by line.

```
>>> def etree2df(root):
...     column_names = []
...     for i in range(0,len(root.getchildren()[0].getchildren())):
...         column_names.append(root.getchildren()[0].getchildren()[i].tag)
...     xml:frame = pd.DataFrame(columns=column_names)
...     for j in range(0, len(root.getchildren())):
...         obj = root.getchildren()[j].getchildren()
...         texts = []
...         for k in range(0, len(column_names)):
...             texts.append(obj[k].text)
...         row = dict(zip(column_names, texts))
...         row_s = pd.Series(row)
...         row_s.name = j
...         xml_frame = pd.concat([xml_frame, row_s.to_frame().T], ignore_index=True)
...     return xml:frame
...
>>> etree2df(root)
            Author           Title      Genre  Price PublishDate
0        Ross, Mark     XML Cookbook  Computer  23.56  2014-22-01
1  Bracket, Barbara  XML for Dummies  Computer  35.95  2014-12-16
```

Reading and Writing Data on Microsoft Excel Files

In the previous section, you saw how data can be easily read from CSV files. It is not uncommon, however, that data are collected in tabular form in an Excel spreadsheet.

pandas provides specific functions for this type of format:

- `to_excel()`

- `read_excel()`

The `read_excel()` function can read Excel 2003 (`.xls`) files and Excel 2007 (`.xlsx`) files. This is possible thanks to the integration of the internal module, `xlrd`.

However, to carry out these readings and format conversions from Excel files, Python uses a special library called `openpyxl`. To continue with the examples, you first have to install it in your virtual environment. If you are using the Anaconda platform, you can open the CMD.exe Prompt console and use this command:

```
conda install openpyxl
```

If you prefer, you can install it graphically by selecting this package in Anaconda Navigator and requesting its installation. See Figure 5-3.

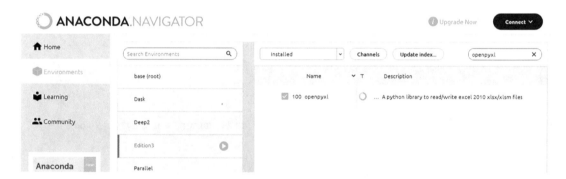

Figure 5-3. *Anaconda Navigator offers a graphical way to install packages like openpyxl*

If you're not using the Anaconda platform, the easiest way to install `openpyxl` is via PyPI.

```
pip install openpyxl
```

Once the package is installed in the proper virtual environment, open an Excel file and enter the data as shown in Figure 5-4. Copy the data in sheet1 and sheet2. Then save the file as `ch05_data.xlsx`.

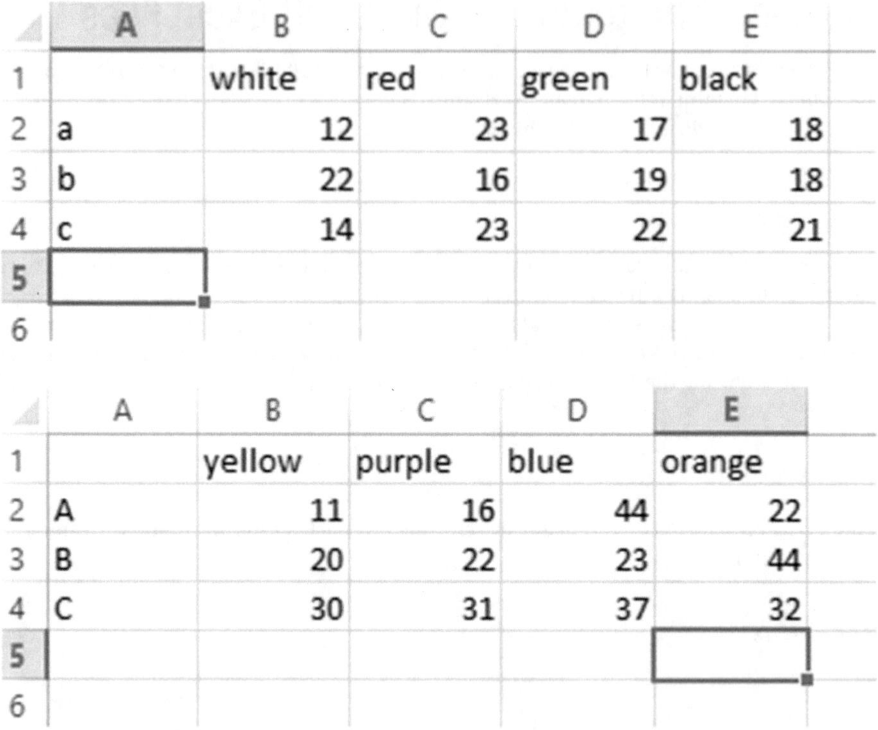

	A	B	C	D	E
1		white	red	green	black
2	a	12	23	17	18
3	b	22	16	19	18
4	c	14	23	22	21
5					
6					

	A	B	C	D	E
1		yellow	purple	blue	orange
2	A	11	16	44	22
3	B	20	22	23	44
4	C	30	31	37	32
5					
6					

Figure 5-4. *The two datasets in sheet1 and sheet2 of an Excel file*

To read the data contained in the XLS file and convert it into a dataframe, you only have to use the `read_excel()` function.

```
>>> pd.read_excel('ch05_data.xlsx')
   white  red  green  black
a     12   23     17     18
b     22   16     19     18
c     14   23     22     21
```

As you can see, by default, the returned dataframe is composed of the data tabulated in the first spreadsheets. If, however, you need to load the data in the second spreadsheet, you must then specify the name of the sheet or the number of the sheet (index) as the second argument.

```
>>> pd.read_excel('ch05_data.xlsx','Sheet2')
   yellow  purple  blue  orange
A      11      16    44      22
B      20      22    23      44
C      30      31    37      32
>>> pd.read_excel('ch05_data.xlsx',1)
   yellow  purple  blue  orange
A      11      16    44      22
B      20      22    23      44
C      30      31    37      32
```

The same applies to writing. To convert a dataframe into an Excel spreadsheet, you have to write the following:

```
>>> frame = pd.DataFrame(np.random.random((4,4)),
...                      index = ['exp1','exp2','exp3','exp4'],
...                      columns = ['Jan2015','Fab2015','Mar2015','Apr2005'])
>>> frame
        Jan2015    Fab2015    Mar2015    Apr2005
exp1   0.030083   0.065339   0.960494   0.510847
exp2   0.531885   0.706945   0.964943   0.085642
exp3   0.981325   0.868894   0.947871   0.387600
exp4   0.832527   0.357885   0.538138   0.357990
>>> frame.to_excel('ch05_data2.xlsx')
```

In the working directory, you will find a new Excel file containing the data, as shown in Figure 5-5.

	A	B	C	D	E
1		**Jan2015**	**Fab2015**	**Mar2015**	**Apr2005**
2	**exp1**	0,030083	0,065339	0,960494	0,510847
3	**exp2**	0,531885	0,706945	0,964943	0,085642
4	**exp3**	0,981325	0,868894	0,947871	0,3876
5	**exp4**	0,832527	0,357885	0,538138	0,35799
6					

Figure 5-5. The dataframe in the Excel file

JSON Data

JSON (JavaScript Object Notation) has become one of the most common standard formats, especially for the transmission of data on the web. So it is normal to work with this data format if you want to use data on the web.

The special feature of this format is its great flexibility, although its structure is far from being the one to which you are well accustomed, that is, tabular structure.

In the first part of this section, you learn how to use the read_json() and to_json() functions to stay within the I/O API functions discussed in this chapter. In the second part of this section, you see another example in which you have to deal with structured data in JSON format, which is much more related to real cases.

In my opinion, a useful online application for checking the JSON format is JSON Viewer, available at http://jsonviewer.stack.hu/. This web application, once you enter or copy data in JSON format, allows you to see if the format you entered is valid. Moreover, it displays the tree structure so that you can better understand its structure (see Figure 5-6).

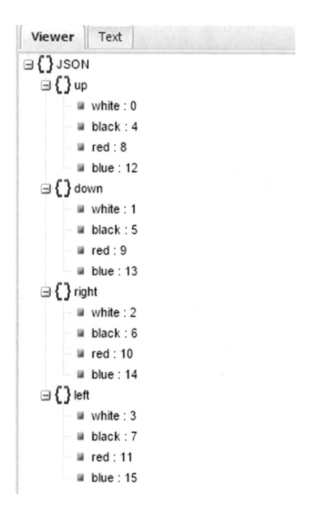

Figure 5-6. *JSON Viewer*

Let's begin with the more useful case, that is, when you have a dataframe and you need to convert it into a JSON file. So, define a dataframe and then call the to_json() function on it, passing as an argument the name of the file that you want to create.

```
>>> frame = pd.DataFrame(np.arange(16).reshape(4,4),
...                      index=['white','black','red','blue'],
...                      columns=['up','down','right','left'])
>>> frame.to_json('frame.json')
```

In the working directory, you will find a new JSON file (see Listing 5-12) containing the dataframe data translated into JSON format.

Listing 5-12. frame.json

```
{"up":{"white":0,"black":4,"red":8,"blue":12},"down":{"white":1,"black":5,"red":9,"blue":
13},"right":{"white":2,"black":6,"red":10,"blue":14},"left":{"white":3,"black":7,"red":11,
"blue":15}}
```

The converse is possible, using the read_json() with the name of the file passed as an argument.

```
>>> pd.read_json('frame.json')
       up  down  right  left
white   0     1      2     3
black   4     5      6     7
red     8     9     10    11
blue   12    13     14    15
```

The example you have seen is a fairly simple case in which the JSON data were in tabular form (since the frame.json file comes from a dataframe). Generally, however, the JSON files do not have a tabular structure. Thus, you need to somehow convert the structure dict file into tabular form. This process is called *normalization*.

The library pandas provides a function, called json_normalize(), that converts a dict or a list in a table. First you have to import the function:

```
>>> from pandas.io.json import json_normalize
```

Then you write a JSON file as described in Listing 5-13 with any text editor. Save it in the working directory as books.json.

Listing 5-13. books.json

```
[{"writer": "Mark Ross",
 "nationality": "USA",
 "books": [
         {"title": "XML Cookbook", "price": 23.56},
         {"title": "Python Fundamentals", "price": 50.70},
         {"title": "The NumPy library", "price": 12.30}
            ]
},
{"writer": "Barbara Bracket",
 "nationality": "UK",
 "books": [
         {"title": "Java Enterprise", "price": 28.60},
         {"title": "HTML5", "price": 31.35},
         {"title": "Python for Dummies", "price": 28.00}
            ]
}]
```

As you can see, the file structure is no longer tabular, but more complex. Then the approach with the read_json() function is no longer valid. As you learn from this example, you can still get the data in tabular form from this structure. First you have to load the contents of the JSON file and convert it into a string.

```
>>> import pandas.io.json as json
>>> file = open('books.json','r')
>>> text = file.read()
>>> text = json.loads(text)
```

Now you are ready to apply the `json_normalize()` function. From a quick look at the contents of the data within the JSON file, for example, you might want to extract a table that contains all the books. Then write the books key as the second argument.

```
>>> pd.json_normalize(text,'books')
                title  price
0         XML Cookbook  23.56
1  Python Fundamentals  50.70
2    The NumPy library  12.30
3      Java Enterprise  28.60
4                HTML5  31.35
5   Python for Dummies  28.00
```

The function reads the contents of all the elements that have books as the key. All properties will be converted into nested column names while the corresponding values will fill the dataframe. For the indexes, the function assigns a sequence of increasing numbers.

However, you get a dataframe containing only some internal information. It would be useful to add the values of other keys on the same level. In this case, you can add other columns by inserting a key list as the third argument of the function.

```
>>> pd.json_normalize(text,'books',['writer','nationality'])
                title  price           writer nationality
0         XML Cookbook  23.56        Mark Ross         USA
1  Python Fundamentals  50.70        Mark Ross         USA
2    The NumPy library  12.30        Mark Ross         USA
3      Java Enterprise  28.60  Barbara Bracket          UK
4                HTML5  31.35  Barbara Bracket          UK
5   Python for Dummies  28.00  Barbara Bracket          UK
```

Now as a result you get a dataframe from a starting tree structure. In Figure 5-7, you can see the dataframe as it is displayed on Jupyter Notebook.

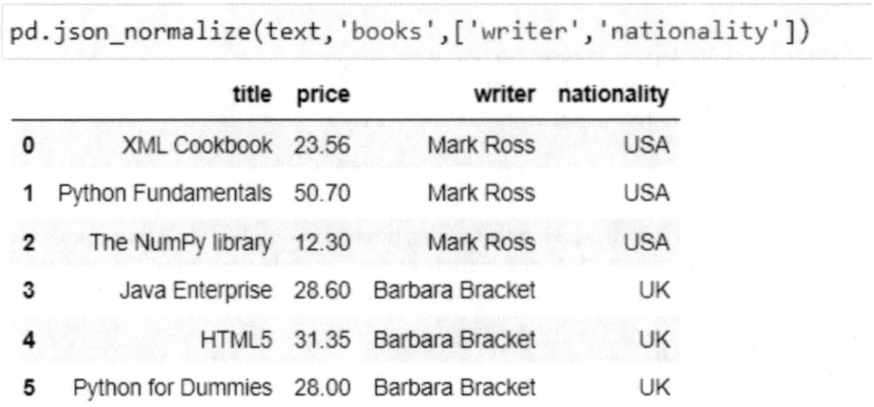

```
pd.json_normalize(text,'books',['writer','nationality'])
```

	title	price	writer	nationality
0	XML Cookbook	23.56	Mark Ross	USA
1	Python Fundamentals	50.70	Mark Ross	USA
2	The NumPy library	12.30	Mark Ross	USA
3	Java Enterprise	28.60	Barbara Bracket	UK
4	HTML5	31.35	Barbara Bracket	UK
5	Python for Dummies	28.00	Barbara Bracket	UK

Figure 5-7. *The dataframe obtained from a starting tree structure in Jupyter Notebook*

The HDF5 Format

So far you have seen how to write and read data in text format. When your data analysis involves large amounts of data, it is preferable to use them in binary format. There are several tools in Python to handle binary data. A library that is having some success in this area is the HDF5 library.

The HDF term stands for *hierarchical data format*, and in fact this library is concerned with reading and writing HDF5 files containing a structure with nodes and the possibility to store multiple datasets.

This library, fully developed in C, however, has also interfaces with other types of languages like Python, MATLAB, and Java. It is very efficient, especially when using this format to save huge amounts of data. Compared to other formats that work more simply in binary, HDF5 supports compression in real time, thereby taking advantage of repetitive patterns in the data structure to compress the file size.

At present, the possible choices in Python are PyTables and h5py. These two forms differ in several respects and therefore their choice depends very much on the needs of those who use them.

h5py provides a direct interface with the high-level APIs HDF5, while PyTables makes abstract many of the details of HDF5 to provide more flexible data containers, indexed tables, querying capabilities, and other media on the calculations.

In this case, you see the PyTables library in action, which must be installed in your virtual environment. If you work on Anaconda, you can easily install it with Anaconda Navigator or from the console with this command:

```
conda install pytables
```

Otherwise you can install it via PyPI.

```
pip install pytables
```

pandas has a class-like dict called HDFStore, using PyTables to store pandas objects. So before working with the format HDF5, you must import the HDFStore class:

```
>>> from pandas.io.pytables import HDFStore
```

Now you're ready to store the data of a dataframe within an .h5 file. First, create a dataframe.

```
>>> frame = pd.DataFrame(np.arange(16).reshape(4,4),
...                      index=['white','black','red','blue'],
...                      columns=['up','down','right','left'])
```

Now create a HDF5 file and call it mydata.h5, then enter the data inside of the dataframe.

```
>>> store = HDFStore('mydata.h5')
>>> store['obj1'] = frame
```

From here, you can guess how to store multiple data structures in the same HDF5 file, specifying for each of them a label.

```
>>> frame
       up  down  right  left
white   0     1      2     3
black   4     5      6     7
red     8     9     10    11
blue   12    13     14    15
>>> store['obj2'] = frame
```

So with this type of format, you can store multiple data structures in a single file, represented by the `store` variable.

```
>>> store
<class 'pandas.io.pytables.HDFStore'>
File path: mydata.h5
```

Even the reverse process is very simple. Taking into account having an HDF5 file containing various data structures, objects inside can be called in the following way:

```
>>> store['obj2']
        up  down  right  left
white    0     1      2     3
black    4     5      6     7
red      8     9     10    11
blue    12    13     14    15
```

Pickle—Python Object Serialization

The `pickle` module implements a powerful algorithm for serialization and deserialization of a data structure implemented in Python. *Pickling* is the process in which the hierarchy of an object is converted into a stream of bytes.

This allows an object to be transmitted and stored, and then to be rebuilt by the receiver itself retaining all the original features.

In Python, the picking operation is carried out by the `pickle` module, but currently there is a module called `_pickle`, which is the result of an enormous amount of work optimizing the `pickle` module (written in C). This module can be in fact 1,000 times faster than the `pickle` module. However, regardless of which module you use, the interfaces of the two modules are almost the same.

Before moving to explicitly mention the I/O functions of pandas that operate on this format, let's look in more detail at the `_pickle` module and see how to use it.

Serialize a Python Object with cPickle

The data format used by the `pickle` (or `cPickle`) module is specific to Python. By default, an ASCII representation is used to represent it, in order to be readable from the human point of view. Then, by opening a file with a text editor, you may be able to understand its contents. To use this module, you must first import it:

```
>>> import _pickle as pickle
```

Then create an object sufficiently complex to have an internal data structure, for example a `dict` object.

```
>>> data = { 'color': ['white','red'], 'value': [5, 7]}
```

Now perform a serialization of the data object through the `dumps()` function of the `cPickle` module.

```
>>> pickled_data = pickle.dumps(data)
```

Now, to see how it serialized the `dict` object, you need to look at the contents of the `pickled_data` variable.

```
>>> print(pickled_data)
b'\x80\x04\x95/\x00\x00\x00\x00\x00\x00\x00}\x94(\x8c\x05color\x94]\x94(\x8c\x05white\x94\
x8c\x03red\x94e\x8c\x05value\x94]\x94(K\x05K\x07eu.'
```

Once you have serialized data, they can easily be written on a file or sent over a socket, pipe, and so on.

After the data are transmitted, it is possible to reconstruct the serialized object (deserialization) with the `loads()` function of the _pickle module.

```
>>> nframe = pickle.loads(pickled_data)
>>> nframe
{'color': ['white', 'red'], 'value': [5, 7]}
```

Pickling with pandas

When it comes to pickling (and unpickling) with the pandas library, everything is much easier. There is no need to import the `cPickle` module in the Python session; the whole operation is performed implicitly.

Also, the serialization format used by pandas is not completely in ASCII.

```
>>> frame = pd.DataFrame(np.arange(16).reshape(4,4), index = ['up','down','left','right'])
>>> frame.to_pickle('frame.pkl')
```

There is a new file called `frame.pkl` in your working directory that contains all the information about the `frame` dataframe.

To open a PKL file and read the contents, simply use this command:

```
>>> pd.read_pickle('frame.pkl')
          0    1    2    3
up        0    1    2    3
down      4    5    6    7
left      8    9   10   11
right    12   13   14   15
```

As you can see, all the implications on the operation of pickling and unpickling are completely hidden from the pandas user, making the job as easy and understandable as possible, for those who must deal specifically with data analysis.

■ **Note** When you use this format, make sure that the file you open is safe. Indeed, the pickle format was not designed to be protected against erroneous and maliciously constructed data.

Interacting with Databases

In many applications, the data rarely come from text files, given that this is certainly not the most efficient way to store data.

The data are often stored in an SQL-based relational database, and also in many alternative NoSQL databases that have become very popular in recent times.

Loading data from SQL in a dataframe is sufficiently simple and pandas has some functions to simplify the process.

The `pandas.io.sql` module provides a unified interface independent of the DB, called `sqlalchemy`. This interface simplifies the connection mode, since regardless of the DB, the commands will always be the same. To make a connection, you use the `create_engine()` function. With this feature you can configure all the properties necessary to use the driver, as a user, password, port, and database instance.

In order to install `sqlalchemy` on Anaconda, you can easily install it with Anaconda Navigator or from the console with the command:

```
conda install sqlalchemy
```

Otherwise you can install it via PyPI.

```
pip install sqlalchemy
```

Here is a list of examples for the various types of databases:

```
>>> from sqlalchemy import create_engine
```

For PostgreSQL:

```
>>> engine = create_engine('postgresql://scott:tiger@localhost:5432/mydatabase')
For MySQL
>>> engine = create_engine('mysql+mysqldb://scott:tiger@localhost/foo')
For Oracle
>>> engine = create_engine('oracle://scott:tiger@127.0.0.1:1521/sidname')
For MSSQL
>>> engine = create_engine('mssql+pyodbc://mydsn')
For SQLite
>>> engine = create_engine('sqlite:///foo.db')
```

Loading and Writing Data with SQLite3

As a first example, you use an SQLite database using the driver's built-in Python `sqlite3`. SQLite3 is a tool that implements a DBMS SQL in a very simple and lightweight way, so it can be incorporated into any application implemented with the Python language. In fact, this practical software allows you to create an embedded database in a single file.

This makes it the perfect tool for anyone who wants to have the functions of a database without having to install a real database. SQLite3 can be the right choice for anyone who wants to practice before going on to a real database, or for anyone who needs to use the functions of a database to collect data, but remaining within a single program, without having to interface with a database.

Create a dataframe that you will use to create a new table on the SQLite3 database.

```
>>> frame = pd.DataFrame( np.arange(20).reshape(4,5),
...                       columns=['white','red','blue','black','green'])
>>> frame
   white  red  blue  black  green
0      0    1     2      3      4
1      5    6     7      8      9
```

```
2     10    11    12    13    14
3     15    16    17    18    19
```

Now it's time to implement the connection to the SQLite3 database.

```
>>> engine = create_engine('sqlite:///foo.db')
```

Convert the dataframe in a table within the database.

```
>>> frame.to_sql('colors',engine)
```

Instead, to read the database, you have to use the read_sql() function with the name of the table and the engine.

```
>>> pd.read_sql('colors',engine)
   index  white  red  blue  black  green
0      0      0    1     2      3      4
1      1      5    6     7      8      9
2      2     10   11    12     13     14
3      3     15   16    17     18     19
```

As you can see, even in this case, the writing operation on the database has become very simple, thanks to the I/O APIs available in the pandas library.

Now you'll see instead the same operations, but not using the I/O API. This can be useful to get an idea of how pandas proves to be an effective tool for reading and writing data to a database.

First, you must establish a connection to the DB and create a table by defining the corrected data types, so as to accommodate the data to be loaded.

```
>>> import sqlite3
>>> query = """
... CREATE TABLE test
... (a VARCHAR(20), b VARCHAR(20),
...  c REAL,       d INTEGER
... );"""
>>> con = sqlite3.connect(':memory:')
>>> con.execute(query)
<sqlite3.Cursor object at 0x0000000009E7D730>
>>> con.commit()
```

Now you can enter data using the SQL INSERT statement.

```
>>> data = [('white','up',1,3),
...         ('black','down',2,8),
...         ('green','up',4,4),
...         ('red','down',5,5)]
>>> stmt = "INSERT INTO test VALUES(?,?,?,?)"
>>> con.executemany(stmt, data)
<sqlite3.Cursor object at 0x0000000009E7D8F0>
>>> con.commit()
```

Now that you've seen how to load the data on a table, it is time to see how to query the database to get the data you just recorded. This is possible using an SQL SELECT statement.

```
>>> cursor = con.execute('select * from test')
>>> cursor
<sqlite3.Cursor object at 0x0000000009E7D730>
>>> rows = cursor.fetchall()
>>> rows
[('white', 'up', 1.0, 3),
 ('black', 'down', 2.0, 8),
 ('green', 'up', 4.0, 4),
 ('red', 'down', 5.0, 5)]
```

You can pass the list of tuples to the constructor of the dataframe, and if you need the name of the columns, you can find them within the description attribute of the cursor.

```
>>> cursor.description
(('a', None, None, None, None, None, None),
 ('b', None, None, None, None, None, None),
 ('c', None, None, None, None, None, None),
 ('d', None, None, None, None, None, None))
>>> pd.DataFrame(rows, columns=zip(*cursor.description)[0])
       a     b  c  d
0  white    up  1  3
1  black  down  2  8
2  green    up  4  4
3    red  down  5  5
```

As you can see, this approach is quite laborious.

Loading and Writing Data with PostgreSQL in a Docker Container

To run this example, you must have installed on your system a PostgreSQL database. In my case, I created a database instance using the Docker system with its containers.

Docker is an application developed by Solomon Hykes in 2013 that allows you to have, generate, and uninstall server applications quickly and cleanly through the container mechanism. An application such as a PostgreSQL database is generally installed on physical servers, and in any case, even if it's installed locally, it requires a lot of time and abundant resources. Installing such applications to perform simple tests or for small uses is very impractical. Furthermore, their complete removal is equally complicated, and being able to install and uninstall this kind of application continuously can become infuriating. Docker allows you to cage complex applications such as PostgreSQL (but also all other databases, web servers, or whatever) in containers. These packages can be installed, activated on specific ports, deactivated, and removed, all with a simple click, as shown in Figure 5-8.

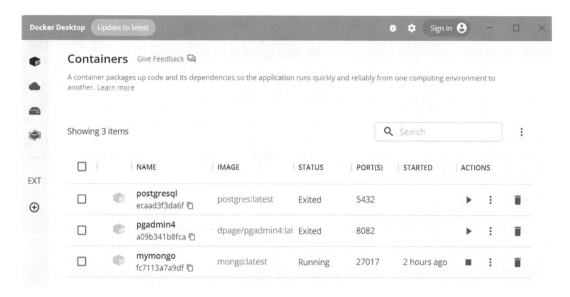

Figure 5-8. *The Docker application with some databases in containers*

Docker installation is very simple. Go to the official site (`www.docker.com/`), where the button with the specific release for your system will appear, as shown in Figure 5-9.

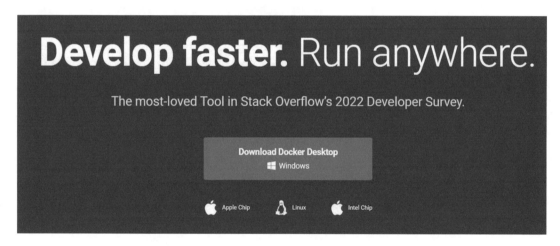

Figure 5-9. *The main page of Docker where you download the application*

Follow the installation instructions. Once Docker is installed, you can proceed with the installation of the container with the latest release of PostgreSQL database. Open the command console and enter the following command:

```
docker pull postgres
```

With this command, you downloaded the container containing a PostgreSQL database into your system (you can find all the information about it at `https://hub.docker.com/_/postgres`). However, the container has not been activated yet. To activate it, run the following command:

```
docker run --name postgresql -e POSTGRES_USER=myusername -e POSTGRES_PASSWORD=mypassword -p
5432:5432 -v /data:/var/lib/postgresql/data -d postgres
```

In this way, you have activated a PostgreSQL database in your system listening on port 5432 and with the login credentials defined as myusername/mypassword. Now you can install another container containing the web application for database management, called pgAdmin. To do this, first download the container corresponding to its latest version by using the following command.

```
docker pull dpage/pgadmin4:latest
```

As you did for the database, you also activate psAdmin in the Docker application by using the following command:

```
docker run --name "pgadmin4" -p 8082:80 -e "PGADMIN_DEFAULT_EMAIL=user@domain.com"
-e "PGADMIN_DEFAULT_PASSWORD=SuperSecret" -d dpage/pgadmin4
```

The application has been activated on port 8080 and the access credentials have been defined. If you look on Docker, you will therefore find a situation like the one shown in Figure 5-10.

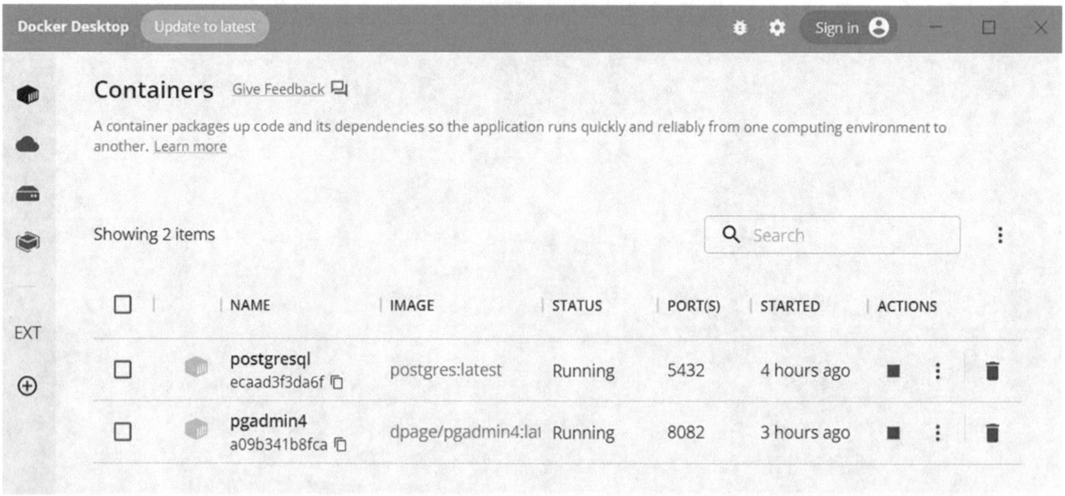

Figure 5-10. *Docker with the two active containers related to the PostgreSQL database and its management application*

Even though the services are up and running, you're not done yet. Open the browser and load the pgAdmin log page by entering the `https://localhost:8082` address. A screen like the one shown in Figure 5-11 will appear.

Figure 5-11. *The login web page of pgAdmin*

Enter the credentials you used to activate the pgAdmin container and access the PostgreSQL database management page, as shown in Figure 5-12.

Figure 5-12. *The pgAdmin administrator page*

Click the Add New Server button and a dialog box with server settings will open. On the General panel, enter the database name as postgres. On the Connection panel, enter 172.17.0.2 as the IP address of the server, myusername as the user, and mypassword as the password (see Figure 5-13). Click the Save button to activate the server.

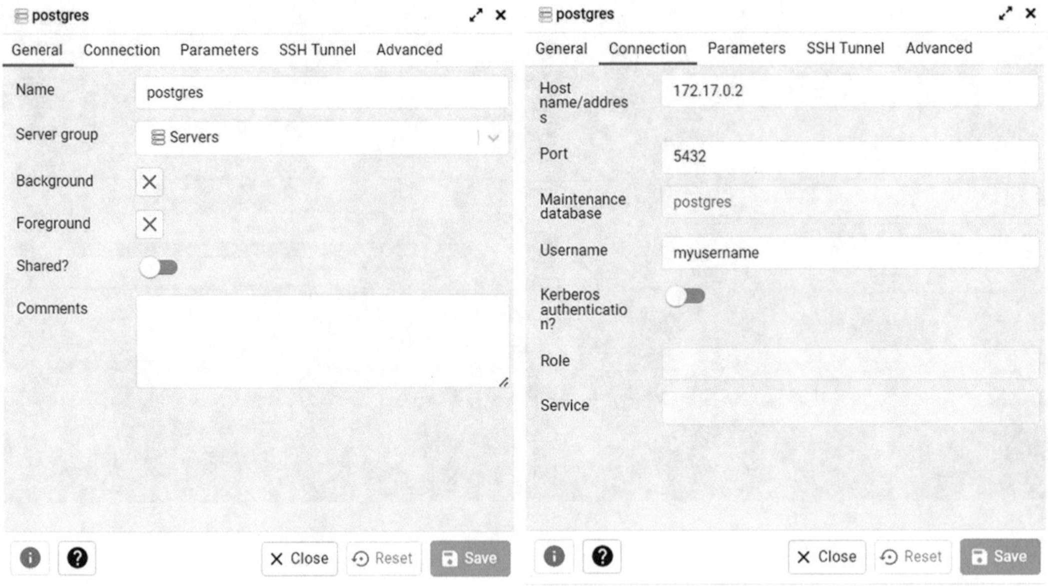

Figure 5-13. *The settings of the PostgreSQL database as a server*

If the IP address does not match, check it directly from the database container by writing this command on the console:

```
docker inspect postgresql
```

A list of all container settings will appear in the output. You can find the IP address by choosing Network Settings ➤ IP Address.

Now that you have a database as a service listening on port 5432, the first thing to do is install the psycopg2 library, which is designed to manage and handle the connection with the databases.

With Anaconda:

```
conda install psycopg2
```

Or if you are using PyPI:

```
pip install psycopg2
```

Now you can establish a connection with the database:

```
>>> import psycopg2
>>> from sqlalchemy import create_engine
>>> engine = create_engine('postgresql://myusername:mypassword@localhost:5432/postgres')
```

Create a dataframe object:

```
>>> import pandas as pd
>>> import numpy as np
>>> frame = pd.DataFrame(np.random.random((4,4)),
                index=['exp1','exp2','exp3','exp4'],
                columns=['feb','mar','apr','may']);
```

Now you see how easily you can transfer this data to a table. With to_sql(), you will record the data in a table called dataframe.

```
>>> frame.to_sql('dataframe',engine)
4
```

If you know the SQL language well, a more classic way to see the new created table and its contents is by using a psql session.

```
psql -U postgres
```

In this case, since you are working on a database in a Docker container, you have to run the following command:

```
>docker exec -it postgresql bash
root@ecaad3f3da6f:/#
```

A command console will open directly inside the container, where you can execute manual commands, such as activate the PostgreSQL shell using psql.

```
root@ecaad3f3da6f:/# psql -h localhost -U myusername
psql (15.3 (Debian 15.3-1.pgdg110+1))
Type "help" for help.

myusername=#
```

Switch to the database that interests you:

```
myusername=# \c postgres
```

Once you're connected to the database, perform an SQL query on the newly created table.

```
postgres=# SELECT * FROM DATAFRAME;
index|       feb        |       mar        |       apr        |       may
-----+------------------+------------------+------------------+------------------
exp1 |0.757871296789076|0.422582915331819|0.979085739226726|0.332288515791064
exp2 |0.124353978978927|0.273461421503087|0.049433776453223|0.0271413946693556
exp3 |0.538089036334938|0.097041417119426|0.905979807772598|0.123448718583967
exp4 |0.736585422687497|0.982331931474687|0.958014824504186|0.448063967996436
(4 righe)
```

Returning to the Python shell, even the conversion of a table in a dataframe is a trivial operation. Even here, there is a `read_sql_table()` function that reads directly on the database and returns a dataframe.

```
>>> pd.read_sql_table('dataframe',engine)
   index      feb       mar       apr       may
0  exp1  0.757871  0.422583  0.979086  0.332289
1  exp2  0.124354  0.273461  0.049434  0.027141
2  exp3  0.538089  0.097041  0.905980  0.123449
3  exp4  0.736585  0.982332  0.958015  0.448064
```

When you want to read data in a database, the conversion of a whole and single table into a dataframe is not the most useful operation. In fact, those who work with relational databases prefer to use the SQL language to choose which data and in what form to export the data by inserting an SQL query.

The text of an SQL query can be integrated in the `read_sql_query()` function.

```
>>> pd.read_sql_query('SELECT index,apr,may FROM DATAFRAME WHERE apr > 0.5',engine)
   index      apr       may
0  exp1  0.979086  0.332289
1  exp3  0.905980  0.123449
2  exp4  0.958015  0.448064
```

Reading and Writing Data with a NoSQL Database: MongoDB

Among all the NoSQL databases (BerkeleyDB, Tokyo Cabinet, and MongoDB), MongoDB is becoming the most widespread. Given its diffusion in many systems, it seems appropriate to consider the possibility of reading and writing data produced with the pandas library during data analysis.

As you have seen with PostgreSQL database, it is possible to take advantage of a Docker container also for a MongoDB database. It's to use this solution instead of installing the database on your system. At the end of the examples in this chapter, you will be free to delete the database, removing the container from Docker by clicking a simple Trash button and leaving your computer clean and unaltered.

Download the Docker container from the web and then activate it by entering the following commands:

```
docker pull mongo
docker run --name mymongo -p 27017:27017 -d mongo
```

Now that the service is listening on port 27017, you can connect to this database using the official driver for MongoDB: pymongo. This package also needs to be installed in your virtual environment. If you work with Anaconda, you can use Anaconda Navigator or enter the console command:

```
conda install pymongo
```

Otherwise, the easiest method is to install it from PyPI.

```
pip install pymongo
```

Once the package is installed, you can proceed directly with the code, importing the MongoClient method at the beginning. This method is important to create a connection with the database.

```
>>> from pymongo import MongoClient
>>> client = MongoClient('localhost',27017)
```

A single instance of MongoDB can support multiple databases at the same time. You need to point to a specific database.

```
>>> db = client.mydatabase
>>> db
Database(MongoClient(host=['localhost:27017'], document_class=dict, tz_aware=False,
connect=True), 'mydatabase')
```

In order to refer to this object, you can also use the following:

```
>>> client['mydatabase']
Database(MongoClient(host=['localhost:27017'], document_class=dict, tz_aware=False,
connect=True), 'mydatabase')
```

Now that you have defined the database, you have to define the collection. The collection is a group of documents stored in MongoDB and can be considered the equivalent of the tables in an SQL database.

```
>>> collection = db.mycollection
>>> db['mycollection']
Collection(Database(MongoClient(host=['localhost:27017'], document_class=dict, tz_
aware=False, connect=True), 'mydatabase'), 'mycollection')
>>> collection
Collection(Database(MongoClient(host=['localhost:27017'], document_class=dict, tz_
aware=False, connect=True), 'mydatabase'), 'mycollection')
```

Now it is the time to load the data in the collection. Create a dataframe.

```
>>> import pandas as pd
>>> import numpy as np
>>> frame = pd.DataFrame( np.arange(20).reshape(4,5),
...                       columns=['white','red','blue','black','green'])
>>> frame
   white  red  blue  black  green
0      0    1     2      3      4
1      5    6     7      8      9
2     10   11    12     13     14
3     15   16    17     18     19
```

Before being added to a collection, this dataframe must be converted into a JSON format. The conversion process is not as direct as you might imagine; this is because you need to set the data to be recorded on the DB in order to be re-extracted as a dataframe as fairly and as simply as possible.

```
>>> import json
>>> record = json.loads(frame.T.to_json()).values()
>>> record
dict_values([{'white': 0, 'red': 1, 'blue': 2, 'black': 3, 'green': 4}, {'white': 5,
'red': 6, 'blue': 7, 'black': 8, 'green': 9}, {'white': 10, 'red': 11, 'blue': 12,
'black': 13, 'green': 14}, {'white': 15, 'red': 16, 'blue': 17, 'black': 18, 'green': 19}])
```

Now you are finally ready to insert a document in the collection, and you can do this using the `insert()` function.

```
>>> collection.mydocument.insert_many(record)
<pymongo.results.InsertManyResult at 0x2f54ac43540>
```

As you can see, you have an object for each line recorded. Now that the data has been loaded into the document within the MongoDB database, you can execute the reverse process—reading data in a document and then converting them to a dataframe.

```
>>> result = collection['mydocument'].find()
>>> df = pd.DataFrame(list(result))>>> del df['_id']
>>> df
   white  red  blue  black  green
0      0    1     2      3      4
1      5    6     7      8      9
2     10   11    12     13     14
3     15   16    17     18     19
```

You have removed the column containing the ID numbers for the internal reference of MongoDB.

Conclusions

In this chapter, you saw how to use the features of the I/O API of the pandas library in order to read and write data to files and databases while preserving the structure of the dataframes. In particular, several modes of writing and reading data according to the type of format were illustrated.

In the last part of the chapter, you saw how to interface to the most popular models of databases to record and/or read data into it directly as a dataframe ready to be processed with the pandas tools.

In the next chapter, you see the most advanced features of the library pandas. Complex instruments like groupby and other forms of data processing are discussed in detail.

CHAPTER 6

■ ■ ■

pandas in Depth: Data Manipulation

In the previous chapter, you saw how to acquire data from data sources such as databases and files. Once you have the data in the dataframe format, they are ready to be manipulated. It's important to prepare the data so that they can be more easily subjected to analysis and manipulation. Especially in preparation for the next phase, the data must be ready for visualization.

This chapter goes in depth into the functionality that the pandas library offers for this stage of data analysis. The three phases of data manipulation are treated individually, illustrating the various operations with a series of examples and explaining how best to use the functions of this library to carry out such operations. The three phases of data manipulation are:

- Data preparation
- Data transformation
- Data aggregation

Data Preparation

Before you start manipulating data, it is necessary to prepare the data and assemble them in the form of data structures so that they can be manipulated later with the tools made available by the pandas library. The different procedures for data preparation are listed here.

- Loading
- Assembling
  - Merging
  - Concatenating
  - Combining
- Reshaping (pivoting)
- Removing

© Fabio Nelli 2023

F. Nelli, *Python Data Analytics*, https://doi.org/10.1007/978-1-4842-9532-8_6

The previous chapter covered loading. In the loading phase, there is also that part of the preparation that concerns the conversion from many different formats into a data structure such as a dataframe. But even after you have the data, probably from different sources and formats, and unified it into a dataframe, you need to perform further operations of preparation. In this chapter, and in particular in this section, you see how to perform all the operations necessary to get the data into a unified data structure.

The data contained in the pandas objects can be assembled in different ways:

- *Merging*—The pandas.merge() function connects the rows in a dataframe based on one or more keys. This mode is very familiar to those who are confident with the SQL language, since it also implements join operations.

- *Concatenating*—The pandas.concat() function concatenates the objects along an axis.

- *Combining*—The pandas.DataFrame.combine_first() function allows you to connect overlapped data in order to fill in missing values in a data structure by taking data from another structure.

Furthermore, part of the preparation process is also pivoting, which consists of the exchange between rows and columns.

Merging

The merging operation, which corresponds to the JOIN operation for those who are familiar with SQL, consists of a combination of data through the connection of rows using one or more keys.

In fact, anyone working with relational databases usually uses the JOIN query with SQL to get data from different tables using some reference values (keys) shared between them. On the basis of these keys, it is possible to obtain new data in a tabular form as the result of the combination of other tables. This operation with the library pandas is called *merging*, and the merge() function performs this kind of operation.

First, you have to import the pandas library and define two dataframes that will serve as examples for this section.

```
>>> import numpy as np
>>> import pandas as pd
>>> frame1 = pd.DataFrame( {'id':['ball','pencil','pen','mug','ashtray'],
...                         'price': [12.33,11.44,33.21,13.23,33.62]})
>>> frame1
        id  price
0      ball  12.33
1    pencil  11.44
2       pen  33.21
3       mug  13.23
4   ashtray  33.62
>>> frame2 = pd.DataFrame( {'id':['pencil','pencil','ball','pen'],
...                         'color': ['white','red','red','black']})
>>> frame2
   color      id
0  white  pencil
1    red  pencil
2    red    ball
3  black     pen
```

Carry out the merging by applying the merge() function to the two dataframe objects.

```
>>> pd.merge(frame1,frame2)
       id  price  color
0     ball  12.33    red
1   pencil  11.44  white
2   pencil  11.44    red
3      pen  33.21  black
```

As you can see from the result, the returned dataframe consists of all rows that have an ID in common. In addition to the common column, the columns from the first and the second dataframe are added.

In this case, you used the merge() function without specifying any column explicitly. In fact, in most cases you need to decide which column on which to base the merging.

To do this, add the on option with the column name as the key for the merging.

```
>>> frame1 = pd.DataFrame( {'id':['ball','pencil','pen','mug','ashtray'],
...                         'color': ['white','red','red','black','green'],
...                         'brand': ['OMG','ABC','ABC','POD','POD']})
>>> frame1
        id   color  brand
0     ball   white    OMG
1   pencil     red    ABC
2      pen     red    ABC
3      mug   black    POD
4  ashtray   green    POD
>>> frame2 = pd.DataFrame( {'id':['pencil','pencil','ball','pen'],
...                         'brand': ['OMG','POD','ABC','POD']})
>>> frame2
       id brand
0  pencil   OMG
1  pencil   POD
2    ball   ABC
3     pen   POD
```

Now, in this case, you have two dataframes with columns of the same name. So if you launch a merge, you do not get any results.

```
>>> pd.merge(frame1,frame2)
Empty DataFrame
Columns: [id, color, brand]
Index: []
```

It is necessary to explicitly define the criteria for merging that pandas must follow, specifying the name of the key column in the on option.

```
>>> pd.merge(frame1,frame2,on='id')
       id  color brand_x brand_y
0     ball  white     OMG     ABC
1   pencil    red     ABC     OMG
2   pencil    red     ABC     POD
3      pen    red     ABC     POD
```

```
>>> pd.merge(frame1,frame2,on='brand')
      id_x  color brand     id_y
0     ball  white   OMG   pencil
1   pencil    red   ABC     ball
2      pen    red   ABC     ball
3      mug  black   POD   pencil
4      mug  black   POD      pen
5  ashtray  green   POD   pencil
6  ashtray  green   POD      pen
```

As expected, the results vary considerably depending on the criteria of merging.

Often, however, the opposite problem arises, that is, you have two dataframes in which the key columns do not have the same name. To remedy this situation, you have to use the left_on and right_on options, which specify the key column for the first and for the second dataframe. Here is an example.

```
>>> frame2.columns = ['sid','brand']
>>> frame2
      sid brand
0  pencil   OMG
1  pencil   POD
2    ball   ABC
3     pen   POD
>>> pd.merge(frame1, frame2, left_on='id', right_on='sid')
       id  color brand_x     sid brand_y
0    ball  white     OMG    ball     ABC
1  pencil    red     ABC  pencil     OMG
2  pencil    red     ABC  pencil     POD
3     pen    red     ABC     pen     POD
```

By default, the merge() function performs an *inner join;* the keys in the result are the result of an intersection.

Other possible options are the *left join,* the *right join,* and the *outer join.* The outer join produces the union of all keys, combining the effect of a left join with a right join. To select the type of join you have to use the how option.

```
>>> frame2.columns = ['id','brand']
>>> pd.merge(frame1,frame2,on='id')
       id  color brand_x brand_y
0    ball  white     OMG     ABC
1  pencil    red     ABC     OMG
2  pencil    red     ABC     POD
3     pen    red     ABC     POD
>>> pd.merge(frame1,frame2,on='id',how='outer')
        id  color brand_x brand_y
0     ball  white     OMG     ABC
1   pencil    red     ABC     OMG
2   pencil    red     ABC     POD
3      pen    red     ABC     POD
4      mug  black     POD     NaN
5  ashtray  green     POD     NaN
```

```
>>> pd.merge(frame1,frame2,on='id',how='left')
        id  color brand_x brand_y
0      ball  white     OMG     ABC
1    pencil    red     ABC     OMG
2    pencil    red     ABC     POD
3       pen    red     ABC     POD
4       mug  black     POD     NaN
5   ashtray  green     POD     NaN
>>> pd.merge(frame1,frame2,on='id',how='right')
        id  color brand_x brand_y
0    pencil    red     ABC     OMG
1    pencil    red     ABC     POD
2      ball  white     OMG     ABC
3       pen    red     ABC     POD
```

To merge multiple keys, you simply add a list to the on option.

```
>>> pd.merge(frame1,frame2,on=['id','brand'],how='outer')
        id  color brand
0      ball  white   OMG
1    pencil    red   ABC
2       pen    red   ABC
3       mug  black   POD
4   ashtray  green   POD
5    pencil    NaN   OMG
6    pencil    NaN   POD
7      ball    NaN   ABC
8       pen    NaN   POD
```

Merging on an Index

In some cases, instead of considering the columns of a dataframe as keys, indexes could be used as keys for merging. Then in order to decide which indexes to consider, you set the left_index or right_index options to True to activate them, with the ability to activate them both.

```
>>> pd.merge(frame1,frame2,right_index=True, left_index=True)
      id_x  color brand_x    id_y brand_y
0     ball  white     OMG  pencil     OMG
1   pencil    red     ABC  pencil     POD
2      pen    red     ABC    ball     ABC
3      mug  black     POD     pen     POD
```

But the dataframe objects have a join() function, which is much more convenient when you want to do the merging by indexes. It can also be used to combine many dataframe objects having the same indexes but with no columns overlapping.

In fact, if you launch the following:

```
>>> frame1.join(frame2)
```

You will get an error code because some columns in frame1 have the same name as frame2. You need to rename the columns in frame2 before launching the join() function.

```
>>> frame2.columns = ['id2','brand2']
>>> frame1.join(frame2)
        id  color brand      id2 brand2
0      ball  white   OMG  pencil    OMG
1    pencil    red   ABC  pencil    POD
2       pen    red   ABC    ball    ABC
3       mug  black   POD     pen    POD
4   ashtray  green   POD     NaN    NaN
```

Here you've performed a merge, but based on the values of the indexes instead of the columns. This time there is also the index 4 that was present only in frame1, but the values corresponding to the columns of frame2 report NaN as a value.

Concatenating

Another type of data combination is referred to as *concatenation*. NumPy provides a concatenate() function to do this kind of operation with arrays.

```
>>> array1 = np.arange(9).reshape((3,3))
>>> array1
array([[0, 1, 2],
       [3, 4, 5],
       [6, 7, 8]])
>>> array2 = np.arange(9).reshape((3,3))+6
>>> array2
array([[ 6,  7,  8],
       [ 9, 10, 11],
       [12, 13, 14]])
>>> np.concatenate([array1,array2],axis=1)
array([[ 0,  1,  2,  6,  7,  8],
       [ 3,  4,  5,  9, 10, 11],
       [ 6,  7,  8, 12, 13, 14]])
>>> np.concatenate([array1,array2],axis=0)
array([[ 0,  1,  2],
       [ 3,  4,  5],
       [ 6,  7,  8],
       [ 6,  7,  8],
       [ 9, 10, 11],
       [12, 13, 14]])
```

With the pandas library and its data structures like series and dataframes, having labeled axes allows you to further generalize the concatenation of arrays. The concat() function is provided by pandas for this kind of operation.

```
>>> ser1 = pd.Series(np.random.rand(4), index=[1,2,3,4])
>>> ser1
1    0.636584
2    0.345030
```

```
3     0.157537
4     0.070351
dtype: float64
>>> ser2 = pd.Series(np.random.rand(4), index=[5,6,7,8])
>>> ser2
5     0.411319
6     0.359946
7     0.987651
8     0.329173
dtype: float64
>>> pd.concat([ser1,ser2])
1     0.636584
2     0.345030
3     0.157537
4     0.070351
5     0.411319
6     0.359946
7     0.987651
8     0.329173
dtype: float64
```

By default, the concat() function works on axis = 0, having as a returned object a series. If you set axis = 1, the result will be a dataframe.

```
>>> pd.concat([ser1,ser2],axis=1)
          0         1
1  0.636584       NaN
2  0.345030       NaN
3  0.157537       NaN
4  0.070351       NaN
5       NaN  0.411319
6       NaN  0.359946
7       NaN  0.987651
8       NaN  0.329173
```

The problem with this kind of operation is that the concatenated parts are not identifiable in the result. For example, you want to create a hierarchical index on the axis of concatenation. To do this, you have to use the keys option.

```
>>> pd.concat([ser1,ser2], keys=[1,2])
1  1     0.636584
   2     0.345030
   3     0.157537
   4     0.070351
2  5     0.411319
   6     0.359946
   7     0.987651
   8     0.329173
dtype: float64
```

In the case of combinations between series along `axis = 1`, the keys become the column headers of the dataframe.

```
>>> pd.concat([ser1,ser2], axis=1, keys=[1,2])
          1         2
1  0.636584       NaN
2  0.345030       NaN
3  0.157537       NaN
4  0.070351       NaN
5       NaN  0.411319
6       NaN  0.359946
7       NaN  0.987651
8       NaN  0.329173
```

So far you have seen the concatenation applied to the series, but the same logic can be applied to the dataframe.

```
>>> frame1 = pd.DataFrame(np.random.rand(9).reshape(3,3),
...                       index=[1,2,3],
...                       columns=['A','B','C'])

>>> frame2 = pd.DataFrame(np.random.rand(9).reshape(3,3),
...                       index=[4,5,6],
...                       columns=['A','B','C'])
>>> pd.concat([frame1, frame2])
          A         B         C
1  0.400663  0.937932  0.938035
2  0.202442  0.001500  0.231215
3  0.940898  0.045196  0.723390
4  0.568636  0.477043  0.913326
5  0.598378  0.315435  0.311443
6  0.619859  0.198060  0.647902
>>> pd.concat([frame1, frame2], axis=1)
          A         B         C         A         B         C
1  0.400663  0.937932  0.938035       NaN       NaN       NaN
2  0.202442  0.001500  0.231215       NaN       NaN       NaN
3  0.940898  0.045196  0.723390       NaN       NaN       NaN
4       NaN       NaN       NaN  0.568636  0.477043  0.913326
5       NaN       NaN       NaN  0.598378  0.315435  0.311443
6       NaN       NaN       NaN  0.619859  0.198060  0.647902
```

Combining

There is another situation in which there is combination of data that cannot be obtained either with merging or with concatenation. Take the case in which you want the two datasets to have indexes that overlap in their entirety or at least partially.

One applicable function to series is `combine_first()`, which performs this kind of operation along with data alignment.

```
>>> ser1 = pd.Series(np.random.rand(5),index=[1,2,3,4,5])
>>> ser1
1    0.942631
2    0.033523
3    0.886323
4    0.809757
5    0.800295
dtype: float64
>>> ser2 = pd.Series(np.random.rand(4),index=[2,4,5,6])
>>> ser2
2    0.739982
4    0.225647
5    0.709576
6    0.214882
dtype: float64
>>> ser1.combine_first(ser2)
1    0.942631
2    0.033523
3    0.886323
4    0.809757
5    0.800295
6    0.214882
dtype: float64
>>> ser2.combine_first(ser1)
1    0.942631
2    0.739982
3    0.886323
4    0.225647
5    0.709576
6    0.214882
dtype: float64
```

Instead, if you want a partial overlap, you can specify only the portion of the series you want to overlap.

```
>>> ser1[:3].combine_first(ser2[:3])
1    0.942631
2    0.033523
3    0.886323
4    0.225647
5    0.709576
dtype: float64
```

Pivoting

In addition to assembling the data to unify the values collected from different sources, another fairly common operation is *pivoting*. In fact, arrangement of the values by row or by column is not always suited to your goals. Sometimes you might want to rearrange the data by column values on rows or vice versa.

Pivoting with Hierarchical Indexing

You have already seen that a dataframe can support hierarchical indexing. This feature can be exploited to rearrange the data in a dataframe. In the context of pivoting, you have two basic operations:

- *Stacking*—Rotates or pivots the data structure, converting columns to rows

- *Unstacking*—Converts rows into columns

```
>>> frame1 = pd.DataFrame(np.arange(9).reshape(3,3),
...                       index=['white','black','red'],
...                       columns=['ball','pen','pencil'])
>>> frame1
        ball  pen  pencil
white     0    1       2
black     3    4       5
red       6    7       8
```

Using the stack() function on the dataframe, you will pivot the columns in rows, thus producing a series:

```
>>> ser5 = frame1.stack()
>>> ser5
white  ball      0
       pen       1
       pencil    2
black  ball      3
       pen       4
       pencil    5
red    ball      6
       pen       7
       pencil    8
dtype: int32
```

From this hierarchically indexed series, you can reassemble the dataframe into a pivoted table by use of the unstack() function.

```
>>> ser5.unstack()
        ball  pen  pencil
white     0    1       2
black     3    4       5
red       6    7       8
```

You can also do the unstack on a different level, specifying the number of levels or its name as the argument of the function.

```
>>> ser5.unstack(0)
        white  black  red
ball        0      3    6
pen         1      4    7
pencil      2      5    8
```

Pivoting from "Long" to "Wide" Format

The most common way to store datasets is produced by the punctual registration of data that will fill a line of the text file, for example, CSV, or a table of a database. This happens especially when you have instrumental readings, calculation results iterated over time, or the simple manual input of a series of values. A similar case of these files is for example the logs file, which is filled line by line by accumulating data in it.

The peculiar characteristic of this type of dataset is to have entries on various columns, often duplicated in subsequent lines. Always remaining in tabular format, this data is referred to as *long* or *stacked* format.

To get a clearer idea about that, consider the following dataframe.

```
>>> longframe = pd.DataFrame({ 'color':['white','white','white',
...                                     'red','red','red',
...                                     'black','black','black'],
...                     'item':['ball','pen','mug',
...                             'ball','pen','mug',
...                             'ball','pen','mug'],
...                     'value': np.random.rand(9)})
>>> longframe
   color  item     value
0  white  ball  0.091438
1  white   pen  0.495049
2  white   mug  0.956225
3    red  ball  0.394441
4    red   pen  0.501164
5    red   mug  0.561832
6  black  ball  0.879022
7  black   pen  0.610975
8  black   mug  0.093324
```

This mode of data recording has some disadvantages. One, for example, is the multiplicity and repetition of some fields. Considering the columns as keys, the data in this format will be difficult to read, especially in fully understanding the relationships between the key values and the rest of the columns.

Instead of the long format, there is another way to arrange the data in a table that is called *wide*. This mode is easier to read, allowing easy connection with other tables, and it occupies much less space. So in general it is a more efficient way of storing the data, although less practical, especially if during the filling of the data.

As a criterion, select a column, or a set of them, as the primary key; then, the values contained in it must be unique.

In this regard, pandas gives you a function that allows you to transform a dataframe from the long type to the wide type. This function is pivot() and it accepts as arguments the column, or columns, which will assume the role of key.

Starting from the previous example, you choose to create a dataframe in wide format by choosing the color column as the key, and item as a second key, the values of which will form the new columns of the dataframe.

```
>>> wideframe = longframe.pivot(index='color',columns='item')
>>> wideframe
          value
item       ball       mug       pen
color
black   0.879022  0.093324  0.610975
red     0.394441  0.561832  0.501164
white   0.091438  0.956225  0.495049
```

As you can now see, in this format, the dataframe is much more compact and the data contained in it are much more readable.

Removing

The last stage of data preparation is the removal of columns and rows. You have already seen this part in Chapter 4. However, for completeness, the description is reiterated here. Define a dataframe by way of example.

```
>>> frame1 = pd.DataFrame(np.arange(9).reshape(3,3),
...                        index=['white','black','red'],
...                        columns=['ball','pen','pencil'])
>>> frame1
        ball  pen  pencil
white     0    1     2
black     3    4     5
red       6    7     8
```

In order to remove a column, you simply apply the del command to the dataframe with the column name specified.

```
>>> del frame1['ball']
>>> frame1
        pen  pencil
white     1     2
black     4     5
red       7     8
```

Instead, to remove an unwanted row, you have to use the drop() function with the label of the corresponding index as an argument.

```
>>> frame1.drop('white')
        pen  pencil
black     4     5
red       7     8
```

Data Transformation

So far you have seen how to prepare data for analysis. This process in effect represents a reassembly of the data contained in a dataframe, with possible additions by other dataframe and removal of unwanted parts.

Now you begin the second stage of data manipulation: the *data transformation*. After you arrange the form of data and their disposal within the data structure, it is important to transform their values. In fact, in this section, you see some common issues and the steps required to overcome them using functions of the pandas library.

Some of these operations involve the presence of duplicate or invalid values, with possible removal or replacement. Other operations relate instead by modifying the indexes. Other steps include handling and processing the numerical values of the data and strings.

Removing Duplicates

Duplicate rows might be present in a dataframe for various reasons. In dataframes of enormous size, the detection of these rows can be very problematic. In this case, pandas provides a series of tools to analyze the duplicate data present in large data structures.

First, create a simple dataframe with some duplicate rows.

```
>>> dframe = pd.DataFrame({ 'color': ['white','white','red','red','white'],
...                         'value': [2,1,3,3,2]})
>>> dframe
   color  value
0  white      2
1  white      1
2    red      3
3    red      3
4  white      2
```

The duplicated() function applied to a dataframe can detect the rows that appear to be duplicated. It returns a series of Booleans where each element corresponds to a row, with True if the row is duplicated (i.e., only the other occurrences, not the first), and with False if there are no duplicates in the previous elements.

```
>>> dframe.duplicated()
0    False
1    False
2    False
3     True
4     True
dtype: bool
```

Having a Boolean series as a return value can be useful in many cases, especially for filtering. In fact, if you want to know which rows are duplicated, just type the following:

```
>>> dframe[dframe.duplicated()]
   color  value
3    red      3
4  white      2
```

Generally, all duplicated rows are to be deleted from the dataframe; to do that, pandas provides the drop_duplicates() function, which returns the dataframes without the duplicate rows.

```
>>> dframe[dframe.duplicated()]
   color  value
3    red      3
4  white      2
```

Mapping

The pandas library provides a set of functions which, as you see in this section, exploit mapping to perform some operations. Mapping is nothing more than the creation of a list of matches between two different values, with the ability to bind a value to a particular label or string.

To define mapping, there is no better object than dict objects.

```
map = {
   'label1' : 'value1,
   'label2' : 'value2,
   ...
}
```

The functions that you see in this section perform specific operations, but they all accept a dict object.

- replace()—Replaces values

- map()—Creates a new column

- rename()—Replaces the index values

Replacing Values via Mapping

Often in the data structure that you have assembled there are values that do not meet your needs. For example, the text may be in a foreign language, or may be a synonym of another value, or may not be expressed in the desired shape. In such cases, a replace operation of various values is often a necessary process.

Define, as an example, a dataframe containing various objects and colors, including two colors that are not in English. Assembly operations are likely to keep maintaining data with values in an undesirable form.

```
>>> frame = pd.DataFrame({ 'item':['ball','mug','pen','pencil','ashtray'],
...                        'color':['white','rosso','verde','black','yellow'],
                           'price':[5.56,4.20,1.30,0.56,2.75]})
>>> frame
      item   color  price
0     ball   white   5.56
1      mug   rosso   4.20
2      pen   verde   1.30
3   pencil   black   0.56
4  ashtray  yellow   2.75
```

To replace the incorrect values with new values, it is necessary to define a mapping of correspondences, containing as a key the new values.

```
>>> newcolors = {
...     'rosso': 'red',
...     'verde': 'green'
... }
```

Now the only thing you can do is use the replace() function with the mapping as an argument.

```
>>> frame.replace(newcolors)
      item   color  price
0     ball   white   5.56
1      mug     red   4.20
2      pen   green   1.30
3   pencil   black   0.56
4  ashtray  yellow   2.75
```

As you can see from the result, the two colors have been replaced with the correct values within the dataframe. A common case, for example, is the replacement of NaN values with another value, for example 0. You can use replace(), which performs its job very well.

```
>>> ser = pd.Series([1,3,np.nan,4,6,np.nan,3])
>>> ser
0    1.0
1    3.0
2    NaN
3    4.0
4    6.0
5    NaN
6    3.0
dtype: float64
>>> ser.replace(np.nan,0)
0    1.0
1    3.0
2    0.0
3    4.0
4    6.0
5    0.0
6    3.0
dtype: float64
```

Adding Values via Mapping

In the previous example, you saw how to substitute values by mapping correspondences. In this case you continue to exploit the mapping of values with another example. In this case you are exploiting mapping to add values in a column depending on the values contained in another column. The mapping will always be defined separately.

```
>>> frame = pd.DataFrame({ 'item':['ball','mug','pen','pencil','ashtray'],
...                        'color':['white','red','green','black','yellow']})
>>> frame
      item    color
0     ball    white
1      mug      red
2      pen    green
3   pencil    black
4  ashtray   yellow
```

Suppose you want to add a column to indicate the price of the item shown in the dataframe. Before you do this, it is assumed that you have a price list available somewhere, in which the price for each type of item is described. Define then a dict object that contains a list of prices for each type of item.

```
>>> prices = {
...     'ball' : 5.56,
...     'mug' : 4.20,
...     'bottle' : 1.30,
...     'scissors' : 3.41,
...     'pen' : 1.30,
...     'pencil' : 0.56,
...     'ashtray' : 2.75
... }
```

The map() function, when applied to a series or to a column of a dataframe, accepts a function or an object containing a dict with mapping. So in this case, you can apply the mapping of the prices on the column item, making sure to add a column to the price dataframe.

```
>>> frame['price'] = frame['item'].map(prices)
>>> frame
      item    color  price
0     ball    white   5.56
1      mug      red   4.20
2      pen    green   1.30
3   pencil    black   0.56
4  ashtray   yellow   2.75
```

Rename the Indexes of the Axes

In a manner very similar to what you saw for the values contained within the series and the dataframe, even the axis label can be transformed in a very similar way using mapping. So to replace the label indexes, pandas provides the rename() function, which takes the mapping as an argument, that is, a dict object.

```
>>> frame
      item   color  price
0     ball   white   5.56
1      mug     red   4.20
2      pen   green   1.30
3   pencil   black   0.56
4  ashtray  yellow   2.75
>>> reindex = {
...     0: 'first',
...     1: 'second',
...     2: 'third',
...     3: 'fourth',
...     4: 'fifth'}
>>> frame.rename(reindex)
           item   color  price
first      ball   white   5.56
second      mug     red   4.20
third       pen   green   1.30
fourth   pencil   black   0.56
fifth   ashtray  yellow   2.75
```

As you can see, by default, the indexes are renamed. If you want to rename columns, you must use the columns option. This time you assign various mapping explicitly to the two index and columns options.

```
>>> recolumn = {
...     'item':'object',
...     'price': 'value'}
>>> frame.rename(index=reindex, columns=recolumn)
         object   color  value
first      ball   white   5.56
second      mug     red   4.20
third       pen   green   1.30
fourth   pencil   black   0.56
fifth   ashtray  yellow   2.75
```

Also here, for the simplest cases in which you have a single value to be replaced, you can avoid having to write and assign many variables.

```
>>> frame.rename(index={1:'first'}, columns={'item':'object'})
        object   color  price
0         ball   white   5.56
first      mug     red   4.20
2          pen   green   1.30
3       pencil   black   0.56
4      ashtray  yellow   2.75
```

So far you have seen that the `rename()` function returns a dataframe with the changes, leaving unchanged the original dataframe. If you want the changes to take effect on the object on which you call the function, set the `inplace` option to `True`.

```
>>> frame.rename(columns={'item':'object'}, inplace=True)
>>> frame
     object   color  price
0      ball   white   5.56
1       mug     red   4.20
2       pen   green   1.30
3    pencil   black   0.56
4   ashtray  yellow   2.75
```

Discretization and Binning

A more complex process of transformation that you see in this section is *discretization*. Sometimes it can be used, especially in some experimental cases, to handle large quantities of data generated in sequence. To carry out an analysis of the data, however, it is necessary to transform this data into discrete categories, for example, by dividing the range of values of such readings into smaller intervals and counting the occurrence or statistics in them. Another case might be when you have a huge number of samples due to precise readings on a population. Even here, to facilitate analysis of the data, it is necessary to divide the range of values into categories and then analyze the occurrences and statistics related to each.

In this case, for example, you may have a reading of an experimental value between 0 and 100. These data are collected in a list.

```
>>> results = [12,34,67,55,28,90,99,12,3,56,74,44,87,23,49,89,87]
```

You know that the experimental values have a range from 0 to 100; therefore you can uniformly divide this interval, for example, into four equal parts, that is, *bins*. The first contains the values between 0 and 25, the second between 26 and 50, the third between 51 and 75, and the last between 76 and 100.

To do this binning with pandas, first you have to define an array containing the values of separation of bin:

```
>>> bins = [0,25,50,75,100]
```

Then you use a special function called `cut()` and apply it to the array of results, also passing the bins.

```
>>> cat = pd.cut(results, bins)
>>> cat
[(0, 25], (25, 50], (50, 75], (50, 75], (25, 50], ..., (75, 100], (0, 25], (25, 50], (75, 100], (75, 100]]
Length: 17
Categories (4, interval[int64, right]): [(0, 25] < (25, 50] < (50, 75] < (75, 100]]
```

The object returned by the `cut()` function is a special object of *Categorical* type. You can consider it as an array of strings indicating the name of the bin. Internally it contains a `categories` array indicating the names of the different internal categories and a `codes` array that contains a list of numbers equal to the elements of `results` (i.e., the array subjected to binning). The number corresponds to the bin to which the corresponding element of `results` is assigned.

```
>>> cat.categories
IntervalIndex([(0, 25], (25, 50], (50, 75], (75, 100]], dtype='interval[int64, right]')
>>> cat.codes
array([0, 1, 2, 2, 1, 3, 3, 0, 0, 2, 2, 1, 3, 0, 1, 3, 3], dtype=int8)
```

Finally, to know the occurrences for each bin, that is, how many results fall into each category, you have to use the value_counts() function.

```
>>> pd.value_counts(cat)
(75, 100]    5
(0, 25]      4
(25, 50]     4
(50, 75]     4
dtype: int64
```

As you can see, each class has the lower limit indicated with a bracket and the upper limit indicated with a parenthesis. This notation is consistent with mathematical notation that is used to indicate the intervals. If the bracket is square, the number belongs to the range (limit closed), and if it is round, the number does not belong to the interval (limit open).

You can give names to various bins by calling them first in an array of strings and then assigning to the labels options inside the cut() function that you have used to create the Categorical object.

```
>>> bin_names = ['unlikely','less likely','likely','highly likely']
>>> pd.cut(results, bins, labels=bin_names)
['unlikely', 'less likely', 'likely', 'likely', 'less likely', ..., 'highly likely',
'unlikely', 'less likely', 'highly likely', 'highly likely']
Length: 17
Categories (4, object): ['unlikely' < 'less likely' < 'likely' < 'highly likely']
```

If the cut() function is passed as an argument to an integer instead of explicating the bin edges, this will divide the range of values of the array into the number of intervals you specify.

The limits of the interval will be taken by the minimum and maximum of the sample data, namely, the array subjected to binning.

```
>>> pd.cut(results, 5)

[(2.904, 22.2], (22.2, 41.4], (60.6, 79.8], (41.4, 60.6], (22.2, 41.4], ..., (79.8, 99.0],
(22.2, 41.4], (41.4, 60.6], (79.8, 99.0], (79.8, 99.0]]
Length: 17
Categories (5, interval[float64, right]): [(2.904, 22.2] < (22.2, 41.4] < (41.4, 60.6] <
(60.6, 79.8] < (79.8, 99.0]]
)
```

In addition to cut(), pandas provides another method for binning: qcut(). This function divides the sample directly into quintiles. In fact, depending on the distribution of the data sample, by using cut(), you will have a different number of occurrences for each bin. Instead, qcut() ensures that the number of occurrences for each bin is equal, but the edges of each bin vary.

```
>>> quintiles = pd.qcut(results, 5)
>>> quintiles
```

```
[(2.999, 24.0], (24.0, 46.0], (62.6, 87.0], (46.0, 62.6], (24.0, 46.0], ..., (62.6, 87.0],
(2.999, 24.0], (46.0, 62.6], (87.0, 99.0], (62.6, 87.0]]
Length: 17
Categories (5, interval[float64, right]): [(2.999, 24.0] < (24.0, 46.0] < (46.0, 62.6]
< (62.6, 87.0] < (87.0, 99.0]]
>>> pd.value_counts(quintiles)
(2.999, 24.0]    4
(62.6, 87.0]     4
(24.0, 46.0]     3
(46.0, 62.6]     3
(87.0, 99.0]     3
dtype: int64
```

As you can see, in the case of quintiles, the intervals bounding the bin differ from those generated by the cut() function. Moreover, if you look at the occurrences for each bin, you will find that qcut() tried to standardize the occurrences for each bin, but in the case of quintiles, the first two bins have an occurrence in more because the number of results is not divisible by five.

Detecting and Filtering Outliers

During data analysis, the need to detect the presence of abnormal values in a data structure often arises. By way of example, create a dataframe with three columns of 1,000 completely random values:

```
>>> randframe = pd.DataFrame(np.random.randn(1000,3))
```

With the describe() function, you can see the statistics for each column.

```
>>> randframe.describe()
                 0            1            2
count  1000.000000  1000.000000  1000.000000
mean      0.021609    -0.022926    -0.019577
std       1.045777     0.998493     1.056961
min      -2.981600    -2.828229    -3.735046
25%      -0.675005    -0.729834    -0.737677
50%       0.003857    -0.016940    -0.031886
75%       0.738968     0.619175     0.718702
max       3.104202     2.942778     3.458472
```

For example, you might consider outliers those that have a value greater than three times the standard deviation. To have only the standard deviation of each column of the dataframe, use the std() function.

```
>>> randframe.std()
0    1.045777
1    0.998493
2    1.056961
dtype: float64
```

Now you apply the filtering of all the values of the dataframe, applying the corresponding standard deviation for each column. Thanks to the any() function, you can apply the filter to each column.

```
>>> randframe[(np.abs(randframe) > (3*randframe.std())).any(axis=1)]
              0          1          2
69   -0.442411  -1.099404   3.206832
576  -0.154413  -1.108671   3.458472
907   2.296649   1.129156  -3.735046
```

Permutation

The operations of permutation (random reordering) on a series or on the rows of a dataframe are easy to do using the numpy.random.permutation() function.

For this example, create a dataframe containing integers in ascending order.

```
>>> nframe = pd.DataFrame(np.arange(25).reshape(5,5))
>>> nframe
    0   1   2   3   4
0   0   1   2   3   4
1   5   6   7   8   9
2  10  11  12  13  14
3  15  16  17  18  19
4  20  21  22  23  24
```

Now create an array of five integers from 0 to 4, arranged in random order with the permutation() function. This will be the new order in which to set the values of a row of the dataframe.

```
>>> new_order = np.random.permutation(5)
>>> new_order
array([2, 3, 0, 1, 4])
```

Now apply it to the dataframe on all lines, using the take() function.

```
>>> nframe.take(new_order)
    0   1   2   3   4
2  10  11  12  13  14
3  15  16  17  18  19
0   0   1   2   3   4
1   5   6   7   8   9
4  20  21  22  23  24
```

As you can see, the order of the rows has changed; now the indices follow the same order as indicated in the new_order array.

You can submit even a portion of the entire dataframe to a permutation. It generates an array that has a sequence limited to a certain range, for example, in this case from 2 to 4.

```
>>> new_order = [3,4,2]
>>> nframe.take(new_order)
    0   1   2   3   4
3  15  16  17  18  19
4  20  21  22  23  24
2  10  11  12  13  14
```

Random Sampling

You have just seen how to extract a portion of the dataframe determined by subjecting it to permutation. Sometimes, when you have a huge dataframe, you may need to sample it randomly, and the quickest way to do that is by using the `np.random.randint()` function.

```
>>> sample = np.random.randint(0, len(nframe), size=3)
>>> sample
array([1, 4, 4])
>>> nframe.take(sample)
    0   1   2   3   4
1   5   6   7   8   9
4  20  21  22  23  24
4  20  21  22  23  24
```

As you can see from this random sampling, you can get the same sample even more often.

String Manipulation

Python is a popular language thanks to its ease of use in the processing of strings and text. Most operations can easily be made by using built-in functions provided by Python. For more complex cases of matching and manipulation, it is necessary to use regular expressions.

Built-in Methods for String Manipulation

In many cases, you have composite strings in which you want to separate the various parts and then assign them to the correct variables. The `split()` function allows you to separate parts of the text, taking as a reference point a separator, for example, a comma.

```
>>> text = '16 Bolton Avenue , Boston'
>>> text.split(',')
['16 Bolton Avenue ', 'Boston']
```

As you can see in the first element, you have a string with a space character at the end. To overcome this common problem, you have to use the `split()` function along with the `strip()` function, which trims the whitespace (including newlines).

```
>>> tokens = [s.strip() for s in text.split(',')]
>>> tokens
['16 Bolton Avenue', 'Boston']
```

The result is an array of strings. If the number of elements is small and always the same, a very interesting way to make assignments may be this:

```
>>> address, city = [s.strip() for s in text.split(',')]
>>> address
'16 Bolton Avenue'
>>> city
'Boston'
```

So far you have seen how to split text into parts, but often you also need the opposite, namely concatenate various strings to form longer text.

The most intuitive and simplest way is to concatenate the various parts of the text with the + operator.

```
>>> address + ',' + city
'16 Bolton Avenue, Boston'
```

This can be useful when you have only two or three strings to be concatenated. If you have many parts to be concatenated, a more practical approach is to use the join() function assigned to the separator character, with which you want to join the various strings.

```
>>> strings = ['A+','A','A-','B','BB','BBB','C+']
>>> ';'.join(strings)
'A+;A;A-;B;BB;BBB;C+'
```

Another category of operations that can be performed on the string is searching for pieces of text in them, that is, substrings. Python provides the keyword that represents the best way of detecting substrings.

```
>>> 'Boston' in text
True
```

However, there are two functions that can serve this purpose: index() and find().

```
>>> text.index('Boston')
19
>>> text.find('Boston')
19
```

In both cases, the function returns the number of the corresponding characters in the text where you have the substring. The difference in the behavior of these two functions can be seen, however, when the substring is not found:

```
>>> text.index('New York')
Traceback (most recent call last):
  File "<stdin>", line 1, in <module>
ValueError: substring not found
>>> text.find('New York')
-1
```

In fact, the index() function returns an error message, and find() returns -1 if the substring is not found. In the same area, you can know how many times a character or combination of characters (substring) occurs within the text. The count() function provides you with this number.

```
>>> text.count('e')
2
>>> text.count('Avenue')
1
```

Another operation that can be performed on strings is replacing or eliminating a substring (or a single character). In both cases, you use the replace() function, where if you are prompted to replace a substring with a blank character, the operation is equivalent to the elimination of the substring from the text.

```
>>> text.replace('Avenue','Street')
'16 Bolton Street , Boston'
>>> text.replace('1', '')'16 Bolton Avenue, Boston'
```

Regular Expressions

Regular expressions provide a very flexible way to search and match string patterns within text. A single expression, generically called *regex*, is a string formed according to the regular expression language. There is a built-in Python module called re, which is responsible for the operation of the regex.

First of all, when you want to use regular expressions, you need to import the module:

```
>>> import re
```

The re module provides a set of functions that can be divided into three categories:

- Pattern matching

- Substitution

- Splitting

Let's start with a few examples. For example, the regex for expressing a sequence of one or more whitespace characters is \s+. In the previous section, to split text into parts through a separator character, you used split(). There is a split() function even for the re module that performs the same operations, but it can accept a regex pattern as the criteria of separation, which makes it considerably more flexible.

```
>>> text = "This is        an\t odd  \n text!"
>>> re.split('\s+', text)
['This', 'is', 'an', 'odd', 'text!']
```

Let's analyze more deeply the mechanism of re module. When you call the re.split() function, the regular expression is first compiled, then it subsequently calls the split() function on the text argument. You can compile the regex function with the re.compile() function, thus obtaining a reusable object regex and so gaining CPU cycles.

This is especially true in the operations of iterative search of a substring in a set or an array of strings.

```
>>> regex = re.compile('\s+')
```

If you create a regex object with the compile() function, you can apply split() directly to it in the following way.

```
>>> regex.split(text)
['This', 'is', 'an', 'odd', 'text!']
```

To match a regex pattern to any other business substrings in the text, you can use the findall() function. It returns a list of all the substrings in the text that meet the requirements of the regex.

For example, if you want to find in a string all the words starting with "A" uppercase, or for example, with "a" regardless of whether it's upper- or lowercase, you need to enter the following:

```
>>> text = 'This is my address: 16 Bolton Avenue, Boston'
>>> re.findall('A\w+',text)
['Avenue']
>>> re.findall('[A,a]\w+',text)
['address', 'Avenue']
```

There are two other functions related to the findall() function—match() and search(). While findall() returns all matches within a list, the search() function returns only the first match. Furthermore, the object returned by this function is a particular object:

```
>>> re.search('[A,a]\w+',text)
<_sre.SRE_Match object; span=(11, 18), match='address'>
```

This object does not contain the value of the substring that responds to the regex pattern, but it returns its start and end positions within the string.

```
>>> search = re.search('[A,a]\w+',text)
>>> search.start()
11
>>> search.end()
18
>>> text[search.start():search.end()]
'address'
```

The match() function performs matching only at the beginning of the string; if there is no match to the first character, it goes no farther in research within the string. If you do not find a match, then it will not return any objects.

```
>>> re.match('[A,a]\w+',text)
```

If match() has a response, it returns an object identical to what you saw for the search() function.

```
>>> re.match('T\w+',text)
<_sre.SRE_Match object; span=(0, 4), match='This'>
>>> match = re.match('T\w+',text)
>>> text[match.start():match.end()]
'This'
```

Data Aggregation

The last stage of data manipulation is data aggregation. Data aggregation involves a transformation that produces a single integer from an array. In fact, you have already made many operations of data aggregation, for example, when you calculated the sum(), mean(), and count(). In fact, these functions operate on a set of data and perform a calculation with a consistent result consisting of a single value. However, a more formal manner and the one with more control in data aggregation is that which includes the categorization of a set.

The categorization of a set of data carried out for grouping is often a critical stage in the process of data analysis. It is a process of transformation since, after dividing the data into different groups, you apply a function that converts or transforms the data in some way, depending on the group they belong to. Very often the two phases of grouping and applying a function are performed in a single step.

Also for this part of the data analysis, pandas provides a tool that's very flexible and high performance: GroupBy.

Again, as in the case of join, those familiar with relational databases and the SQL language can find similarities. Nevertheless, languages such as SQL are quite limited when applied to operations on groups. In fact, given the flexibility of a programming language like Python, with all the libraries available, especially pandas, you can perform very complex operations on groups.

GroupBy

This section analyzes in detail the process of GroupBy and how it works. Generally, it refers to its internal mechanism as a process called *split-apply-combine*. In its pattern of operation you may conceive this process as divided into three phases expressed by three operations:

- *Splitting*—Division into groups of datasets

- *Applying*—Application of a function on each group

- *Combining*—Combination of all the results obtained by different groups

Analyze the three different phases (see Figure 6-1). In the first phase, that of splitting, the data contained within a data structure, such as a series or a dataframe, are divided into several groups, according to given criteria, which is often linked to indexes or to certain values in a column. In the jargon of SQL, values contained in this column are reported as keys. Furthermore, if you are working with two-dimensional objects such as a dataframe, the grouping criterion may be applied both to the line (axis = 0) for that column (axis = 1).

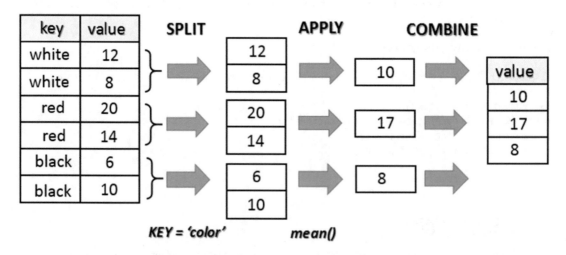

Figure 6-1. *The split-apply-combine mechanism*

The second phase, that of applying, consists of applying a function, or better a calculation expressed precisely by a function, which produces a new and single value that's specific to that group.

The last phase, that of combining, collects all the results obtained from each group and combines them to form a new object.

A Practical Example

You have just seen that the process of data aggregation in pandas is divided into various phases called split-apply-combine. With these phases, pandas are not expressed explicitly with the functions as you would expect, but by a groupby() function that generates a GroupBy object, which is the core of the whole process.

To better understand this mechanism, let's switch to a practical example. First define a dataframe containing numeric and string values.

```
>>> frame = pd.DataFrame({ 'color': ['white','red','green','red','green'],
...                        'object': ['pen','pencil','pencil','ashtray','pen'],
...                        'price1' : [5.56,4.20,1.30,0.56,2.75],
...                        'price2' : [4.75,4.12,1.60,0.75,3.15]})
>>> frame
    color   object  price1  price2
0   white      pen    5.56    4.75
1     red   pencil    4.20    4.12
2   green   pencil    1.30    1.60
3     red  ashtray    0.56    0.75
4   green      pen    2.75    3.15
```

Suppose you want to calculate the average of the price1 column using group labels listed in the color column. There are several ways to do this. You can for example access the price1 column and call the groupby() function with the color column.

```
>>> group = frame['price1'].groupby(frame['color'])
>>> group
<pandas.core.groupby.generic.SeriesGroupBy object at 0x000001942131D110>
```

The object that you get is a GroupBy object. In the operation that you just did, there was not really any calculation; there was just a collection of all the information needed to calculate the average. What you have done is group all the rows with the same value of color into a single item.

To analyze in detail how the dataframe was divided into groups of rows, you call the attribute groups' GroupBy object.

```
>>> group.groups
{'green': [2, 4], 'red': [1, 3], 'white': [0]}
```

As you can see, each group is listed and explicitly specifies the rows of the dataframe assigned to each of them. Now it is sufficient to apply the operation on the group to obtain the results for each individual group.

```
>>> group.mean()
color
green    2.025
red      2.380
white    5.560
Name: price1, dtype: float64
>>> group.sum()
color
green    4.05
red      4.76
white    5.56
Name: price1, dtype: float64
```

Hierarchical Grouping

You have seen how to group the data according to the values of a column as a key choice. The same thing can be extended to multiple columns, that is, make a grouping of multiple keys hierarchical.

```
>>> ggroup = frame['price1'].groupby([frame['color'],frame['object']])
>>> ggroup.groups
{('green', 'pen'): [4], ('green', 'pencil'): [2], ('red', 'ashtray'): [3], ('red',
'pencil'): [1], ('white', 'pen'): [0]}
>>> ggroup.sum()
color   object
green   pen          2.75
        pencil       1.30
red     ashtray      0.56
        pencil       4.20
white   pen          5.56
Name: price1, dtype: float64
```

So far you have applied the grouping to a single column of data, but in reality it can be extended to multiple columns or to the entire dataframe. If you do not need to reuse the object GroupBy several times, it is convenient to combine in a single passing all of the grouping and calculation to be done, without defining any intermediate variable.

```
>>> frame[['price1','price2']].groupby(frame['color']).mean()
       price1  price2
color
green   2.025   2.375
red     2.380   2.435
white   5.560   4.750
>>> frame.groupby(frame['color']).mean(numeric_only=True)       price1  price2
color
green   2.025   2.375
red     2.380   2.435
white   5.560   4.750
```

Group Iteration

The GroupBy object supports an iteration to generate a sequence of two tuples containing the name of the group together with the data portion.

```
>>> for name, group in frame.groupby('color'):
...     print(name)
...     print(group)
...
green
   color  object  price1  price2
2  green  pencil    1.30    1.60
4  green     pen    2.75    3.15
```

```
red
   color   object  price1  price2
1    red   pencil    4.20    4.12
3    red  ashtray    0.56    0.75
white
   color object  price1  price2
0  white    pen    5.56    4.75
```

This example only applied the `print` variable for illustration. In fact, you replace the printing operation of a variable with the function to be applied to it.

Chain of Transformations

From these examples, you have seen that for each grouping, when subjected to some function calculation or other operations in general, regardless of how it was obtained and the selection criteria, the result will be a data structure series (if you selected a single column data) or a dataframe, which then retains the index system and the name of the columns.

```
>>> result1 = frame['price1'].groupby(frame['color']).mean()
>>> type(result1)
pandas.core.series.Series
>>> result2 = frame.groupby(frame['color']).mean(numeric_only=True)
>>> type(result2)
pandas.core.frame.DataFrame
```

It is therefore possible to select a single column at any point in the various phases of this process. Here are three cases in which the selection of a single column in three different stages of the process applies. This example illustrates the great flexibility of this system of grouping provided by pandas.

```
>>> frame['price1'].groupby(frame['color']).mean()
color
green    2.025
red      2.380
white    5.560
Name: price1, dtype: float64
>>> frame.groupby(frame['color'])['price1'].mean()
color
green    2.025
red      2.380
white    5.560
Name: price1, dtype: float64
>>> (frame.groupby(frame['color']).mean(numeric_only=True))['price1']
color
green    2.025
red      2.380
white    5.560
Name: price1, dtype: float64
```

In addition, after an operation of aggregation, the names of some columns may not be very meaningful. In fact it is often useful to add a prefix to the column name that describes the type of business combination. Adding a prefix, instead of completely replacing the name, is very useful for keeping track of the source data from which they derive aggregate values. This is important if you apply a process of transformation chain (a series or dataframe is generated from another), because it is important to keep some reference with the source data.

```
>>> means = frame.groupby('color').mean(numeric_only=True).add_prefix('mean_')>>> means
       mean_price1  mean_price2
color
green        2.025        2.375
red          2.380        2.435
white        5.560        4.750
```

Functions on Groups

Although many methods have not been implemented specifically for use with GroupBy, they actually work correctly with data structures as the series. You saw in the previous section how easy it is to get the series by a GroupBy object, by specifying the name of the column and then by applying the method to make the calculation. For example, you can use the calculation of quantiles with the quantiles() function.

```
>>> group = frame.groupby('color')
>>> group['price1'].quantile(0.6)
color
green    2.170
red      2.744
white    5.560
Name: price1, dtype: float64
```

You can also define your own aggregation functions. Define the function separately and then pass it as an argument to the mark() function. For example, you can calculate the range of the values of each group.

```
>>> def range(series):
...        return series.max() - series.min()
...
>>> group['price1'].agg(range)
color
green    1.45
red      3.64
white    0.00
Name: price1, dtype: float64
```

You can also use more aggregate functions at the same time, with the mark() function passing an array containing the list of operations to be done, which will become the new columns.

```
>>> group['price1'].agg(['mean','std',range])
          mean       std  range
color
green    2.025  1.025305   1.45
red      2.380  2.573869   3.64
white    5.560       NaN   0.00
```

Advanced Data Aggregation

This section introduces the `transform()` and `apply()` functions, which allow you to perform many kinds of group operations, some of which are very complex.

Now suppose you want to bring together in the same dataframe the following: the dataframe of origin (the one containing the data) and that obtained by the calculation of group aggregation, for example, the sum.

```
>>> frame = pd.DataFrame({ 'color':['white','red','green','red','green'],
...                        'price1':[5.56,4.20,1.30,0.56,2.75],
...                        'price2':[4.75,4.12,1.60,0.75,3.15]})
>>> frame
    color  price1  price2
0   white    5.56    4.75
1     red    4.20    4.12
2   green    1.30    1.60
3     red    0.56    0.75
4   green    2.75    3.15
>>> sums = frame.groupby('color').sum().add_prefix('tot_')
>>> sums
        tot_price1  tot_price2
color
green         4.05        4.75
red           4.76        4.87
white         5.56        4.75
>>> pd.merge(frame,sums,left_on='color',right_index=True)
    color  price1  price2  tot_price1  tot_price2
0   white    5.56    4.75        5.56        4.75
1     red    4.20    4.12        4.76        4.87
3     red    0.56    0.75        4.76        4.87
2   green    1.30    1.60        4.05        4.75
4   green    2.75    3.15        4.05        4.75
```

Thanks to `merge()`, you can add the results of the aggregation in each line of the dataframe to start. But there is another way to do this type of operation. That is by using `transform()`. This function performs aggregation as you have seen before, but at the same time, it shows the values calculated based on the key value on each line of the dataframe to start.

```
>>> frame.groupby('color').transform(np.sum).add_prefix('tot_')
    tot_price1  tot_price2
0         5.56        4.75
1         4.76        4.87
2         4.05        4.75
3         4.76        4.87
4         4.05        4.75
```

As you can see, the `transform()` method is a more specialized function that has very specific requirements: the function passed as an argument must produce a single scalar value (aggregation) to be broadcasted.

The method to cover more general GroupBy is applicable to apply(). This method applies in its entirety the split-apply-combine scheme. In fact, this function divides the object into parts in order to be manipulated, invokes the passage of functions on each piece, and then tries to chain together the various parts.

```
>>> frame = pd.DataFrame( { 'color':['white','black','white','white','black','black'],
...                          'status':['up','up','down','down','down','up'],
...                          'value1':[12.33,14.55,22.34,27.84,23.40,18.33],
...                          'value2':[11.23,31.80,29.99,31.18,18.25,22.44]})
>>> frame
   color status  value1  value2
0  white     up   12.33   11.23
1  black     up   14.55   31.80
2  white   down   22.34   29.99
3  white   down   27.84   31.18
4  black   down   23.40   18.25
>>> frame.groupby(['color','status']).apply( lambda x: x.max())
                  color status  value1  value2
color status
black down        black   down   23.40   18.25
      up          black     up   18.33   31.80
white down        white   down   27.84   31.18
      up          white     up   12.33   11.23
5  black     up   18.33   22.44
>>> frame.rename(index=reindex, columns=recolumn)
         color   object  value
first    white     ball   5.56
second     red      mug   4.20
third    green      pen   1.30
fourth   black   pencil   0.56
fifth   yellow  ashtray   2.75
>>> temp = pd.date_range('1/1/2015', periods=10, freq= 'H')
>>> temp
DatetimeIndex(['2015-01-01 00:00:00', '2015-01-01 01:00:00',
               '2015-01-01 02:00:00', '2015-01-01 03:00:00',
               '2015-01-01 04:00:00', '2015-01-01 05:00:00',
               '2015-01-01 06:00:00', '2015-01-01 07:00:00',
               '2015-01-01 08:00:00', '2015-01-01 09:00:00'],
              dtype='datetime64[ns]', freq='H')
Length: 10, Freq: H, Timezone: None
>>> timeseries = pd.Series(np.random.rand(10), index=temp)
>>> timeseries
2015-01-01 00:00:00    0.368960
2015-01-01 01:00:00    0.486875
2015-01-01 02:00:00    0.074269
2015-01-01 03:00:00    0.694613
2015-01-01 04:00:00    0.936190
2015-01-01 05:00:00    0.903345
2015-01-01 06:00:00    0.790933
2015-01-01 07:00:00    0.128697
2015-01-01 08:00:00    0.515943
2015-01-01 09:00:00    0.227647
```

```
Freq: H, dtype: float64
>>> timetable = pd.DataFrame( {'date': temp, 'value1' : np.random.rand(10),
...                                        'value2' : np.random.rand(10)})
>>> timetable
                 date     value1    value2
0 2015-01-01 00:00:00   0.545737  0.772712
1 2015-01-01 01:00:00   0.236035  0.082847
2 2015-01-01 02:00:00   0.248293  0.938431
3 2015-01-01 03:00:00   0.888109  0.605302
4 2015-01-01 04:00:00   0.632222  0.080418
5 2015-01-01 05:00:00   0.249867  0.235366
6 2015-01-01 06:00:00   0.993940  0.125965
7 2015-01-01 07:00:00   0.154491  0.641867
8 2015-01-01 08:00:00   0.856238  0.521911
9 2015-01-01 09:00:00   0.307773  0.332822
```

You then add to the dataframe a column that represents a set of text values that you will use as key values.

```
>>> timetable['cat'] = ['up','down','left','left','up','up','down','right','right','up']
>>> timetable
                 date     value1    value2    cat
0 2015-01-01 00:00:00   0.545737  0.772712     up
1 2015-01-01 01:00:00   0.236035  0.082847   down
2 2015-01-01 02:00:00   0.248293  0.938431   left
3 2015-01-01 03:00:00   0.888109  0.605302   left
4 2015-01-01 04:00:00   0.632222  0.080418     up
5 2015-01-01 05:00:00   0.249867  0.235366     up
6 2015-01-01 06:00:00   0.993940  0.125965   down
7 2015-01-01 07:00:00   0.154491  0.641867  right
8 2015-01-01 08:00:00   0.856238  0.521911  right
9 2015-01-01 09:00:00   0.307773  0.332822     up
```

The example shown here, however, has duplicate key values.

Conclusions

In this chapter, you saw the three basic parts that divide the data manipulation phase: preparation, processing, and data aggregation. Thanks to a series of examples, you learned about a set of library functions that allow pandas to perform these operations.

You saw how to apply these functions on simple data structures so that you can become familiar with how they work and understand their applicability to more complex cases. You now have the knowledge you need to prepare a dataset for the next phase of data analysis: data visualization.

The next chapter presents the Python library `matplotlib`, which can convert data structures in any chart.

■ ■ ■

## CHAPTER 7

Data Visualization with matplotlib and Seaborn

The previous chapters covered the Python libraries that are responsible for data processing, and this chapter covers the libraries that take care of visualization. You'll first get a broad overview of the matplotlib library, and then the chapter concludes with the seaborn library, which extends matplotlib with the representation of statistical graphic elements.

Data visualization is often underestimated in data analysis, but it is a very important factor because incorrect or inefficient data representation can ruin an otherwise excellent analysis.

The matplotlib Library

matplotlib is a Python library specializing in the development of two-dimensional charts (including 3D charts). In recent years, it has been widespread in scientific and engineering circles (https:// matplotlib.org/).

Among all its features that make it the most used tool to represent data graphically, there are a few that stand out:

- Extreme simplicity in its use

- Gradual development and interactive data visualization

- Expressions and text in LaTeX

- Greater control over graphic elements

- Ability to export it to many formats, such as PNG, PDF, SVG, and EPS

matplotlib is designed to reproduce as much as possible an environment similar to MATLAB in terms of both graphical view and syntactic form. This approach has proved successful, as it can exploit the experience of software (MATLAB) that has been on the market for several years and is now widespread in all professional technical and scientific circles. Not only is matplotlib based on a scheme known and quite familiar to most experts in the field, but it also exploits those optimizations that over the years have led to a deducibility and simplicity in its use. That makes this library an excellent choice for those approaching data visualization for the first time, especially those without any experience with applications such as MATLAB.

In addition to its simplicity and deducibility, the matplotlib library inherited *interactivity* from MATLAB as well. That is, the analyst can insert command after command to control the gradual development of a graphical representation of data. This mode is well suited to the more interactive approaches of Python as the IPython QtConsole and Jupyter Notebook (see Chapter 2), thus providing an environment for data analysis that has little to envy from other tools such as Mathematica, IDL, or MATLAB.

© Fabio Nelli 2023
F. Nelli, *Python Data Analytics*, https://doi.org/10.1007/978-1-4842-9532-8_7

The genius of those who developed this beautiful library was to use and incorporate the good things currently available and in use in science. This is not only limited, as you have seen, to the operating mode of MATLAB and similar, but also to models of textual formatting of scientific expressions and symbols represented by LaTeX. Because of its great capacity for display and presentation of scientific expressions, LaTeX has been an irreplaceable element in scientific publications and documentations, where the need to visually represent expressions like integrals, summations, and derivatives is mandatory. Therefore, `matplotlib` integrates this remarkable instrument in order to improve the representative capacity of charts.

In addition, you must not forget that `matplotlib` is not a separate application but a library of a programming language like Python. It therefore takes full advantage of the potential that programming languages offer. `matplotlib` looks like a graphics library that allows you to programmatically manage the graphic elements that make up a chart so that the graphical display can be controlled in its entirety. The ability to program the graphical representation allows you to manage the reproducibility of the data representation across multiple environments, especially when you make changes or when the data is updated.

Moreover, because `matplotlib` is a Python library, it allows you to exploit the full potential of other libraries available to any developer who implements this language. In fact, with regard to data analysis, `matplotlib` normally cooperates with a set of other libraries such as NumPy and pandas, but many other libraries can be integrated without any problem.

Finally, graphical representations obtained through encoding with this library can be exported in the most common graphic formats (such as PNG and SVG) and then be used in other applications, documentation, web pages, and so on.

Installation

There are many options for installing the `matplotlib` library. If you choose to use the Anaconda distribution, installing the `matplotlib` package is very simple. You can do this graphically, using Anaconda Navigator. Activate the virtual environment you need to work on and then look for `matplotlib` among the distribution's packages to install, as shown in the Figure 7-1 with points 1, 2, 3, and 4. Then select the two packages— `matplotlib` and `matplotlib-base`—in the list of available ones. Finally, click the Apply button at the bottom right to start the installation.

Figure 7-1. *Installing matplotlib with Anaconda Navigator*

Soon after, a dialog will appear in which the dependencies will be shown to you and will ask you to confirm the installation. Click Apply again and complete the installation.

If instead you prefer to use the command console, on Anaconda you can open a session via the CMD. exe Prompt and then enter the following command:

```
conda install matplotlib
```

If instead you have decided not to use the Anaconda platform on your system, you can install matplotlib directly from the command console of your system using the pip command.

```
pip install matplotlib
```

In the output, you will see all the dependencies needed for the installation. You will then be asked in YES-NO form for confirmation to continue the installation. Press y to complete the installation.

The matplotlib Architecture

One of the key tasks that matplotlib must take on is provide a set of functions and tools that allow representation and manipulation of a *Figure* (the main object), along with all internal objects of which it is composed. However, matplotlib not only deals with graphics but also provides all the tools for the event handling and the ability to animate graphics. So, thanks to these additional features, matplotlib can produce interactive charts based on the events triggered by pressing a key on the keyboard or upon mouse movement.

The architecture of matplotlib is logically structured into three layers, which are placed at three different levels (see Figure 7-2). The communication is unidirectional, that is, each layer can communicate with the underlying layer, while the lower layers cannot communicate with the top ones.

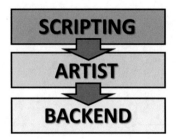

Figure 7-2. *The three layers of the matplotlib architecture*

The three layers are as follows:

- Scripting
- Artist
- Backend

Backend Layer

In the diagram of the `matplotlib` architecture in Figure 7-2, the layer that works at the lowest level is the *Backend* layer. This layer contains the `matplotlib` APIs, a set of classes that implement the graphic elements at a low level.

- `FigureCanvas` is the object that embodies the concept of a drawing area.
- `Renderer` is the object that draws on `FigureCanvas`.
- `Event` is the object that handles user inputs (keyboard and mouse events).

Artist Layer

As an intermediate layer, there is a layer called *Artist*. All the elements that make up a chart, such as the title, axis labels, markers, and so on, are instances of the Artist object. Each of these instances plays its role within a hierarchical structure (as shown in Figure 7-3).

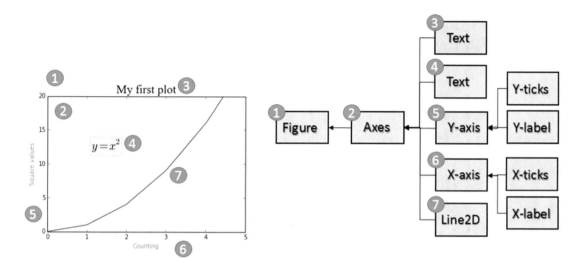

Figure 7-3. *Each element of a chart corresponds to an instance of an Artist structured in a hierarchy*

There are two Artist classes: primitive and composite.

- The *primitive artists* are individual objects that constitute the basic elements that form a graphical representation in a plot, for example a Line2D, or as a geometric figure such as a Rectangle or Circle, or even pieces of text.

- The *composite artists* are graphic elements that are composed of several base elements, namely, the primitive artists. Composite artists are for example the Axis, Ticks, Axes, and Figures (see Figure 7-4).

Generally, working at this level, you will have to deal often with objects in the higher hierarchy as Figure, Axes, and Axis. So it is important to fully understand what these objects are and what role they play within the graphical representation. Figure 7-4 shows the three main Artist objects (composite artists) that are generally used in all implementations performed at this level.

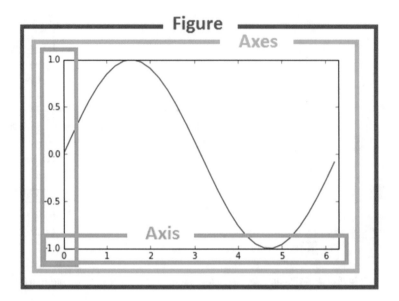

Figure 7-4. *The three main artist objects in the hierarchy of the Artist layer*

- *Figure* is the object with the highest level in the hierarchy. It corresponds to the entire graphical representation and generally can contain many Axes.

- *Axes* is generally what you mean as a plot or chart. Each Axis object belongs to only one Figure and is characterized by two Artist Axis (three in the three-dimensional case). Other objects, such as the title, the x label, and the y label, belong to this composite artist.

- *Axis* objects that take into account the numerical values to be represented on Axes define the limits and manage the ticks (the mark on the axes) and tick labels (the label text represented on each tick). The position of the tick is adjusted by an object called a *Locator,* while the formatting tick label is regulated by an object called a *Formatter*.

Scripting Layer (pyplot)

Artist classes and their related functions (the matplotlib API) are particularly suitable to all developers, especially for those who work on web application servers or develop the GUI. But for purposes of calculation, and in particular for the analysis and visualization of data, the scripting layer is best. This layer consists of an interface called *pyplot*.

pylab and pyplot

In general there is talk of *pylab* and *pyplot*. But what is the difference between these two packages? pylab is a module that is installed along with matplotlib, while pyplot is an internal module of matplotlib. Often you will find references to one or the other approach.

```
from pylab import *
```

and

```
import matplotlib.pyplot as plt
import numpy as np
```

pylab combines the functionality of pyplot with the capabilities of NumPy in a single namespace, and therefore you do not need to import NumPy separately. Furthermore, if you import pylab, pyplot and NumPy functions can be called directly without any reference to a module (namespace), making the environment more similar to MATLAB.

```
plot(x,y)
array([1,2,3,4])
```

Instead of

```
plt.plot()
np.array([1,2,3,4]
```

The pyplot package provides the classic Python interface for programming the matplotlib library, has its own namespace, and requires the import of the NumPy package separately. This approach is the one chosen for this book; it is the main topic of this chapter; and it is used for the rest of the book. This approach is shared and approved by most Python developers.

pyplot

The pyplot module is a collection of command-style functions that allow you to use matplotlib much like MATLAB. Each pyplot function will operate or make some changes to the Figure object, for example, the creation of the Figure itself, the creation of a plotting area, representation of a line, decoration of the plot with a label, and so on.

pyplot also is *stateful*, in that it tracks the status of the current figure and its plotting area. The called functions act on the current figure.

The Plotting Window

To get familiar with the matplotlib library and in a particular way with pyplot, you will start creating a simple interactive chart. Using matplotlib, this operation is very simple; in fact, you can achieve it using only three lines of code.

First, you can start using matplotlib from a simple command console, where you enter one command at a time.

The first thing to do is import the pyplot package and rename it as plt.

```
>>> import matplotlib.pyplot as plt
```

In Python, the constructors generally are not necessary; everything is already implicitly defined. In fact when you import the package, the plt object with all its graphics capabilities has already been instantiated and is ready to use. In fact, you can simply use the plot() function to pass the values to be plotted.

Thus, you can simply pass the values that you want to represent as a sequence of integers.

```
>>> plt.plot([1,2,3,4])
```

As you can see from the output, a Line2D object has been generated. The object is a line that represents the linear trend of the points included in the chart.

```
[<matplotlib.lines.Line2D object at 0x000001FE04AE4510>]
```

Now it is all set. You just have to give the command to show the plot using the show() function.

```
>>> plt.show()
```

The result will be the one shown in Figure 7-5. A window, called the *plotting window,* with a toolbar and the plot represented within it, will appear next to the command console (in a very similar way to what happens when working with MATLAB).

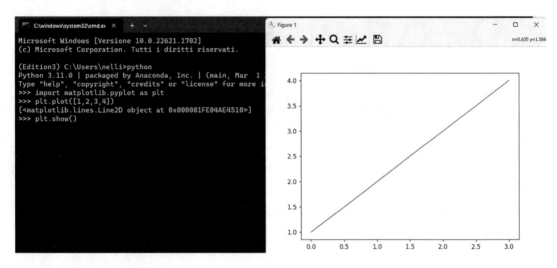

Figure 7-5. *The plotting window*

The plotting window is characterized by a toolbar at the top in which there is a series of buttons.

- Resets the original view

- Goes to the previous/next view

- Pans axes with left mouse, zoom with right

- Zooms to rectangle

- Configures subplots

- 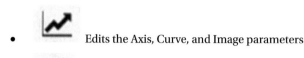 Edits the Axis, Curve, and Image parameters

- 💾 Saves/exports the figure

As you can soon see, this window allows you to manage the produced image autonomously and independently, as a separate application.

Data Visualization with Jupyter Notebook

A more convenient way to work interactively and professionally with matplot-generated charts is to use Jupyter Notebook.

In fact, by opening a Notebook on Jupyter and entering the previous code, you will find that the chart will be displayed directly integrated into the Notebook, without directly invoking the show() command (see Figure 7-6).

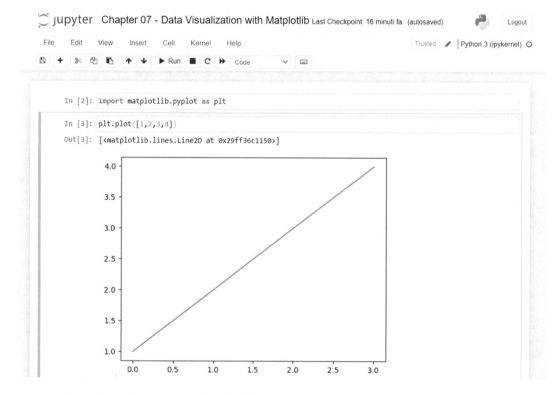

Figure 7-6. *Viewing a chart in Jupyter Notebook*

If you pass only a list of numbers or an array to the plt.plot() function, matplotlib assumes it is the sequence of y values of the chart, and it associates them to the natural sequence of values x: 0,1,2,3, ...

Generally, a plot represents value pairs (x, y), so if you want to define a chart correctly, you must define two arrays, the first containing the values on the x-axis and the second containing the values on the y-axis. You then add a second array with the values of the points on the y-axis. Now the points of the first array will correspond to the values of the points on the x-axis.

You write the following code in a new cell or edit the previous cell. In the first case, you will have a new chart completely separate from the first; in the second case, the existing chart will be updated based on the new values.

```
plt.plot([1,2,3,4],[1,4,9,16])
```

In both cases, the result is similar to the one shown in Figure 7-7.

```
In [6]:  plt.plot([1,2,3,4],[1,4,9,16])

Out[6]:  [<matplotlib.lines.Line2D at 0x29ff5014a90>]
```

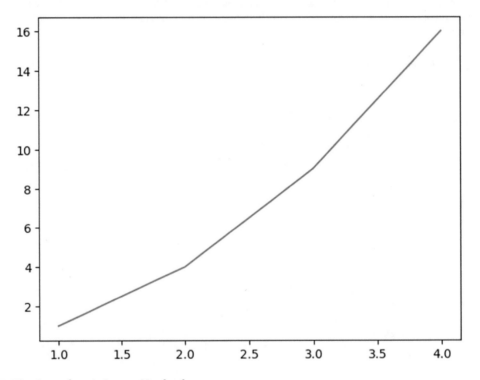

Figure 7-7. *Viewing a chart in Jupyter Notebook*

Set the Properties of the Plot

The plot() function can accept a third argument, which describes the specifics of how you want the point to be represented on the chart. As you can see in Figure 7-7, the points are represented by a blue line. In fact, if you do not specify otherwise, the plot is represented taking into account a default configuration of the plt. plot() function:

- The size of the axes matches perfectly with the range of the input data

- There is neither a title nor axis labels

- There is no legend

- A blue line connecting the points is drawn

Therefore, you may want to change this representation to have a real plot in which each pair of values (x, y) is represented by a red dot. Then you must add a third argument that defines some characteristics to the graph to override the default ones. In this case, you can pass 'ro' as the third argument to the plot() function.

```
plt.plot([1,2,3,4],[1,4,9,16],'ro')
```

Running the cell of the Notebook, you will get the chart shown in Figure 7-8.

```
In [7]: plt.plot([1,2,3,4],[1,4,9,16], 'ro')

Out[7]: [<matplotlib.lines.Line2D at 0x29ff508d810>]
```

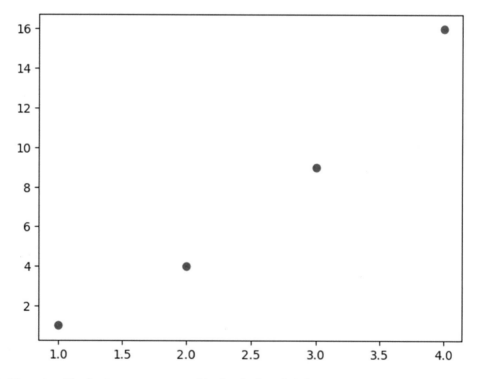

Figure 7-8. *The pairs of (x,y) values are represented in the plot by red circles*

Another possible change to the default behavior would be to define a range both on the x-axis and on the y-axis within a list [xmin, xmax, ymin, ymax] and then pass it as an argument to the axis() function.

In addition to the axis ranges, it is possible to define many other features of the graph. For example, you can also add a title using the `title()` function. Then, write the following code in a cell of the Notebook:

```
plt.axis([0,5,0,20])
plt.title('My first plot')
plt.plot([1,2,3,4],[1,4,9,16],'ro')
```

In Figure 7-9, you can see how the new settings make the plot more readable. In fact, the end points of the dataset are now represented within the plot rather than at the edges. Also the title of the plot is now visible at the top.

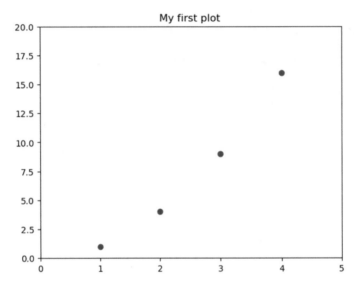

Figure 7-9. *The plot after the properties have been set*

matplotlib and NumPy

Even the `matplot` library, despite being a fully graphical library, has as its foundation the NumPy library. In fact, you have seen so far how to pass lists as arguments, both to represent the data and to set the extremes of the axes. These lists have been converted internally to NumPy arrays.

Therefore, you can directly enter NumPy arrays as input data. This array of data, which has been processed by pandas, can be directly used with `matplotlib` without further processing.

As an example, you see how it is possible to plot three different trends in the same plot (see Figure 7-10). You can choose for this example the `sin()` function belonging to the `math` module. You will need to import it. To generate points following a sinusoidal trend, you use the NumPy library. Generate a series of points on the x-axis using the `arange()` function, while for the values on the y-axis you will use the `map()` function to apply the `sin()` function on all the items of the array (without using a `for` loop).

```
import math
import numpy as np
t = np.arange(0,2.5,0.1)
y1 = np.sin(math.pi*t)
y2 = np.sin(math.pi*t+math.pi/2)
y3 = np.sin(math.pi*t-math.pi/2)
plt.plot(t,y1,'b*',t,y2,'g^',t,y3,'ys')
```

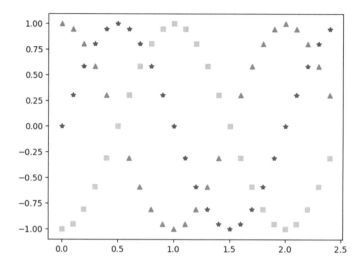

Figure 7-10. *Three sinusoidal trends phase-shifted by π / 4 represented by markers*

As you can see in Figure 7-10, the plot represents the three different temporal trends with three different colors and markers. In these cases, when the trend of a function is so obvious, the plot is perhaps not the most appropriate representation, but it is better to use the lines (see Figure 7-10). To differentiate the three trends with something other than color, you can use the pattern composed of different combinations of dots and dashes (- and .).

```
plt.plot(t,y1,'b--',t,y2,'g',t,y3,'r-.')
```

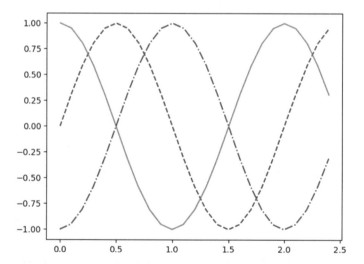

Figure 7-11. *This chart represents the three sinusoidal patterns with colored lines*

Using kwargs

The objects that make up a chart have many attributes that characterize them. These attributes are all default values, but can be set through the use of *keyword args,* often referred as *kwargs.*

These keywords are passed as arguments to functions. In reference documentation of the various functions of the `matplotlib` library, you will always find them referred to as *kwargs* in the last position. For example, the `plot()` function that you are using in these examples is referred to in the following way.

```
matplotlib.pyplot.plot(*args, **kwargs)
```

For a practical example, the thickness of a line can be changed if you set the `linewidth` keyword (see Figure 7-12).

```
plt.plot([1,2,4,2,1,0,1,2,1,4],linewidth=2.0)
```

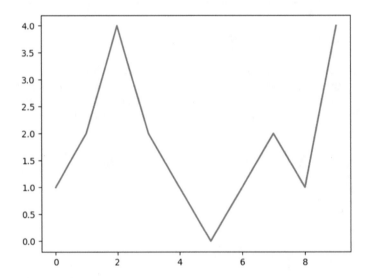

Figure 7-12. *The thickness of a line can be set directly from the plot() function*

Working with Multiple Figures and Axes

So far you have seen how all pyplot commands are routed to the display of a single figure. Actually, `matplotlib` allows you to manage multiple figures simultaneously, and within each figure, it offers the ability to view different plots defined as subplots.

So when you are working with pyplot, you must always keep in mind the concept of the current Figure and current Axes (that is, the plot shown within the figure).

Now you will see an example where two subplots are represented in a single figure. The `subplot()` function, in addition to subdividing the figure in different drawing areas, is used to focus the commands on a specific subplot.

The argument passed to the subplot() function sets the mode of subdivision and determines which is the current subplot. The current subplot will be the only figure that will be affected by the commands. The argument of the subplot() function is composed of three integers. The first number defines how many parts the figure is split into vertically. The second number defines how many parts the figure is divided into horizontally. The third number selects which is the current subplot on which you can direct commands.

Now you will display two sinusoidal trends (sine and cosine) and the best way to do that is to divide the canvas vertically in two horizontal subplots (as shown in Figure 7-13). So the numbers to pass as arguments are 211 and 212.

```
t = np.arange(0,5,0.1)
y1 = np.sin(2*np.pi*t)
y2 = np.sin(2*np.pi*t)
plt.subplot(211)
plt.plot(t,y1,'b-.')
plt.subplot(212)
plt.plot(t,y2,'r--')
```

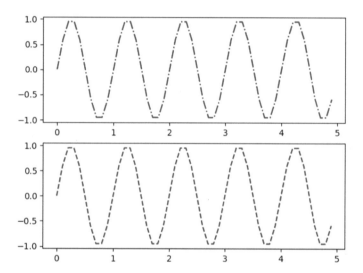

Figure 7-13. *The figure has been divided into two horizontal subplots*

You do the same thing by dividing the figure into two vertical subplots. The numbers to be passed as arguments to the subplot() function are 121 and 122 (as shown in Figure 7-14).

```
t = np.arange(0.,1.,0.05)
y1 = np.sin(2*np.pi*t)
y2 = np.cos(2*np.pi*t)
plt.subplot(121)
plt.plot(t,y1,'b-.')
plt.subplot(122)
plt.plot(t,y2,'r--')
```

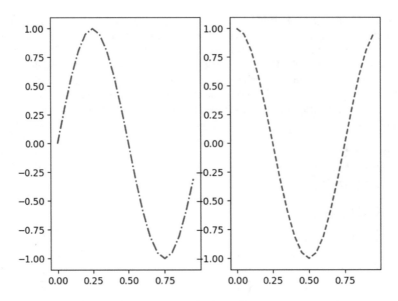

Figure 7-14. *The figure has been divided into two vertical subplots*

Adding Elements to the Chart

In order to make a chart more informative, many times it is not enough to represent the data using lines or markers and assign the range of values using two axes. In fact, there are many other elements that can be added to a chart in order to enrich it with additional information.

In this section, you see how to add elements, such as text labels, a legend, and so on, to a chart.

Adding Text

You've already seen how you can add a title to a chart with the title() function. Two other textual indications you can add are the *axis labels.* This is possible through the use of two other specific functions, called xlabel() and ylabel(). These functions take as an argument a string, which will be the shown text. Now add two axis labels to the chart. They describe which kind of value is assigned to each axis (as shown in Figure 7-15).

```
plt.axis([0,5,0,20])
plt.title('My first plot')
plt.xlabel('Counting')
plt.ylabel('Square values')
plt.plot([1,2,3,4],[1,4,9,16],'ro')
```

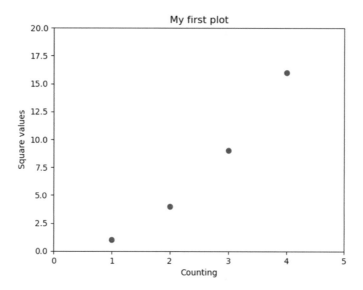

Figure 7-15. *A plot is more informative when it has axis labels*

Thanks to the keywords, you can change the characteristics of the text. For example, you can modify the title by changing the font and increasing the size of the characters. You can also modify the color of the axis labels to accentuate the title of the plot (as shown in Figure 7-16).

```
plt.axis([0,5,0,20])
plt.title('My first plot',fontsize=20,fontname='Times New Roman')
plt.xlabel('Counting',color='gray')
plt.ylabel('Square values',color='gray')
plt.plot([1,2,3,4],[1,4,9,16],'ro')
```

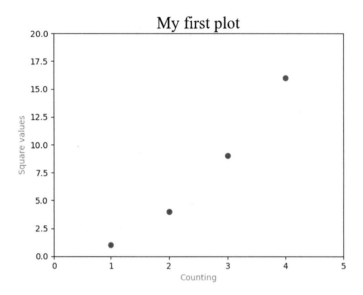

Figure 7-16. *The text can be modified by setting the keywords*

But `matplotlib` is not limited to this: pyplot allows you to add text to any position within a chart. This feature is performed by a specific function called `text()`.

```
text(x,y,s, fontdict=None, **kwargs)
```

The first two arguments are the coordinates of the location where you want to place the text. `s` is the string of text to be added, and `fontdict` (optional) is the font that you want to use. Finally, you can add the keywords.

Add the label to each point of the plot. Because the first two arguments to the `text()` function are the coordinates of the graph, you have to use the coordinates of the four points of the plot shifted slightly on the y-axis.

```
plt.axis([0,5,0,20])
plt.title('My first plot',fontsize=20,fontname='Times New Roman')
plt.xlabel('Counting',color='gray')
plt.ylabel('Square values',color='gray')
plt.text(1,1.5,'First')
plt.text(2,4.5,'Second')
plt.text(3,9.5,'Third')
plt.text(4,16.5,'Fourth')
plt.plot([1,2,3,4],[1,4,9,16],'ro')
```

As you can see in Figure 7-17, each point of the plot now has a label.

Figure 7-17. *Every point of the plot has an informative label*

Because matplotlib is a graphics library designed to be used in scientific circles, it must be able to exploit the full potential of scientific language, including mathematical expressions. matplotlib offers the possibility to integrate LaTeX expressions, thereby allowing you to insert mathematical expressions within the chart.

To do this, you can add a LaTeX expression to the text, enclosing it between two $ characters. The interpreter will recognize them as LaTeX expressions and convert them into the corresponding graphic, which can be a mathematical expression, a formula, mathematical characters, or just Greek letters. Generally you have to precede the string containing LaTeX expressions with an r, which indicates raw text, in order to avoid unintended escape sequences.

Here, you can also use the keywords to further enrich the text to be shown in the plot. Therefore, as an example, you can add the formula describing the trend followed by the point of the plot and enclose it in a colored bounding box (see Figure 7-18).

```
plt.axis([0,5,0,20])
plt.title('My first plot',fontsize=20,fontname='Times New Roman')
plt.xlabel('Counting',color='gray')
plt.ylabel('Square values',color='gray')
plt.text(1,1.5,'First')
plt.text(2,4.5,'Second')
plt.text(3,9.5,'Third')
plt.text(4,16.5,'Fourth')
plt.text(1.1,12,r'$y = x^2$',fontsize=20,bbox={'facecolor':'yellow','alpha':0.2})
plt.plot([1,2,3,4],[1,4,9,16],'ro')
```

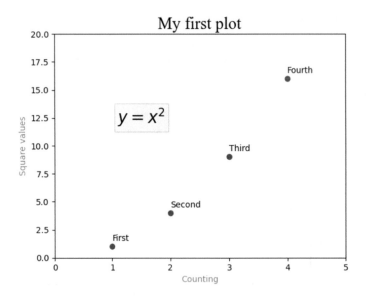

Figure 7-18. *Any mathematical expression can be seen in the context of a chart*

To get a complete view of the potential offered by LaTeX, consult Appendix A of this book.

Adding a Grid

Another element you can add to a plot is a grid. Often its addition is necessary in order to better understand the position occupied by each point on the chart.

Adding a grid to a chart is a very simple operation: just add the grid() function, passing True as an argument (see Figure 7-19).

```
plt.axis([0,5,0,20])
plt.title('My first plot',fontsize=20,fontname='Times New Roman')
plt.xlabel('Counting',color='gray')
plt.ylabel('Square values',color='gray')
plt.text(1,1.5,'First')
plt.text(2,4.5,'Second')
plt.text(3,9.5,'Third')
plt.text(4,16.5,'Fourth')
plt.text(1.1,12,r'$y = x^2$',fontsize=20,bbox={'facecolor':'yellow','alpha':0.2})
plt.grid(True)
plt.plot([1,2,3,4],[1,4,9,16],'ro')
```

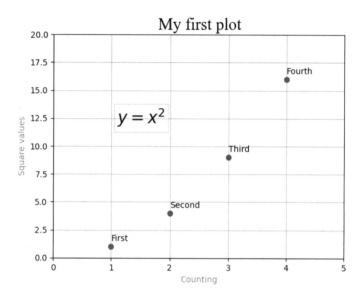

Figure 7-19. *A grid makes it easier to read the values of the data points represented on a chart*

Adding a Legend

Another very important component that should be present in any chart is the legend. pyplot also provides a specific function for this type of object: legend().

Add a legend to your chart with the legend() function and a string indicating the words for the legend. In this example, you assign the First series name to the input data array (see Figure 7-20).

```
plt.axis([0,5,0,20])
plt.title('My first plot',fontsize=20,fontname='Times New Roman')
plt.xlabel('Counting',color='gray')
plt.ylabel('Square values',color='gray')
plt.text(2,4.5,'Second')
plt.text(3,9.5,'Third')
plt.text(4,16.5,'Fourth')
plt.text(1.1,12,'$y = x^2$',fontsize=20,bbox={'facecolor':'yellow','alpha':0.2})
plt.grid(True)
plt.plot([1,2,3,4],[1,4,9,16],'ro')
plt.legend(['First series'])
```

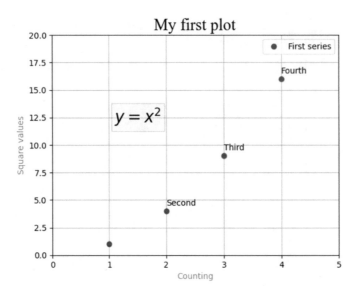

Figure 7-20. The legend is added to the upper-right corner by default

As you can see in Figure 7-19, the legend is added to the upper-right corner by default. If you want to change this behavior, you will need to add a few kwargs. For example, the position occupied by the legend is set by assigning numbers from 0 to 10 to the `loc` kwarg. Each of these numbers characterizes one of the corners of the chart (see Table 7-1). A value of 1 is the default, that is, the upper-right corner. In the next example, you will move the legend to the upper-left corner so it will not overlap with the points represented in the plot.

Table 7-1. The Possible Values for the loc Keyword

Location Code	Location String
0	best
1	upper-right
2	upper-left
3	lower-right
4	lower-left
5	right
6	center-left
7	center-right
8	lower-center
9	upper-center
10	center

Before you begin to modify the code to move the legend, I want to add a small notice. Generally, the legends are used to indicate the definition of a series to the reader via a label associated with a color and/or a marker that distinguishes it in the plot. So far in the examples, you have used a single series that was expressed by a single plot() function. Now, you have to focus on a more general case in which the same plot shows multiple series simultaneously. Each series in the chart will be characterized by a specific color and a specific marker (see Figure 7-21). In terms of code, instead, each series will be characterized by a call to the plot() function and the order in which they are defined will correspond to the order of the text labels passed as an argument to the legend() function.

```python
import matplotlib.pyplot as plt
plt.axis([0,5,0,20])
plt.title('My first plot',fontsize=20,fontname='Times New Roman')
plt.xlabel('Counting',color='gray')
plt.ylabel('Square values',color='gray')
plt.text(1,1.5,'First')
plt.text(2,4.5,'Second')
plt.text(3,9.5,'Third')
plt.text(4,16.5,'Fourth')
plt.text(1.1,12,'$y = x^2$',fontsize=20,bbox={'facecolor':'yellow','alpha':0.2})
plt.grid(True)
plt.plot([1,2,3,4],[1,4,9,16],'ro')
plt.plot([1,2,3,4],[0.8,3.5,8,15],'g^')
plt.plot([1,2,3,4],[0.5,2.5,4,12],'b*')
plt.legend(['First series','Second series','Third series'],loc=2)
```

Figure 7-21. *A legend is necessary in every multiseries chart*

Saving Your Charts

In this section you learn how to save your chart in different ways depending on your needs. If you need to reproduce your chart in different notebooks or Python sessions, or reuse them in future projects, it is a good practice to save the Python code. On the other hand, if you need to make reports or presentations, it can be very useful to save your chart as an image. Moreover, it is possible to save your chart as a HTML page, and this can be very useful when you need to share your work on the web.

Saving the Code

As you can see from the examples in the previous sections, the code concerning the representation of a single chart is growing into a fair number of rows. Once you think you've reached a good point in your development process, you can choose to save all the rows of code in a .py file that you can recall at any time.

You can use the magic command save% followed by the name of the file you want to save, followed by the number of input prompts containing the row of code that you want to save. If all the code is written in only one prompt, as in this case, you have to add only its number; otherwise if you want to save the code written in many prompts, for example from 10 to 20, you have to indicate this range with the two numbers separated by a -, that is, 10-20.

In your case, you would save the Python code underlying the representation of your first chart contained in the input prompt with the number 21.

```
In [21]: import matplotlib.pyplot as plt
...
```

You need to insert the following command in a new cell and then execute it to save the code into a new .py file.

```
%save my_first_chart 171
```

After you launch the command, you will get as output the message that a new file called my_first_chart.py has been created in the working directory and the code contained in it, as shown in Figure 7-22.

```
In [22]: %save my_first_chart 21
         The following commands were written to file `my_first_chart.py`:
         import matplotlib.pyplot as plt
         plt.axis([0,5,0,20])
         plt.title('My first plot',fontsize=20,fontname='Times New Roman')
         plt.xlabel('Counting',color='gray')
         plt.ylabel('Square values',color='gray')
         plt.text(1,1.5,'First')
         plt.text(2,4.5,'Second')
         plt.text(3,9.5,'Third')
         plt.text(4,16.5,'Fourth')
         plt.text(1.1,12,'$y = x^2$',fontsize=20,bbox={'facecolor':'yellow','alpha':0.2})
         plt.grid(True)
         plt.plot([1,2,3,4],[1,4,9,16],'ro')
         plt.plot([1,2,3,4],[0.8,3.5,8,15],'g^')
         plt.plot([1,2,3,4],[0.5,2.5,4,12],'b*')
         plt.legend(['First series','Second series','Third series'],loc=2)
```

Figure 7-22. The code written inside the cells can be saved to a file

In the same way as with saving, it is possible to load the code present in a file into a cell of the Notebook; you do this using the magic command %load.

```
%load my_first_chart.py
```

If you are not interested in the code of a file, but only in its output, you can write inside a cell the magic command %run.

```
%run my_first_chart.py
```

In this example, the result is the chart visualization in the output of the cell.

Saving Your Notebook as an HTML File or as Other File Formats

An interesting aspect of being able to work with Jupyter Notebook is the possibility of converting this Notebook into an HTML page. On the Notebook menu, click File and then Download As to see how many file formats the Notebook can be converted into (see Figure 7-23).

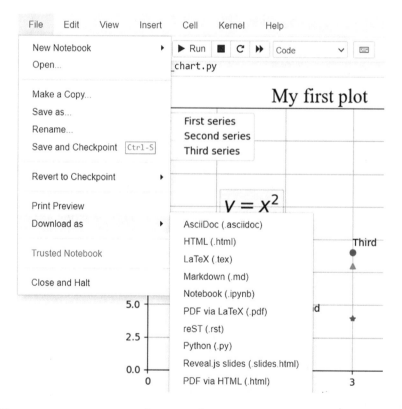

Figure 7-23. *You can save your current session as a web page*

Saving Your Chart Directly as an Image

If you are interested in saving only the figure of a chart as an image file, ignoring all the code you've written during the session, this is also possible. In fact, thanks to the `savefig()` function, you can directly save the chart in PNG format, although you should take care to add this function to the end of the same series of commands (otherwise you'll get a blank PNG file).

```
plt.axis([0,5,0,20])
plt.title('My first plot',fontsize=20,fontname='Times New Roman')
plt.xlabel('Counting',color='gray')
plt.ylabel('Square values',color='gray')
plt.text(1,1.5,'First')
plt.text(2,4.5,'Second')
plt.text(3,9.5,'Third')
plt.text(4,16.5,'Fourth')
plt.text(1.1,12,'$y = x^2$',fontsize=20,bbox={'facecolor':'yellow','alpha':0.2})
plt.grid(True)
 plt.plot([1,2,3,4],[1,4,9,16],'ro')
plt.plot([1,2,3,4],[0.8,3.5,8,15],'g^')
plt.plot([1,2,3,4],[0.5,2.5,4,12],'b*')
plt.legend(['First series','Second series','Third series'],loc=2)
plt.savefig('my_chart.png')
```

When you execute the previous code, a new file will be created in your working directory. This file will be named my_chart.png and will contain the image of your chart.

Handling Date Values

One of the most common problems encountered when doing data analysis is handling data of the date-time type. Displaying that data along an axis (normally the x-axis) can be problematic, especially when managing ticks (see Figure 7-24).

It is very common to use a dataframe with columns similar to the following one as a data source to display:

```
import pandas as pd
import datetime
df = pd.DataFrame({'year': 2015,
                   'month': [1,1,2,2,3,3,4,4],
                   'day': [23,28,3,21,15,24,8,24],
                   'readings': [12,22,25,20,18,15,17,14]})
df['events'] = pd.to_datetime(df[['year','month','day']])
df
```

Running the code, you will get a dataframe like the one in Figure 7-24, in which there is an Events column of date-time type values.

	year	month	day	readings	events
0	2015	1	23	12	2015-01-23
1	2015	1	28	22	2015-01-28
2	2015	2	3	25	2015-02-03
3	2015	2	21	20	2015-02-21
4	2015	3	15	18	2015-03-15
5	2015	3	24	15	2015-03-24
6	2015	4	8	17	2015-04-08
7	2015	4	24	14	2015-04-24

Figure 7-24. *A dataframe containing a date-time data column*

In this case, you might want to use a linear chart to visualize the trend of Readings values over time. It's very simple to do this; you use the plot() function to obtain a chart line like the one shown in Figure 7-25.

```
import matplotlib.pyplot as plt
plt.plot(df['events'],df['readings'])
```

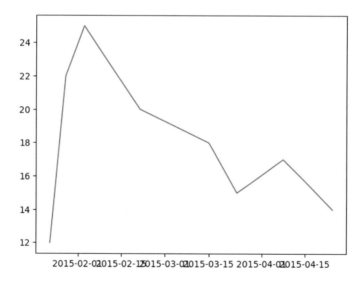

Figure 7-25. *If they aren't handled, displaying date-time values can be problematic*

As you can see in Figure 7-25, automatic management of ticks, and especially the tick labels, can be a disaster. The dates expressed in this way are difficult to read, there are no clear time intervals elapsed between one point and another, and there is also overlap.

To manage dates it is therefore advisable to define a time scale with appropriate objects. First you need to import matplotlib.dates, a module specialized for this type of data. Then you define the scales of the times, as in this case, a scale of days and one of months, through the MonthLocator() and DayLocator() functions. In these cases, the formatting is also very important, and to avoid overlap or unnecessary references, you have to limit the tick labels to the essential, which in this case is year-month. This format can be passed as an argument to the DateFormatter() function.

After you define the two scales, one for the days and one for the months, you can set two different kinds of ticks on the x-axis, using the set_major_locator() and set_minor_locator() functions on the xaxis object. Instead, to set the text format of the tick labels that refer to the months, you have to use the set_major_formatter() function.

Changing all these settings, you finally obtain the plot shown in Figure 7-26.

```
import numpy as np
import matplotlib.pyplot as plt
import matplotlib.dates as mdates
months = mdates.MonthLocator()
days = mdates.DayLocator()
timeFmt = mdates.DateFormatter('%Y-%m')
fig, ax = plt.subplots()
plt.plot(df['events'],df['readings'])
ax.xaxis.set_major_locator(months)
ax.xaxis.set_major_formatter(timeFmt)
ax.xaxis.set_minor_locator(days)
```

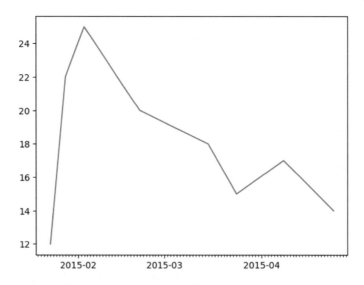

Figure 7-26. *Now the tick labels of the x-axis refer only to the months, making the plot more readable*

Chart Typology

In the previous sections, you saw a number of examples relating to the architecture of the `matplotlib` library. Now that you are familiar with using the main graphic elements in a chart, it is time to see a series of examples treating different types of charts, starting from the most common ones such as linear charts, bar charts, and pie charts, up to a discussion about some that are more sophisticated but commonly used, nonetheless.

This part of the chapter is very important since the purpose of this library is the visualization of the results produced by data analysis. Thus, knowing how to choose the proper type of chart is a fundamental choice. Remember that excellent data analysis represented incorrectly can lead to a wrong interpretation of the experimental results.

Line Charts

Among all the chart types, the linear chart is the simplest. A line chart is a sequence of data points connected by a line. Each data point consists of a pair of values (x,y), which will be reported in the chart according to the scale of values of the two axes (x and y).

By way of example, you can begin to plot the points generated by a mathematical function. Then, you can consider a generic mathematical function such as this:

$$y = sin(3^* x)/x$$

Therefore, if you want to create a sequence of data points, you need to create two NumPy arrays. First you create an array containing the x values to be referred to the x-axis. In order to define a sequence of increasing values, you will use the `np.arange()` function. Because the function is sinusoidal, you should refer to values that are multiples and submultiples of the Greek pi (*np.pi*). Then, using these sequence of values, you can obtain the y values by applying the `np.sin()` function directly to these values (thanks to NumPy!).

After all this, you have to plot them by calling the `plot()` function. You will obtain a line chart, as shown in Figure 7-27.

```
import matplotlib.pyplot as plt
import numpy as np
x = np.arange(-2*np.pi,2*np.pi,0.01)
y = np.sin(3*x)/x
plt.plot(x,y)
```

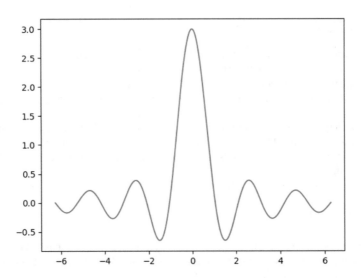

Figure 7-27. *A mathematical function represented in a line chart*

Now you can extend the case in which you want to display a family of functions, such as this:

$$y = sin(n^*x)/x$$

varying the parameter n.

```
import matplotlib.pyplot as plt
import numpy as np
x = np.arange(-2*np.pi,2*np.pi,0.01)
y = np.sin(x)/x
y2 = np.sin(2*x)/x
y3 = np.sin(3*x)/x
plt.plot(x,y)
plt.plot(x,y2)
plt.plot(x,y3)
```

As you can see in Figure 7-28, a different color is automatically assigned to each line. All the plots are represented on the same scale; that is, the data points of each series refer to the same x-axis and y-axis. This is because each call of the plot() function takes into account the previous calls to same function, so the Figure applies the changes, keeping memory of the previous commands until the Figure is not displayed (using show() with Python and Enter with the IPython QtConsole).

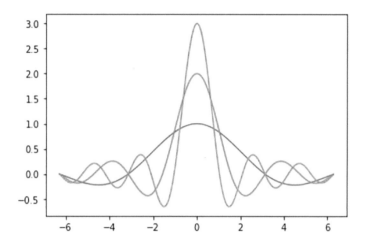

Figure 7-28. *Three different series are drawn with different colors in the same chart*

As you saw in the previous sections, regardless of the default settings, you can select the type of stroke, color, and so on. As the third argument of the plot() function, you can specify some codes that correspond to the color (see Table 7-2) and other codes that correspond to line styles, all included in the same string. Another possibility is to use two kwargs separately—color to define the color and linestyle to define the stroke (see Figure 7-29).

```
import matplotlib.pyplot as plt
import numpy as np
x = np.arange(-2*np.pi,2*np.pi,0.01)
y = np.sin(x)/x
y2 = np.sin(2*x)/x
y3 = np.sin(3*x)/x
plt.plot(x,y,'k--',linewidth=3)
plt.plot(x,y2,'m-.')
plt.plot(x,y3,color='#87a3cc',linestyle='--')
```

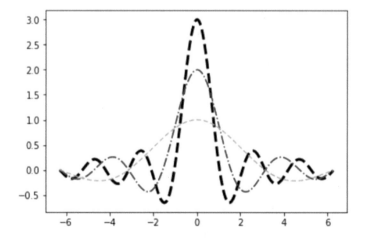

Figure 7-29. *You can define colors and line styles using character codes*

213

Table 7-2. *Color Codes*

Code	Color
b	blue
g	green
r	red
c	cyan
m	magenta
y	yellow
k	black
w	white

You have just defined a range from -2π to 2π on the x-axis, but by default, values on ticks are shown in numerical form. Therefore, you need to replace the numerical values with multiple of π. You can also replace the ticks on the y-axis. To do all this, you have to use the xticks() and yticks() functions, passing to each of them two lists of values. The first list contains values corresponding to the positions where the ticks are to be placed, and the second contains the tick labels. In this particular case, you have to use strings containing LaTeX format in order to correctly display the π symbol. Remember to define them within two $ characters and to add an r as the prefix.

```
import matplotlib.pyplot as plt
import numpy as np
x = np.arange(-2*np.pi,2*np.pi,0.01)
y = np.sin(3*x)/x
y2 = np.sin(2*x)/x
y3 = np.sin(x)/x
plt.plot(x,y,color='b')
plt.plot(x,y2,color='r')
plt.plot(x,y3,color='g')
plt.xticks([-2*np.pi, -np.pi, 0, np.pi, 2*np.pi],
          [r'$-2\pi$',r'$-\pi$',r'$0$',r'$+\pi$',r'$+2\pi$'])
plt.yticks([-1,0,1,2,3],
          [r'$-1$',r'$0$',r'$+1$',r'$+2$',r'$+3$'])
```

In the end, you will get a clean and pleasant line chart showing Greek characters, as shown in Figure 7-30.

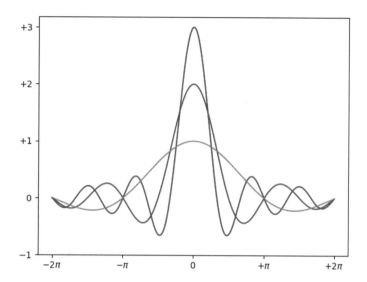

Figure 7-30. *The tick label can be improved by adding text with LaTeX format*

In all the linear charts you have seen so far, you always have the x-axis and y-axis placed at the edge of the figure (corresponding to the sides of the bounding border box). Another way of displaying axes is to have the two axes pass through the origin (0, 0), that is, the two Cartesian axes.

To do this, you must first capture the Axes object through the gca() function. Then through this object, you can select each of the four sides making up the bounding box, specifying for each one its position: right, left, bottom, and top. Crop the sides that do not match any axis (right and bottom) using the set_color() function and indicating none for color. Then, the sides corresponding to the x- and y-axes are moved to pass through the origin (0,0) with the set_position() function.

```
import matplotlib.pyplot as plt
import numpy as np
x = np.arange(-2*np.pi,2*np.pi,0.01)
y = np.sin(3*x)/x
y2 = np.sin(2*x)/x
y3 = np.sin(x)/x
plt.plot(x,y,color='b')
plt.plot(x,y2,color='r')
plt.plot(x,y3,color='g')
plt.xticks([-2*np.pi, -np.pi, 0, np.pi, 2*np.pi],
          [r'$-2\pi$',r'$-\pi$',r'$0$',r'$+\pi$',r'$+2\pi$'])
plt.yticks([-1,0,+1,+2,+3],
          [r'$-1$',r'$0$',r'$+1$',r'$+2$',r'$+3$'])
ax = plt.gca()
ax.spines['right'].set_color('none')
ax.spines['top'].set_color('none')
ax.xaxis.set_ticks_position('bottom')
ax.spines['bottom'].set_position(('data',0))
ax.yaxis.set_ticks_position('left')
ax.spines['left'].set_position(('data',0))
```

Now the chart will show the two axes crossing in the middle of the figure, that is, the origin of the Cartesian axes, as shown in Figure 7-31.

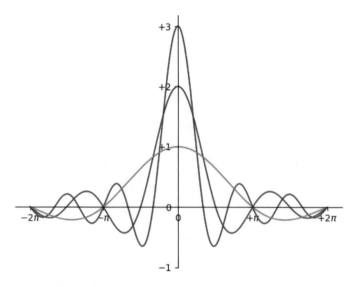

Figure 7-31. *The chart shows two Cartesian axes*

Often, it is very useful to be able to specify a particular point of the line using a notation and optionally add an arrow to better indicate the position of the point. For example, this notation may be a LaTeX expression, such as the formula for the limit of the function sinx/x where x tends to 0.

In this regard, matplotlib provides a function called annotate(), which is especially useful in these cases, even if the numerous kwargs needed to obtain a good result can make its settings quite complex. The first argument is the string to be represented containing the expression in LaTeX; then you can add the various kwargs. The point of the chart to note is indicated by a list containing the coordinates of the point [x, y] passed to the xy kwarg. The distance of the textual notation from the point to be highlighted is defined by the xytext kwarg and represented by means of a curved arrow whose characteristics are defined in the arrowprops kwarg.

```
import matplotlib.pyplot as plt
import numpy as np
x = np.arange(-2*np.pi,2*np.pi,0.01)
y = np.sin(3*x)/x
y2 = np.sin(2*x)/x
y3 = np.sin(x)/x
plt.plot(x,y,color='b')
plt.plot(x,y2,color='r')
plt.plot(x,y3,color='g')
plt.xticks([-2*np.pi, -np.pi, 0, np.pi, 2*np.pi],
           [r'$-2\pi$',r'$-\pi$',r'$0$',r'$+\pi$',r'$+2\pi$'])
plt.yticks([-1,0,+1,+2,+3],
           [r'$-1$',r'$0$',r'$+1$',r'$+2$',r'$+3$'])
plt.annotate(r'$\lim_{x\to 0}\frac{\sin(x)}{x}= 1$',
             xy=[0,1],xycoords='data',
             xytext=[30,30],
             fontsize=16,
```

```
            textcoords='offset points',
            arrowprops=dict(arrowstyle="->",connectionstyle="arc3,rad=.2"))ax = plt.gca()
ax.spines['right'].set_color('none')
ax.spines['top'].set_color('none')
ax.xaxis.set_ticks_position('bottom')
ax.spines['bottom'].set_position(('data',0))
ax.yaxis.set_ticks_position('left')
ax.spines['left'].set_position(('data',0))
```

Running this code, you will get the chart with the mathematical notation of the limit, which is the point shown by the arrow in Figure 7-32.

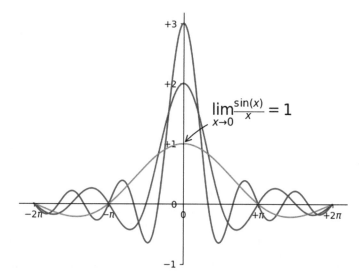

Figure 7-32. Mathematical expressions can be added to a chart with the annotate() function

Line Charts with pandas

Moving to more practical cases, or at least more closely related to data analysis, now is the time to see how easy it is to apply the matplotlib library to the dataframes of the pandas library. It is very easy to visualize the data in a dataframe as a linear chart. It is sufficient to pass the dataframe as an argument to the plot() function to obtain a multiseries linear chart (see Figure 7-33).

```
import matplotlib.pyplot as plt
import numpy as np
import pandas as pd
data = {'series1':[1,3,4,3,5],
                'series2':[2,4,5,2,4],
                'series3':[3,2,3,1,3]}
df = pd.DataFrame(data)
x = np.arange(5)
plt.axis([0,5,0,7])
plt.plot(x,df)
plt.legend(data, loc=2)
```

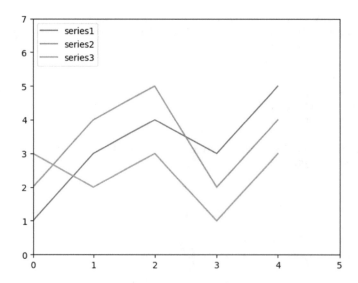

Figure 7-33. *The multiseries line chart displays the data in a pandas dataframe*

Histograms

A *histogram* consists of adjacent rectangles erected on the x-axis, split into discrete intervals called *bins*, and with an area proportional to the frequency of the occurrences for that bin. This kind of visualization is commonly used in statistical studies about distribution of samples.

In order to represent a histogram, pyplot provides a special function called hist(). This graphic function also has a feature that other functions producing charts do not have. The hist() function, in addition to drawing the histogram, returns a tuple of values that are the results of the calculation of the histogram. In fact, the hist() function can also implement the calculation of the histogram, that is, it is sufficient to provide a series of samples of values as an argument and the number of bins in which to be divided, and it will take care of dividing the range of samples in many intervals (bins), and then calculate the occurrences for each bin. The result of this operation, in addition to being shown in graphical form (see Figure 7-34), will be returned in the form of a tuple.

(n, bins, patches)

To understand this operation, a practical example is best. Then you can generate a population of 100 random values from 0 to 100 using the random.randint() function.

```
import matplotlib.pyplot as plt
import numpy as np
pop = np.random.randint(0,100,100)
pop
Out[ ]:
array([53, 11, 34, 75, 40, 89, 97, 78, 19, 50, 65, 30, 11, 38, 27, 11, 33,
       23, 22, 54, 78, 83, 66, 19, 15, 70, 32, 78, 50, 56, 42, 60, 48, 13,
       70, 83, 23, 69,  7, 76, 69,  8, 62,  5, 92, 71, 42, 98, 51, 20, 46,
       32, 53, 64,  7, 22, 84, 14, 82, 39, 17, 27, 73,  5, 78, 41, 90, 73,
       33, 57, 43, 57, 76, 98, 84, 62, 11, 98, 37, 95, 31, 86, 24, 83, 58,
       83, 95, 48, 67, 96, 82,  8, 54, 41, 72, 32, 92, 64,  3,  4])
```

Now, create the histogram of these samples by passing as an argument the hist() function. Say for example, you want to divide the occurrences into 20 bins (if not specified, the default value is ten bins). To do that, you have to use the bin kwarg, as shown in Figure 7-33.

```
n,bins,patches = plt.hist(pop,bins=20)
```

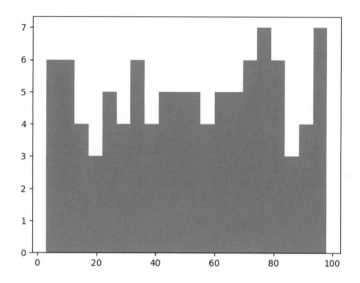

Figure 7-34. *The histogram shows the occurrences in each bin*

Bar Charts

Another very common type of chart is the bar chart. It is very similar to a histogram but in this case the x-axis is not used to reference numerical values but categories. The realization of the bar chart is very simple with matplotlib, using the bar() function.

```
import matplotlib.pyplot as plt
index = [0,1,2,3,4]
values = [5,7,3,4,6]
plt.bar(index,values)
```

With these few lines of code, you obtain the bar chart shown in Figure 7-35.

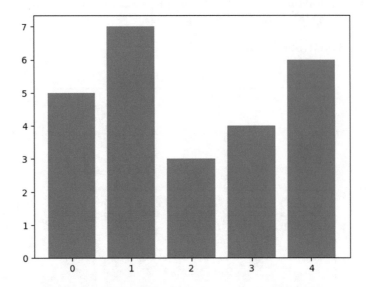

Figure 7-35. *The simplest bar chart with matplotlib*

If you look at Figure 7-34, you can see that the indices are drawn on the x-axis at the beginning of each bar. Actually, because each bar corresponds to a category, it would be better if you specified the categories through the tick label, defined by a list of strings passed to the xticks() function. As for the location of these tick labels, you have to pass a list containing the values corresponding to their positions on the x-axis as the first argument of the xticks() function. At the end you will get a bar chart, as shown in Figure 7-36.

```
import numpy as np
index = np.arange(5)
values1 = [5,7,3,4,6]
plt.bar(index,values1)
plt.xticks(index,['A','B','C','D','E'])
```

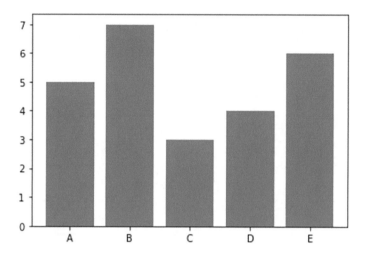

Figure 7-36. *A simple bar chart with categories on the x-axis*

Actually there are many other steps you can take to further refine the bar chart. Each of these finishes is set by adding a specific kwarg as an argument in the bar() function. For example, you can add the standard deviation values of the bar through the yerr kwarg along with a list containing the standard deviations. This kwarg is usually combined with another kwarg called error_kw, which, in turn, accepts other kwargs specialized for representing error bars. Two very specific kwargs used in this case are eColor, which specifies the color of the error bars, and capsize, which defines the width of the transverse lines that mark the ends of the error bars.

Another kwarg that you can use is alpha, which indicates the degree of transparency of the colored bar. Alpha is a value ranging from 0 to 1. When this value is 0, the object is completely transparent to become gradually more significant with the increase of the value, until arriving at 1, at which the color is fully represented.

As usual, the use of a legend is recommended, so in this case you should use a kwarg called label to identify the series that you are representing.

At the end you will get a bar chart with error bars, as shown in Figure 7-37.

```
import numpy as np
index = np.arange(5)
values1 = [5,7,3,4,6]
std1 = [0.8,1,0.4,0.9,1.3]
plt.title('A Bar Chart')
plt.bar(index,values1,yerr=std1,
        error_kw={'ecolor':'0.1','capsize':6},
        alpha=0.7,label='First')plt.xticks(index,['A','B','C','D','E'])
plt.legend(loc=2)
```

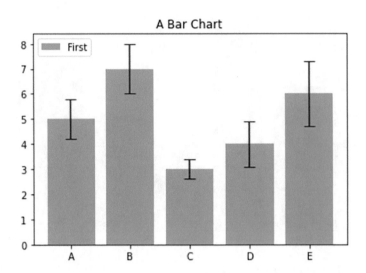

Figure 7-37. *A bar chart with error bars*

Horizontal Bar Charts

So far you have seen the bar chart oriented vertically. There are also bar charts oriented horizontally. This mode is implemented by a special function called barh(). The arguments and the kwargs valid for the bar() function remain the same for this function. The only change that you have to take into account is that the roles of the axes are reversed. Now, the categories are represented on the y-axis and the numerical values are shown on the x-axis (see Figure 7-38).

```
import matplotlib.pyplot as plt
import numpy as np
index = np.arange(5)
values1 = [5,7,3,4,6]
std1 = [0.8,1,0.4,0.9,1.3]
plt.title('A Horizontal Bar Chart')
plt.barh(index,values1,xerr=std1,error_kw={'ecolor':'0.1','capsize':6},alpha=0.7,
label='First')
plt.yticks(index,['A','B','C','D','E'])
plt.legend(loc=5)
```

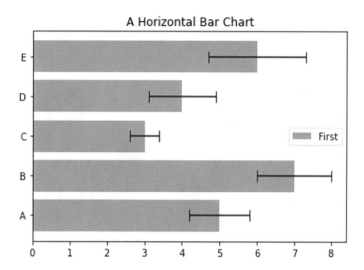

Figure 7-38. *A simple horizontal bar chart*

Multiserial Bar Charts

As line charts, bar charts are generally used to simultaneously display larger series of values. But in this case it is necessary to make some clarifications on how to structure a multiseries bar chart. So far you have defined a sequence of indexes, each corresponding to a bar, to be assigned to the x-axis. These indices should represent categories. In this case, however, you have more bars that must share the same category.

One approach used to overcome this problem is to divide the space occupied by an index (for convenience its width is 1) in as many parts as the bars sharing the index that you want to display. Moreover, it is advisable to add space, which will serve as a gap to separate one category with respect to the next (as shown in Figure 7-39).

```
import matplotlib.pyplot as plt
import numpy as np
index = np.arange(5)+0.5
values1 = [5,7,3,4,6]
values2 = [6,6,4,5,7]
values3 = [5,6,5,4,6]
bw = 0.3
plt.axis([0,5.5,0,8])
plt.title('A Multiseries Bar Chart',fontsize=20)
plt.bar(index,values1,bw,color='b')
plt.bar(index+bw,values2,bw,color='g')
plt.bar(index+2*bw,values3,bw,color='r')
plt.xticks(index+bw,['A','B','C','D','E'])
```

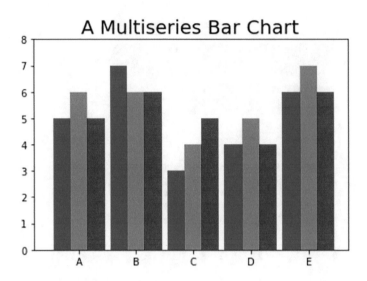

Figure 7-39. *A multiseries bar chart displaying three series*

Regarding the multiseries horizontal bar chart (see Figure 7-40), things are very similar. You have to replace the bar() function with the corresponding barh() function and remember to replace the xticks() function with the yticks() function. You also need to reverse the range of values covered by the axes in the axis() function.

```
import matplotlib.pyplot as plt
import numpy as np
index = np.arange(5)+0.5
values1 = [5,7,3,4,6]
values2 = [6,6,4,5,7]
values3 = [5,6,5,4,6]
bw = 0.3
plt.axis([0,8,0,5.5])
plt.title('A Multiseries Horizontal Bar Chart',fontsize=20)
plt.barh(index,values1,bw,color='b')
plt.barh(index+bw,values2,bw,color='g')
plt.barh(index+2*bw,values3,bw,color='r')
plt.yticks(index+bw,['A','B','C','D','E'])
```

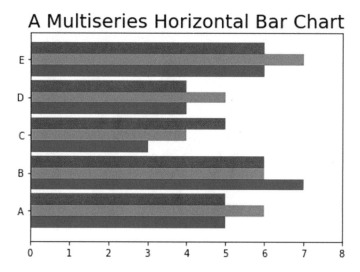

Figure 7-40. A multiseries horizontal bar chart

Multiseries Bar Charts with a pandas Dataframe

As you saw in the line charts, the matplotlib library also provides the ability to directly represent the dataframe objects containing the results of data analysis in the form of bar charts. And even here it does it quickly, directly, and automatically. The only thing you need to do is use the plot() function applied to the dataframe object and specify inside a kwarg called kind, to which you have to assign the type of chart you want to represent, which in this case is bar. Without specifying any other settings, you will get the bar chart shown in Figure 7-41.

```
import matplotlib.pyplot as plt
import numpy as np
import pandas as pd
data = {'series1':[1,3,4,3,5],
        'series2':[2,4,5,2,4],
        'series3':[3,2,3,1,3]}
df = pd.DataFrame(data, index=['A','B','C','D','E'])
df.plot(kind='bar', xlabel='Class',rot=0, ylabel='Value')
```

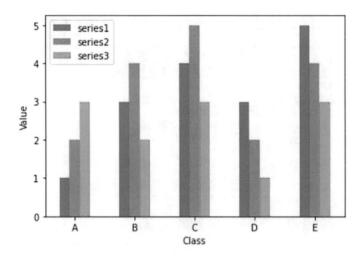

Figure 7-41. *The values in a dataframe can be directly displayed as a bar chart*

However, if you want to get more control, or if your case requires it, you can still extract portions of the dataframe as NumPy arrays and use them as illustrated in the previous examples in this section. That is, by passing them separately as arguments to the matplotlib functions.

Moreover, regarding the horizontal bar chart, the same rules can be applied, but remember to set barh as the value of the kind kwarg.

```
import matplotlib.pyplot as plt
import numpy as np
import pandas as pd
data = {'series1':[1,3,4,3,5],
        'series2':[2,4,5,2,4],
        'series3':[3,2,3,1,3]}
df = pd.DataFrame(data, index=['A','B','C','D','E'])
df.plot(kind='barh',ylabel='Class', xlabel='Value')
```

You'll get a multiseries horizontal bar chart, as shown in Figure 7-42.

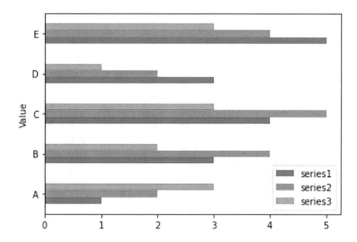

Figure 7-42. *A horizontal bar chart could be a valid alternative to visualize your dataframe values*

Multiseries Stacked Bar Charts

Another way to represent a multiseries bar chart is in the stacked form, in which the bars are stacked one on the other. This is especially useful when you want to show the total value obtained by the sum of all the bars.

To transform a simple multiseries bar chart in a stacked one, you add the bottom kwarg to each bar() function. Each series must be assigned to the corresponding bottom kwarg. At the end, you will obtain the stacked bar chart, as shown in Figure 7-43.

```
import matplotlib.pyplot as plt
import numpy as np
series1 = np.array([3,4,5,3])
series2 = np.array([1,2,2,5])
series3 = np.array([2,3,3,4])
index = np.arange(4)
plt.axis([-0.5,3.5,0,15])
plt.title('A Multiseries Stacked Bar Chart')
plt.bar(index,series1,color='r')
plt.bar(index,series2,color='b',bottom=series1)
plt.bar(index,series3,color='g',bottom=(series2+series1))
plt.xticks(index,['Jan23','Feb23','Mar23','Apr23'])
plt.legend(['series1','series2','series3'], loc='upper left')
```

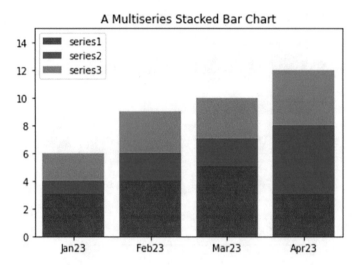

Figure 7-43. *A multiseries stacked bar*

Here too, in order to create the equivalent horizontal stacked bar chart, you need to replace the bar() function with barh() function, being careful to change the other parameters as well. Indeed, the xticks() function should be replaced with the yticks() function because the labels of the categories must now be reported on the y-axis. After making all these changes, you will obtain the horizontal stacked bar chart shown in Figure 7-44.

```
import matplotlib.pyplot as plt
import numpy as np
index = np.arange(4)
series1 = np.array([3,4,5,3])
series2 = np.array([1,2,2,5])
series3 = np.array([2,3,3,4])
plt.axis([0,15,-0.5,3.5])
plt.title('A Multiseries Horizontal Stacked Bar Chart')
plt.barh(index,series1,color='r')
plt.barh(index,series2,color='g',left=series1)
plt.barh(index,series3,color='b',left=(series1+series2))
plt.yticks(index,['Jan23','Feb23','Mar23','Apr23'])
plt.legend(['series1','series2','series3'], loc='lower right')
```

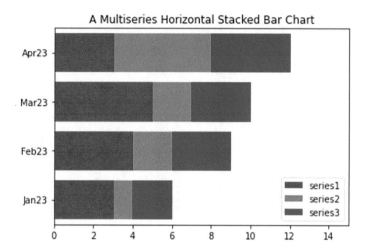

Figure 7-44. *A multiseries horizontal stacked bar chart*

Stacked Bar Charts with a pandas Dataframe

Also with regard to stacked bar charts, it is very simple to directly represent the values contained in the dataframe object by using the plot() function. You need only to add as an argument the stacked kwarg set to True (see Figure 7-45).

```
import matplotlib.pyplot as plt
import pandas as pd

data = {'series1':[1,3,4,3,5],
        'series2':[2,4,5,2,4],
        'series3':[3,2,3,1,3]}
df = pd.DataFrame(data, index=['A','B','C','D','E'])
df.plot(kind='bar', stacked=True, rot=0, xlabel='Class', ylabel='Value')
```

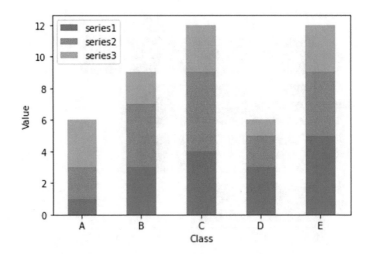

Figure 7-45. *The values of a dataframe can be directly displayed as a stacked bar chart*

Other Bar Chart Representations

Another type of very useful representation is that of a bar chart for comparison, where two series of values sharing the same categories are compared by placing the bars in opposite directions along the y-axis. In order to do this, you have to put the y values of one of the two series in a negative form. Also in this example, you will see the possibility of coloring the inner color of the bars in a different way. In fact, you can do this by setting the two different colors on a specific kwarg: facecolor.

Furthermore, in this example, you will see how to add the y value with a label at the end of each bar. This can increase the readability of the bar chart. You can do this using a for loop in which the text() function will show the y value. You can adjust the label position with the two kwargs called ha and va, which control the horizontal and vertical alignment, respectively. The result will be the chart shown in Figure 7-46.

```
import matplotlib.pyplot as plt
x0 = np.arange(8)
y1 = np.array([1,3,4,6,4,3,2,1])
y2 = np.array([1,2,5,4,3,3,2,1])
plt.ylim(-7,7)
plt.bar(x0,y1,0.9,facecolor='r')
plt.bar(x0,-y2,0.9,facecolor='b')
plt.xticks(())
plt.grid(True)
for x, y in zip(x0, y1):
        plt.text(x + 0.4, y + 0.05, '%d' % y, ha='center', va= 'bottom')
for x, y in zip(x0, y2):
        plt.text(x + 0.4, -y - 0.05, '%d' % y, ha='center', va= 'top')
```

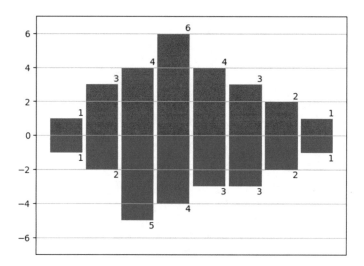

Figure 7-46. *Two series can be compared using this kind of bar chart*

Pie Charts

An alternative way to display data to the bar charts is the pie chart, easily obtainable using the pie() function.

Even for this type of function, you pass as the main argument a list containing the values to be displayed. I chose the percentages (their sum is 100), but you can use any kind of value. It will be up to the pie() function to inherently calculate the percentage occupied by each value.

Also with this type of representation, you need to define some key features using the kwargs. For example, if you want to define the sequence of the colors, which will be assigned to the sequence of input values correspondingly, you have to use the colors kwarg. Therefore, you have to assign a list of strings, each containing the name of the desired color. Another important feature is to add labels to each slice of the pie. To do this, you have to use the labels kwarg to which you assign a list of strings containing the labels to be displayed in sequence.

In addition, in order to draw the pie chart in a perfectly spherical way, you have to add the axis() function to the end, specifying the string 'equal' as an argument. You will get a pie chart as shown in Figure 7-47.

```
import matplotlib.pyplot as plt
labels = ['Nokia','Samsung','Apple','Lumia']
values = [10,30,45,15]
colors = ['yellow','green','red','blue']
plt.pie(values,labels=labels,colors=colors)
plt.axis('equal')
```

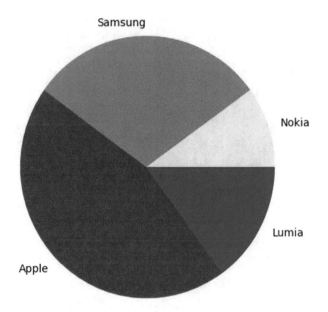

Figure 7-47. *A very simple pie chart*

To add complexity to the pie chart, you can draw it with a slice extracted from the pie. This is usually done when you want to focus on a specific slice. In this case, for example, you would highlight the slice referring to Nokia. In order to do this, there is a special kwarg named explode. It is nothing but a sequence of float values of 0 or 1, where 1 corresponds to the fully extended slice and 0 corresponds to slices completely in the pie. All intermediate values correspond to an intermediate degree of extraction (see Figure 7-48).

You can also add a title to the pie chart with the title() function. You can adjust the angle of rotation of the pie by adding the startangle kwarg, which takes an integer value between 0 and 360, which are the degrees of rotation precisely (0 is the default value).

The modified chart should appear as shown in Figure 7-48.

```
import matplotlib.pyplot as plt
labels = ['Nokia','Samsung','Apple','Lumia']
values = [10,30,45,15]
colors = ['yellow','green','red','blue']
explode = [0.3,0,0,0]
plt.title('A Pie Chart')
plt.pie(values,labels=labels,colors=colors,explode=explode,startangle=180)
plt.axis('equal')
```

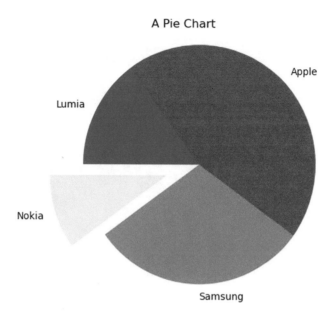

Figure 7-48. *A more advanced pie chart*

But the possible additions that you can insert in a pie chart do not end here. For example, a pie chart does not have axes with ticks, so it is difficult to imagine the perfect percentage represented by each slice. To overcome this, you can use the autopct kwarg, which adds to the center of each slice a text label showing the corresponding value.

If you want to make it an even more appealing image, you can add a shadow with the shadow kwarg and setting it to True. In the end, you will get a pie chart as shown in Figure 7-49.

```
import matplotlib.pyplot as plt
labels = ['Nokia','Samsung','Apple','Lumia']
values = [10,30,45,15]
colors = ['yellow','green','red','blue']
explode = [0.3,0,0,0]
plt.title('A Pie Chart')
plt.pie(values,
        labels=labels,
        colors=colors,
        explode=explode,
        shadow=True,
        autopct='%1.1f%%',
        startangle=180)plt.axis('equal')
```

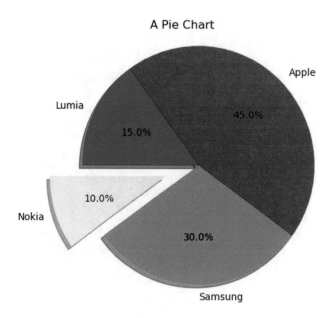

Figure 7-49. *An even more advanced pie chart*

Pie Charts with a pandas Dataframe

Even for the pie chart, you can represent the values contained in a dataframe object. In this case, however, the pie chart can represent only one series at a time, so in this example you display only the values of the first series by specifying df['series1']. You have to specify the type of chart you want to represent through the kind kwarg in the plot() function, which in this case is pie. Furthermore, because you want to represent a pie chart as perfectly circular, it is necessary that you add the figsize kwarg. At the end, you will obtain a pie chart as shown in Figure 7-50.

```
import matplotlib.pyplot as plt
import pandas as pd
data = {'series1':[1,3,4,3,5],
                'series2':[2,4,5,2,4],
                'series3':[3,2,3,1,3]}
df = pd.DataFrame(data)
df['series1'].plot(kind='pie',figsize=(6,6))
```

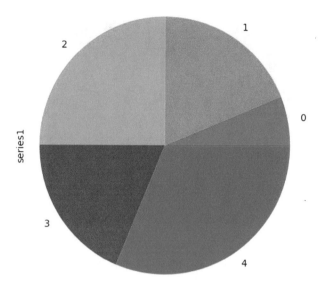

Figure 7-50. *The values in a pandas dataframe can be directly drawn as a pie chart*

Advanced Charts

In addition to the more classical charts such as bar charts and pie charts, you might want to represent your results in alternative ways. On the Internet and in various publications, there are many examples in which many alternative graphics solutions are discussed and proposed, some really brilliant and captivating. This section only shows some graphic representations; a more detailed discussion about this topic is beyond the purpose of this book. You can use this section as an introduction to a world that is constantly expanding: data visualization.

Contour Plots

A quite common type of chart in the scientific world is the *contour plot* or *contour map*. This visualization is in fact suitable for displaying three-dimensional surfaces through a contour map composed of closed curves showing the points on the surface that are located at the same level, or that have the same z value.

Although visually the contour plot is a very complex structure, its implementation is not so difficult, thanks to the `matplotlib` library. First, you need the function $z = f(x, y)$ to generate a three-dimensional surface. Then, once you have defined a range of values x, y that will define the area of the map to be displayed, you can calculate the z values for each pair (x, y), applying the function $f(x, y)$ just defined in order to obtain a matrix of z values. Finally, thanks to the `contour()` function, you can generate the contour map of the surface. It is often desirable to add a color map along with a contour map. That is, the areas delimited by the curves of level are filled with a color gradient, defined by a color map. For example, as in Figure 7-51, you may indicate negative values with increasingly dark shades of blue, and move to yellow and then red with the increase of positive values.

235

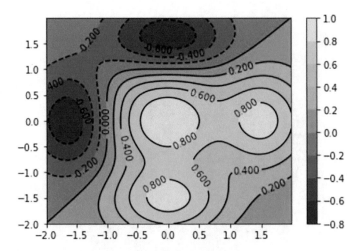

Figure 7-51. *A contour map can describe the z values of a surface*

Furthermore, when you have to deal with this kind of representation, adding a color scale as a reference to the side of the graph is almost a must. This is possible by simply adding the `colorbar()` function to the end of the code.

```
import matplotlib.pyplot as plt
import numpy as np
dx = 0.01; dy = 0.01
x = np.arange(-2.0,2.0,dx)
y = np.arange(-2.0,2.0,dy)
X,Y = np.meshgrid(x,y)
def f(x,y):
            return (1 - y**5 + x**5)*np.exp(-x**2-y**2)
C = plt.contour(X,Y,f(X,Y),8,colors='black')
plt.contourf(X,Y,f(X,Y),8)
plt.clabel(C, inline=1, fontsize=10)
plt.colorbar()
```

The standard color gradient (color map) is represented in Figure 7-51. You can choose among a large number of color maps available just specifying them with the cmap kwarg.

Polar Charts

Another type of advanced chart that is popular is the *polar chart*. This type of chart is characterized by a series of sectors that extend radially; each of these areas will occupy a certain angle. Thus you can display two different values by assigning them to the magnitudes that characterize the polar chart—the extension of the radius r and the angle θ occupied by the sector. These in fact are the polar coordinates (r, θ), an alternative way of representing functions at the coordinate axes. From the graphical point of view, you can imagine it as a kind of chart that has characteristics both of the pie chart and of the bar chart. In fact as the pie chart, the angle of each sector gives percentage information represented by that category with respect to the total. As for the bar chart, the radial extension is the numerical value of that category.

So far you have used the standard set of colors using single characters as the color code (e.g., r to indicate red). In fact you can use any sequence of colors you want. You simply have to define a list of string values that contain RGB codes in the #rrggbb format corresponding to the colors you want.

Oddly, to get a polar chart you have to use the bar() function and pass the list containing the angles θ and a list of the radial extension of each sector. The result will be a polar chart, as shown in Figure 7-52.

```
import matplotlib.pyplot as plt
import numpy as np
N = 8
 theta = np.arange(0.,2 * np.pi, 2 * np.pi / N)
radii = np.array([4,7,5,3,1,5,6,7])
plt.axes([0.025, 0.025, 0.95, 0.95], polar=True)
colors = np.array(['#4bb2c5', '#c5b47f', '#EAA228', '#579575', '#839557', '#958c12',
 '#953579', '#4b5de4'])
 bars = plt.bar(theta, radii, width=(2*np.pi/N), bottom=0.0, color=colors)
```

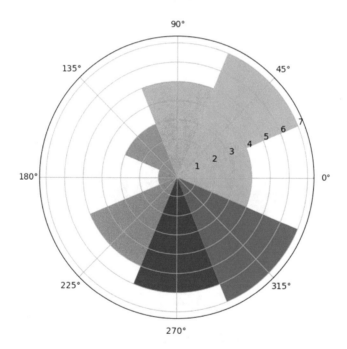

Figure 7-52. *A polar chart*

The mplot3d Toolkit

The mplot3d toolkit is included with all standard installations of matplotlib and allows you to extend the capabilities of visualization to 3D data. If the figure is displayed in a separate window, you can rotate the axes of the three-dimensional representation with the mouse.

With this package you are still using the Figure object, only instead of the Axes object, you will define a new kind of object, called Axes3D, which is introduced in this toolkit. Thus, you need to add a new import to the code if you want to use the Axes3D object.

```
from mpl_toolkits.mplot3d import Axes3D
```

3D Surfaces

In a previous section, you used the contour plot to represent the three-dimensional surfaces through the level lines. Using the mplot3D package, you can draw surfaces directly in 3D. In this example, you use the same function $z = f(x, y)$ you used in the contour map.

Once you have calculated the meshgrid, you can view the surface with the plot_surface() function. A three-dimensional blue surface will appear, as shown in Figure 7-53.

```
from mpl_toolkits.mplot3d import Axes3D
import matplotlib.pyplot as plt
fig = plt.figure()
ax = fig.add_subplot(111, projection="3d")
X = np.arange(-2,2,0.1)
Y = np.arange(-2,2,0.1)
X,Y = np.meshgrid(X,Y)
def f(x,y):
    return (1 - y**5 + x**5)*np.exp(-x**2-y**2)
ax.plot_surface(X,Y,f(X,Y), rstride=1, cstride=1)
```

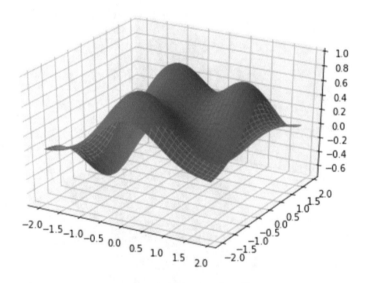

Figure 7-53. *A 3D surface can be represented with the mplot3d toolkit*

A 3D surface stands out most by changing the color map, for example by setting the cmap kwarg. You can also rotate the surface using the view_init() function. In fact, this function adjusts the view point from which you see the surface, changing the two kwargs called elev and azim. Through their combination you can get the surface displayed from any angle. The first kwarg adjusts the height at which the surface is seen, while azim adjusts the angle of rotation of the surface.

For instance, you can change the color map using plt.cm.hot and moving the view point to elev=30 and azim=125. The result is shown in Figure 7-54.

```
from mpl_toolkits.mplot3d import Axes3D
import matplotlib.pyplot as plt
fig = plt.figure()
```

```
ax = fig.add_subplot(111, projection="3d")
X = np.arange(-2,2,0.1)
Y = np.arange(-2,2,0.1)
X,Y = np.meshgrid(X,Y)
def f(x,y):
    return (1 - y**5 + x**5)*np.exp(-x**2-y**2)
ax.plot_surface(X,Y,f(X,Y), rstride=1, cstride=1, cmap=plt.cm.hot)
ax.view_init(elev=30,azim=125)
```

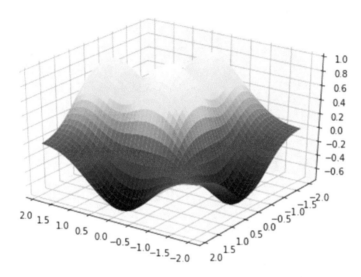

Figure 7-54. *The 3D surface can be rotated and observed from a higher viewpoint*

Scatter Plots in 3D

The mode most used among all 3D views remains the 3D scatter plot. With this type of visualization, you can identify if the points follow particular trends, but above all if they tend to cluster.

In this case, you use the scatter() function as the 2D case but applied to the Axes3D object. By doing this, you can visualize different series, expressed by the calls to the scatter() function, all together in the same 3D representation (see Figure 7-55).

```
import matplotlib.pyplot as plt
import numpy as np
from mpl_toolkits.mplot3d import Axes3D
xs = np.random.randint(30,40,100)
ys = np.random.randint(20,30,100)
zs = np.random.randint(10,20,100)
xs2 = np.random.randint(50,60,100)
ys2 = np.random.randint(30,40,100)
zs2 = np.random.randint(50,70,100)
xs3 = np.random.randint(10,30,100)
ys3 = np.random.randint(40,50,100)
zs3 = np.random.randint(40,50,100)
fig = plt.figure()
```

```
ax = fig.add_subplot(111, projection="3d")
ax.scatter(xs,ys,zs)
ax.scatter(xs2,ys2,zs2,c='r',marker='^')
ax.scatter(xs3,ys3,zs3,c='g',marker='*')
ax.set_xlabel('X Label')
ax.set_ylabel('Y Label')
ax.set_zlabel('Z Label')
```

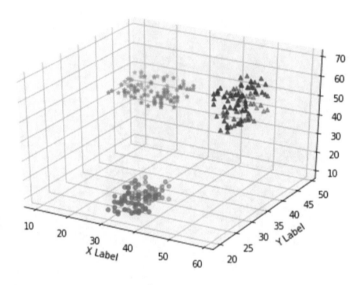

Figure 7-55. *This 3D scatter plot shows three different clusters*

Bar Charts in 3D

Another type of 3D plot widely used in data analysis is the 3D bar chart. Also in this case, you use the bar() function applied to the object Axes3D. If you define multiple series, you can accumulate several calls to the bar() function in the same 3D visualization (see Figure 7-56).

```
import matplotlib.pyplot as plt
import numpy as np
from mpl_toolkits.mplot3d import Axes3D
x = np.arange(8)
y = np.random.randint(0,10,8)
y2 = y + np.random.randint(0,3,8)
y3 = y2 + np.random.randint(0,3,8)
y4 = y3 + np.random.randint(0,3,8)
y5 = y4 + np.random.randint(0,3,8)
clr = ['#4bb2c5', '#c5b47f', '#EAA228', '#579575', '#839557', '#958c12', '#953579', '#4b5de4']
fig = plt.figure()
ax = fig.add_subplot(111, projection="3d")
ax.bar(x,y,0,zdir='y',color=clr)
ax.bar(x,y2,10,zdir='y',color=clr)
```

```
ax.bar(x,y3,20,zdir='y',color=clr)
ax.bar(x,y4,30,zdir='y',color=clr)
 ax.bar(x,y5,40,zdir='y',color=clr)
ax.set_xlabel('X Axis')
ax.set_ylabel('Y Axis')
ax.set_zlabel('Z Axis')
ax.view_init(elev=40)
```

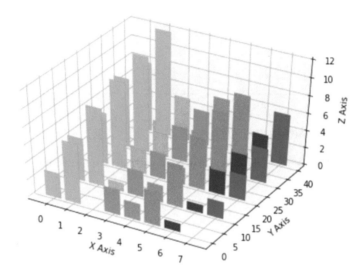

Figure 7-56. *A 3D bar chart*

Multipanel Plots

So far you've seen different ways of representing data through a chart. You saw how to see more charts in the same figure by separating them with subplots. In this section, you deepen your understanding of this topic by analyzing more complex cases.

Display Subplots Within Other Subplots

Now an even more advanced method is explained: the ability to view charts within others, enclosed within frames. Because you are dealing with frames, that is, Axes objects, you need to separate the main Axes (i.e., the general chart) from the frame you want to add that will be another instance of Axes. To do this, you use the figures() function to get the Figure object, on which you will define two different Axes objects using the add_axes() function. See the result of this example in Figure 7-57.

```
import matplotlib.pyplot as plt
fig = plt.figure()
ax = fig.add_axes([0.1,0.1,0.8,0.8])
inner_ax = fig.add_axes([0.6,0.6,0.25,0.25])
```

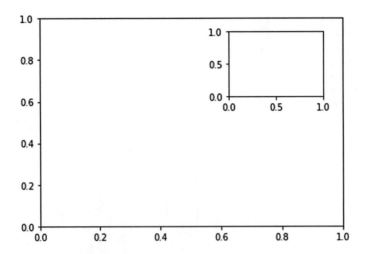

Figure 7-57. *A subplot is displayed within another plot*

To better understand the effect of this mode of display, you can fill the previous Axes with real data, as shown in Figure 7-58.

```
import matplotlib.pyplot as plt
import numpy as np
fig = plt.figure()
ax = fig.add_axes([0.1,0.1,0.8,0.8])
inner_ax = fig.add_axes([0.6,0.6,0.25,0.25])
x1 = np.arange(10)
y1 = np.array([1,2,7,1,5,2,4,2,3,1])
x2 = np.arange(10)
y2 = np.array([1,3,4,5,4,5,2,6,4,3])
ax.plot(x1,y1)
inner_ax.plot(x2,y2)
```

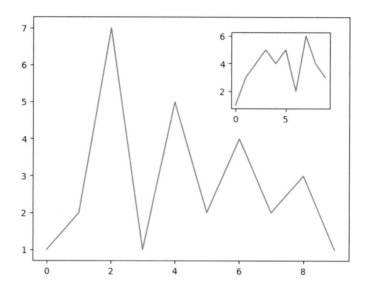

Figure 7-58. *A more realistic visualization of a subplot within another plot*

Grids of Subplots

You have already seen the creation of subplots. It is quite simple to add subplots using the subplots()
function and then dividing a plot into sectors. matplotlib allows you to manage even more complex cases
using another function called GridSpec(). This subdivision allows splitting the drawing area into a grid of
subareas, to which you can assign one or more of them to each subplot, so that in the end you can obtain
subplots with different sizes and orientations, as you can see in Figure 7-59.

```
import matplotlib.pyplot as plt
gs = plt.GridSpec(3,3)
fig = plt.figure(figsize=(6,6))
fig.add_subplot(gs[1,:2])
fig.add_subplot(gs[0,:2])
fig.add_subplot(gs[2,0])
fig.add_subplot(gs[:2,2])
fig.add_subplot(gs[2,1:])
```

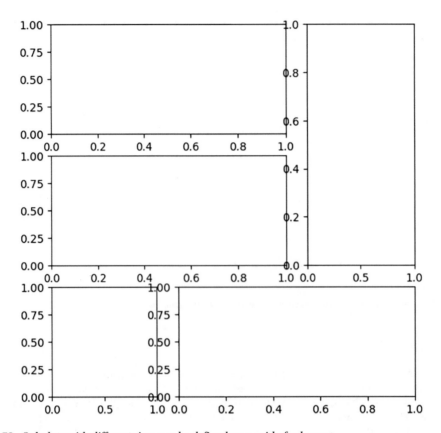

Figure 7-59. *Subplots with different sizes can be defined on a grid of subareas*

Now that it's clear to you how to manage the grid by assigning the various sectors to the subplot, it's time to see how to use these subplots. In fact, you can use the Axes object returned by each `add_subplot()` function to call the `plot()` function and draw the corresponding plot (see Figure 7-60).

```
import matplotlib.pyplot as plt
import numpy as np
gs = plt.GridSpec(3,3)
fig = plt.figure(figsize=(6,6))
x1 = np.array([1,3,2,5])
y1 = np.array([4,3,7,2])
x2 = np.arange(5)
y2 = np.array([3,2,4,6,4])
s1 = fig.add_subplot(gs[1,:2])
s1.plot(x,y,'r')
s2 = fig.add_subplot(gs[0,:2])
s2.bar(x2,y2)
s3 = fig.add_subplot(gs[2,0])
s3.barh(x2,y2,color='g')
s4 = fig.add_subplot(gs[:2,2])
s4.plot(x2,y2,'k')
s5 = fig.add_subplot(gs[2,1:])
s5.plot(x1,y1,'b^',x2,y2,'yo')
```

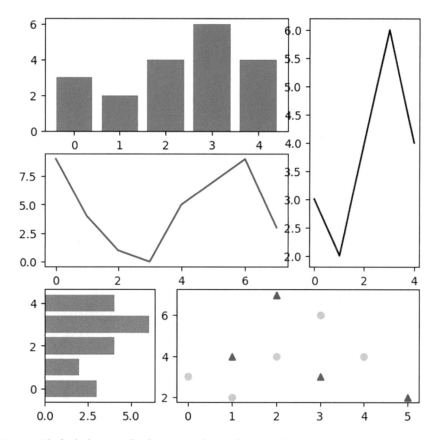

Figure 7-60. *A grid of subplots can display many plots at the same time*

The Seaborn Library

Seaborn is a data visualization library based on the `matplotlib` library. Unlike `matplotlib`, this library already offers graphic solutions prepared with incorporated graphics and statistical calculation elements. This library was specifically designed to work closely with pandas and its data structures.

First you need to install the `seaborn` library inside your virtual environment. If you are using the Anaconda platform, you can simply install it through Anaconda Navigator. Select the package among those available in the virtual environment you are working on and start the installation, as shown in Figure 7-61.

Figure 7-61. *Installing Seaborn with Anaconda Navigator*

Or you can install the `seaborn` library through the CMD.exe Prompt console. Enter the following command:

```
conda install seaborn
```

For those who do not have an Anaconda platform, you can still install the package through the PyPI system with the `pip` command.

```
pip install seaborn
```

Once Seaborn is installed, you can open an IPython session, or rather a Jupyter Notebook, and import the library:

```
import seaborn as sns
```

At this point you can begin to chart with Seaborn. This task is divided into three parts:

- Selecting a theme
- Loading a dataset
- Creating the graph

Let's start with an extremely simple example that introduces these three steps.

First you need to select a theme. By theme, I mean a series of graphic settings (shape, color, grid) that determine the graphic layout and style. In this regard, Seaborn provides a series of ready-made themes, which can be selected by using the `set_theme()` function and passing the name of the theme as an argument. If you want to use the default one, it is sufficient not to set any arguments to the function.

```
sns.set_theme()
```

Otherwise, you can specify the various styles and predefined color palettes according to your needs or tastes, specifying the optional parameters with relative names. As an example:

```
sns.set_theme(style='darkgrid', palette='Set2')
```

This example sets a graphic theme with a background of dark grids and one of the many color palettes available in the library.

At this point, the next step is to load a dataset containing the points and information to be displayed in a graph. Later, you will see how to load data from a pandas dataframe, but for the moment you will use one of the many example datasets available in the library.

```
tips = sns.load_dataset("tips")
tips
```

As a result, you will obtain a dataframe (see Figure 7-62) containing a series of data relating to the tips left to restaurant waiters by different individuals, categorized by gender, by type of meal, and by the cost of the meal.

	total_bill	tip	sex	smoker	day	time	size
0	16.99	1.01	Female	No	Sun	Dinner	2
1	10.34	1.66	Male	No	Sun	Dinner	3
2	21.01	3.50	Male	No	Sun	Dinner	3
3	23.68	3.31	Male	No	Sun	Dinner	2
4	24.59	3.61	Female	No	Sun	Dinner	4
...	...	...	...	...	...	...	...
239	29.03	5.92	Male	No	Sat	Dinner	3
240	27.18	2.00	Female	Yes	Sat	Dinner	2
241	22.67	2.00	Male	Yes	Sat	Dinner	2
242	17.82	1.75	Male	No	Sat	Dinner	2
243	18.78	3.00	Female	No	Thur	Dinner	2

244 rows × 7 columns

Figure 7-62. *The contents of the tip dataframe in the Seaborn library*

Now that you have looked at the content of the dataframe, you might want to create a chart in which the different tips (x-axis) are represented in relation to the bill paid (y-axis). To make a graph with Seaborn, you use the relplot() function. This function accepts a large number of parameters, which will then specify how the different dataframe data will be associated with different characteristics of the chart. In this case, it is very simple, since you only want to relate the values contained in the Tip and Total_Bill columns on the Cartesian axes.

```
sns.relplot(
    data=tips,
    x='total_bill', y='tip',
)
```

In this case, you obtain a scatter plot identical to the one shown in Figure 7-63.

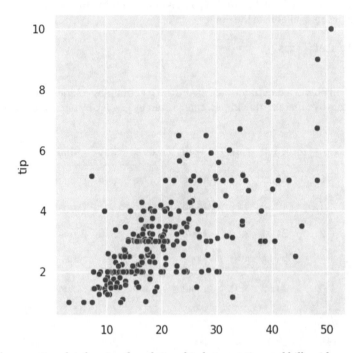

Figure 7-63. *Seaborn scatter plot showing the relationship between tips and bill paid*

You can see the tendency of the points to a certain linear relationship, which tends to define a tip value of 20 percent of the bill paid, even though there is a good probability that the tips are less than this value and a very small probability that they are higher.

You can move forward with the data analysis by adding more information to your graph. Taking a look at the tips dataframe, you can see that each tip is associated with the gender of the payer. Let's see if there is a different tendency between these two genders in giving tips. You need to insert the gender information into the graph, differentiating the appearance of the points represented using the style and hue parameters, which respectively indicate the shape of the point and its color.

```
sns.relplot(
    data=tips,
    x='total_bill', y='tip',
    hue='sex', style='sex',
)
```

When you run the code, you will get a chart identical to the one shown in Figure 7-64.

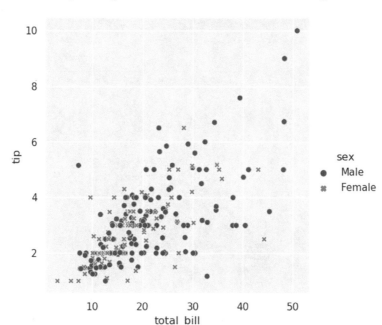

Figure 7-64. *Scatter plot in which the dots separate the data relating to the payer's gender*

In this case, it is clear that there is no distinction in the behavior of the payers, regardless of whether they are men or women.

But Seaborn is not limited only to graphical representations, rather it offers statistical tools that analyze the data to be displayed and add graphical components with statistical information as a result. In the scatter plot example, there is a discrete linear relationship between the amount of tips and the bill paid. Well, Seaborn offers a dedicated function called lmplot(). It makes a scatter plot graph like the one made earlier with relplot(), but with the addition of a graph showing the regression fit on the base scatter plot.

```
sns.lmplot(data=tips, x='total_bill', y='tip')
```

Running this function, you will get a scatter plot like the one shown in Figure 7-65.

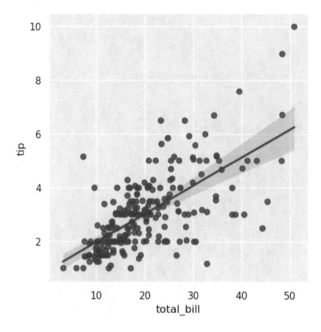

Figure 7-65. *Scatter plot with a linear regression graphic element*

As you can see, what may at first appear to be a 20 percent trend with a quick scan has been proven wrong. Looking at the trend of the straight line, it is clear that this value becomes gradually smaller as the amount paid increases. It can therefore be assumed that there is a 20 percent trend in tipping for small bill expenses like $10, but then as the figure increases, the tips will gradually decrease in percentage (although the tip figure apparently always increases).

Continuing this data analysis approach, you previously made the assumption that the percentage of tips given did not depend on the gender of the payer. You therefore can represent the previous graph taking into account this subdivision by gender.

```
sns.lmplot(data=tips, x='total_bill', y='tip', hue='sex')
```

Running this function, you obtain a scatter plot like the one shown in Figure 7-66.

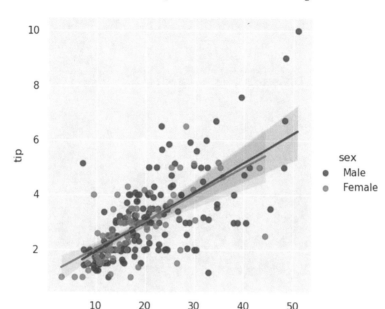

Figure 7-66. *The scatter plot with linear regressions performed by gender*

As you can see, the assumption you made with a quick glance proved to be true. The linear regression lines are virtually identical for men and women.

But what if you wanted to have statistical information in numerical form and not just as graphics? So far you have seen that Seaborn can add elements graphically for statistical evaluations such as linear regression on a set of points. However, to add precise numerical values to be reported on the scatter plot, it is necessary to obtain them using the SciPy library.

If you want to add these numbers to your chart, you have to add a few lines of code to the previous case (the one without distinction of gender, given that the statistics are almost the same).

```
import seaborn as sns
import matplotlib.pyplot as plt
from scipy import stats

g = sns.lmplot(data=tips, x='total_bill', y='tip')

def annotate(data, **kws):
    slope, intercept, r, p, std_err = stats.linregress(data['total_bill'],data['tip'])
    ax = plt.gca()
    ax.text(.05, .8, 'y={0:.1f}x+{1:.1f}, r={2:.2f}, \np={3:.2g}, std_err={4:.3f}'.
    format(slope,intercept,r,p, std_err),
            transform=ax.transAxes)

g.map_dataframe(annotate)
plt.show()
```

The scatter plot is thus obtained with all the statistical information of the linear regression, such as the coefficients of the line, the *rho* and *p* values, and the standard deviation (see Figure 7-67).

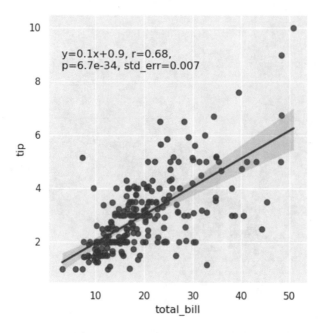

Figure 7-67. *Scatter plot with addition of linear regression statistic values obtained from SciPy*

A further and interesting approach with Seaborn is to graphically study the statistical distributions of a series of samples. Being able to graphically observe how certain values are distributed allows you to acquire a lot of information about the types of values you are studying. For example, using the tips dataframe example, with Seaborn it is possible to visualize the distribution of the values of the tips and the bills paid, to carry out an analysis on the sample taken into consideration. In this regard, the jointplot() function allows you to keep the graph of the scatter plot like the one seen previously, but add the distributions of the values of the points on the x-axis and y-axis as bar plots corresponding to each axis.

```
sns.jointplot(data=tips, x='total_bill', y='tip')
```

By performing this simple function, you obtain a splendid scatter plot like the one shown in Figure 7-68.

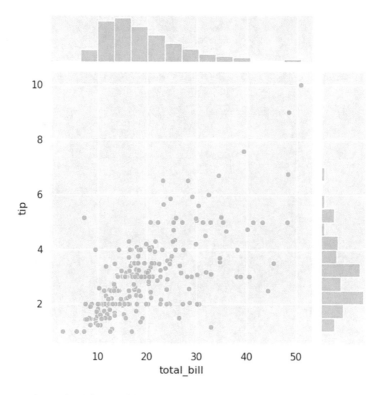

Figure 7-68. *Scatter plot with addition of distributions of values on both axes*

At a quick glance, you can see how the two distributions are very similar, mainly centering with their maximum value on a $15 bill and a $3 tip. It is possible to obtain other useful information on the population of the samples, for example, by reintroducing the gender distinction. In this case, it is sufficient to add the sex value to the hue parameter.

```
sns.jointplot(data=tips, x='total_bill', y='tip', hue='sex')
```

If you run this function, you obtain a scatter plot similar to the one shown in Figure 7-69.

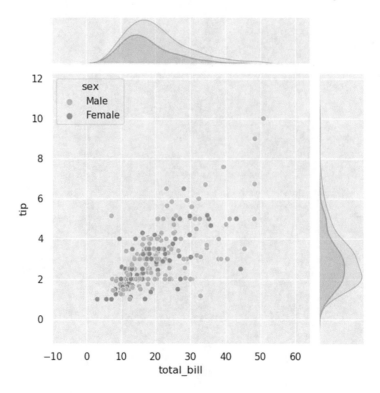

Figure 7-69. *Scatter plot with populations distinguished by gender, with relative distributions of values*

The values present in the dataframe relate more to males than to females. As far as distributions are concerned, it would seem that men tend to pay slightly more expensive bills at restaurants (perhaps because they eat more or perhaps because they could be more gallant… ☺), while no differences in "generosity" can be deduced between the two sexes, as the maximum points are equivalent between the two distributions. In the case of accounts, although the distribution of the male population is greater, the maximum point is shifted toward higher figures.

So far you have used one of the many datasets available within the library, but in real cases, you will have to use dataframes external to it. Consider as an example one of the dataframes you used for `matplotlib` to generate multiseries bar plots.

```
import numpy as np
import pandas as pd
data = {'type':['A','B','C','D','E'],
        'series1':[1,3,4,3,5],
        'series2':[2,4,5,2,4],
        'series3':[3,2,3,1,3]}
df = pd.DataFrame(data)
df
```

Running the code, you will get the dataframe shown in Figure 7-70.

	type	series1	series2	series3
0	A	1	2	3
1	B	3	4	2
2	C	4	5	3
3	D	3	2	1
4	E	5	4	3

Figure 7-70. *Dataframe to be used to generate a multiseries bar plot with Seaborn*

But the dataframe in this form is not congenial to be used with Seaborn. It is appropriate to transform it into a different form. Indeed, in the dataframe, you have three columns of the same type: *series1, series2,* and *series3*. You can concatenate these three columns into one, adding a new column containing the name of these as a value. To do this, you can use *unpivoting*, which with pandas is achieved with the merge() function.

```
df_unpivot = pd.melt(df, id_vars='type', value_vars=['series1', 'series2', 'series3'])
df_unpivot
```

By performing the unpivoting operation, you will get a new dataframe like the one shown in Figure 7-71.

	type	variable	value
0	A	series1	1
1	B	series1	3
2	C	series1	4
3	D	series1	3
4	E	series1	5
5	A	series2	2
6	B	series2	4
7	C	series2	5
8	D	series2	2
9	E	series2	4
10	A	series3	3
11	B	series3	2
12	C	series3	3
13	D	series3	1
14	E	series3	3

Figure 7-71. *The result of unpivoting the starting dataframe*

Now that you have a dataframe of suitable structure to be used with Seaborn, you can use the `barplot()` function to generate the chart.

```
import seaborn as sns
%matplotlib inline
sns.barplot(data=df_unpivot, x='type', y='value', hue='variable')
```

Running the code, you will get a bar plot like the one shown in Figure 7-72.

`<matplotlib.axes._subplots.AxesSubplot at 0x7fb08d4f26d0>`

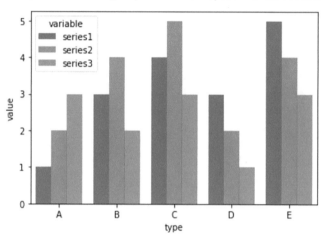

Figure 7-72. *The multiseries bar plot*

What you've seen here is just one of the many graphical possibilities available with Seaborn. For each case study, it is advisable to look for a graphical modality and a relative graphical-statistical approach that is adequate. Given the enormity of possible cases, the best advice is to look in the official documentation (`https://seaborn.pydata.org/index.html`) for the best solution.

Conclusions

In this chapter, you received all the fundamental aspects of the `matplotlib` library, and through a series of examples, you learned about the basic tools for handling data visualization. You have become familiar with various examples of how to develop different types of charts with a few lines of code. The chapter then went on to introduce, with a series of examples, the `seaborn` library, which extends the graphical aspects of `matplotlib` with `stat` elements.

This chapter concludes the part of the book about the libraries that provide the basic tools to perform data analysis. In the next chapter, you begin to learn about topics most closely related to data analysis.

## CHAPTER 8

Machine Learning with scikit-learn

In the chain of processes that make up data analysis, the construction phase of predictive models and their validation are done by a powerful library called scikit-learn. In this chapter, you'll see some examples that illustrate the basic construction of predictive models with some different methods.

The scikit-learn Library

scikit-learn is a Python module that integrates many machine learning algorithms. This library was developed initially by Cournapeu in 2007, but the first real release was in 2010.

This library is part of the SciPy (Scientific Python) group, which is a set of libraries created for scientific computing and especially for data analysis, many of which are discussed in this book. Generally, these libraries are defined as *SciKits*, hence the first part of the name of this library. The second part of the library's name is derived from *machine learning*, the discipline pertaining to this library.

Machine Learning

Machine learning is a discipline that deals with the study of methods for pattern recognition in datasets undergoing data analysis. In particular, it deals with the development of algorithms that learn from data and make predictions. Each methodology is based on building a specific model.

There are many methods that belong to the learning machine, each with its unique characteristics, which are specific to the nature of the data and the predictive model that you want to build. The choice of which method to apply is called a *learning problem*.

The data to be subjected to a pattern in the learning phase can be arrays composed of a single value per element, or of a multivariate value. These values are often referred to as *features* or *attributes*.

Supervised and Unsupervised Learning

Depending on the type of the data and the model to be built, you can separate learning problems into two broad categories—supervised and unsupervised.

© Fabio Nelli 2023
F. Nelli, *Python Data Analytics*, https://doi.org/10.1007/978-1-4842-9532-8_8

Supervised Learning

They are the methods in which the training set contains additional attributes that you want to predict (the *target*). Thanks to these values, you can instruct the model to provide similar values when you have to submit new values (the *test set*).

- *Classification*—The data in the training set belong to two or more classes or categories; then, the data, already being labeled, allow you to teach the system to recognize the characteristics that distinguish each class. When you need to consider a new value unknown to the system, the system can evaluate its class according to its characteristics.

- *Regression*—When the value to be predicted is a continuous variable. The simplest case to understand is when you want to find the line that describes the trend from a series of points represented in a scatterplot.

Unsupervised Learning

These are the methods in which the training set consists of a series of input values, x, without any corresponding target value.

- *Clustering*—The goal of these methods is to discover groups of similar examples in a dataset.

- *Dimensionality reduction*—Reduction of a high-dimensional dataset to one with only two or three dimensions is useful not just for data visualization, but for converting data of very high dimensionality into data of much lower dimensionality such that each of the lower dimensions conveys much more information.

In addition to these two main categories, there is another group of methods that serves to validate and evaluate the models.

Training Set and Testing Set

Machine learning enables the system to learn properties of a model from a dataset and apply these properties to new data. This is because a common practice in machine learning is to evaluate an algorithm. This valuation consists of splitting the data into two parts, one called the *training set*, which is used to learn the properties of the data, and the other called the *testing set,* on which to test these properties.

Supervised Learning with scikit-learn

In this chapter, you see a number of examples of supervised learning.

- Classification, using the Iris Dataset
  - K-Nearest Neighbors Classifier
  - Support Vector Machines (SVC)
- Regression, using the Diabetes Dataset
  - Linear Regression
  - Support Vector Machines (SVR)

Supervised learning consists of learning possible patterns between two or more features reading values from a training set; the learning is possible because the training set contains known results (target or labels). All models in `scikit-learn` are referred to as *supervised estimators,* using the `fit(x, y)` function that does the training. x comprises the features observed, while y indicates the target. Once the estimator has carried out the training, it can predict the value of y for any new observation x not labeled. This operation is completed using the `predict(x)` function.

The Iris Flower Dataset

The *Iris Flower Dataset* is a particular dataset used for the first time by Sir Ronald Fisher in 1936. It is often also called the Anderson Iris Dataset, after the person who collected the data by measuring the size of the various parts of the iris. In this dataset, data from three different species of iris (Iris silky, virginica Iris, and Iris versicolor) are collected. These data correspond to the length and width of the sepals and the length and width of the petals (see Figure 8-1).

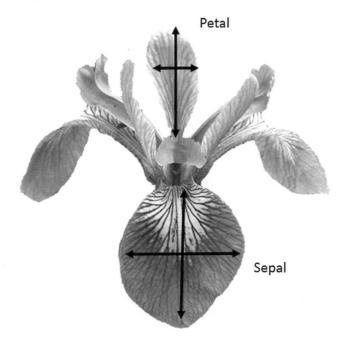

Figure 8-1. *Iris versicolor and the petal and sepal width and length*

This dataset is currently being used as a good example for many types of analysis, in particular for the problems of *classification,* which can be approached by means of machine learning methodologies. It is no coincidence then that this dataset is provided with the `scikit-learn` library as a 150x4 NumPy array.

Now you will study this dataset in detail, importing it into a Jupyter Notebook or into a normal Python session.

```
from sklearn import datasets
iris = datasets.load_iris()
```

In this way, you loaded all the data and metadata concerning the Iris Dataset into the `iris` variable. In order to see the values of the data collected in this variable, it is sufficient to call the `data` attribute of the `iris` variable.

```
iris.data
Out[ ]:
array([[ 5.1,  3.5,  1.4,  0.2],
       [ 4.9,  3. ,  1.4,  0.2],
       [ 4.7,  3.2,  1.3,  0.2],
       [ 4.6,  3.1,  1.5,  0.2],
       ...
```

As you can see, you get an array of 150 elements, each containing four numeric values: the size of sepals and petals respectively.

To determine instead what kind of flower belongs to each item, refer to the `target` attribute.

```
iris.target
Out[ ]:
array([0, 0, 0, 0, 0, 0, 0, 0, 0, 0, 0, 0, 0, 0, 0, 0, 0, 0, 0, 0, 0, 0, 0,
       0, 0, 0, 0, 0, 0, 0, 0, 0, 0, 0, 0, 0, 0, 0, 0, 0, 0, 0, 0, 0, 0,
       0, 0, 0, 0, 1, 1, 1, 1, 1, 1, 1, 1, 1, 1, 1, 1, 1, 1, 1, 1, 1, 1,
       1, 1, 1, 1, 1, 1, 1, 1, 1, 1, 1, 1, 1, 1, 1, 1, 1, 1, 1, 1, 1, 1,
       1, 1, 1, 1, 1, 1, 1, 1, 2, 2, 2, 2, 2, 2, 2, 2, 2, 2, 2, 2, 2, 2,
       2, 2, 2, 2, 2, 2, 2, 2, 2, 2, 2, 2, 2, 2, 2, 2, 2, 2, 2, 2, 2, 2,
       2, 2, 2, 2, 2, 2, 2, 2, 2, 2, 2, 2])
```

You obtain 150 items with three possible integer values (0, 1, and 2), which correspond to the three species of iris. To determine the correspondence between the species and number, you have to call the `target_names` attribute.

```
iris.target_names
Out[ ]:
array(['setosa', 'versicolor', 'virginica'], dtype='<U10')
```

To better understand this dataset, you can use the `matplotlib` library, using the techniques you learned in Chapter 7. Therefore create a scatterplot that displays the three different species in three different colors. The x-axis will represent the length of the sepal, while the y-axis will represent the width of the sepal.

```
import matplotlib.pyplot as plt
import matplotlib.patches as mpatches
from sklearn import datasets
iris = datasets.load_iris()
x = iris.data[:,0]  #X-Axis - sepal length
y = iris.data[:,1]  #Y-Axis - sepal length
species = iris.target      #Species
x_min, x_max = x.min() - .5,x.max() + .5
y_min, y_max = y.min() - .5,y.max() + .5
#SCATTERPLOT WITH MATPLOTLIB
plt.figure()
plt.title('Iris Dataset - Classification By Sepal Sizes')
plt.scatter(x,y, c=species)
```

```
plt.xlabel('Sepal length')
plt.ylabel('Sepal width')
plt.xlim(x_min, x_max)
plt.ylim(y_min, y_max)
plt.xticks(())
plt.yticks(())
plt.show()
```

As a result, you get the scatterplot shown in Figure 8-2. The blue ones are the Iris setosa, the green ones are the Iris versicolor, and red ones are the Iris virginica.

From Figure 8-2 you can see how the Iris setosa features differ from the other two, forming a cluster of blue dots separate from the others.

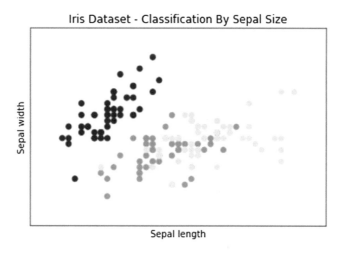

Figure 8-2. *The different species of irises are shown with different colors*

Try to follow the same procedure, but this time using the other two variables—that is, the measure of the length and width of the petal. You can use the same code and change just a few values.

```
import matplotlib.pyplot as plt
import matplotlib.patches as mpatches
from sklearn import datasets
iris = datasets.load_iris()
x = iris.data[:,2]  #X-Axis - petal length
y = iris.data[:,3]  #Y-Axis - petal length
species = iris.target      #Species
x_min, x_max = x.min() - .5,x.max() + .5
y_min, y_max = y.min() - .5,y.max() + .5
#SCATTERPLOT
plt.figure()
plt.title('Iris Dataset - Classification By Petal Sizes', size=14)
plt.scatter(x,y, c=species)
plt.xlabel('Petal length')
plt.ylabel('Petal width')
```

```
plt.xlim(x_min, x_max)
plt.ylim(y_min, y_max)
plt.xticks(())
plt.yticks(())
```

The result is the scatterplot shown in Figure 8-3. In this case, the division between the three species is much more evident. As you can see, you have three different clusters.

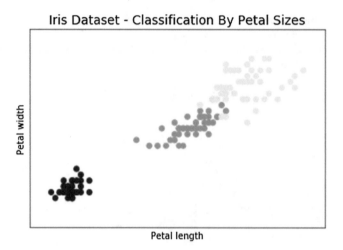

Figure 8-3. *The different species of irises are shown with different colors*

The PCA Decomposition

You have seen how the three species can be characterized, taking into account four measurements of the petal and sepal sizes. It represented two scatterplots, one for the petals and one for sepals, but how can you unify the whole thing? Four dimensions are a problem that even a 3D scatterplot cannot solve.

In this regard, a special technique called *Principal Component Analysis (PCA)* has been developed. This technique allows you to reduce the number of dimensions of a system, keeping all the information for the characterization of the various points. The new dimensions are called *principal components*.

Hence, PCA is employed before applying the machine learning algorithm, as it minimizes the number of variables used. It does this by analyzing the contribution of each of them to the maximum amount of variance in the dataset. It determines which of these features do not contribute significantly (they provide the same information as other features) and discards them, eliminating them from the calculation of the machine learning model. This greatly lightens the model and the calculations related to it.

This technique compresses the number of features involved in the machine learning calculation, creating new ones (or rather transforming the existing ones) in a smaller number, but with the same information as the starting dataset (approximation).

From a strictly mathematical point of view, from the input data matrix X (num_observations x num_features), the covariance matrix Cov(X) is obtained, which is then used to calculate the eigenvectors (principal components) and their corresponding eigenvalues. The k eigenvectors obtained are the principal components (see Figure 8-4). The first k principal components are the eigenvectors corresponding to the k largest eigenvalues.

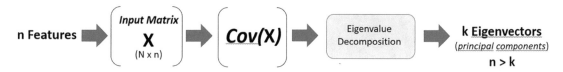

Figure 8-4. *The processes involved in the PCA technique*

Let's put this into practice using the Iris Dataset. You saw in the previous section that this dataset is made up of four features. You can apply the PCA method to analyze this technique in detail. To see if it is possible to reduce this number but still keep as much of the information as possible for the machine learning model, you write the following code:

```
from sklearn.decomposition import PCA
import numpy as np

covar_matrix = PCA(n_components=4)
covar_matrix.fit(iris.data)
variance = covar_matrix.explained_variance_ratio_
var = np.cumsum(np.round(variance, decimals=3)*100)
var
Out[ ]:
array([ 92.5, 97.8, 99.5, 100. ])
```

This code first imports the PCA() function from the sklearn.decomposition module and uses it to derive the covariance matrix. In this case, the technique considers all four features in the calculation (therefore, n = k) by imposing the n_components parameter equal to 4. Then it calculates the eigenvalue decomposition on this matrix using the fit() method. At this point, you can evaluate the contribution of each of the four features to the maximum value of the dataset variance. You extract the variance and then, with the NumPy function cumsum(), you obtain the cumulative contribution as a percentage made by each added feature. From the result, you can see that even if you reduced the size of the input dataset to just one feature, you would still retain 92 percent of the information. It can therefore be said that the information of a single feature could already be almost sufficient to obtain a good machine learning model. You can see it all graphically through matplotlib.

```
plt.ylabel('% Variance explained')
plt.xlabel('# of Features')
plt.title('PCA Analysis')
plt.ylim(90, 100.5)
plt.xticks([0, 1, 2, 3], [1,2,3,4])
plt.axvline(2, linestyle='--', c='#bbbbbb' )
plt.plot(var)
```

Running this code produces a plot like the one shown in Figure 8-5.

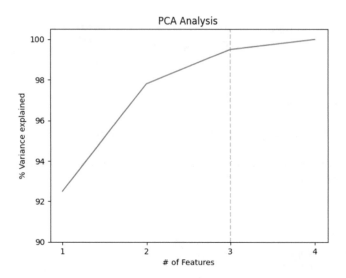

Figure 8-5. *The contribution of each feature to the total variance*

As you can clearly see from Figure 8-5, the use of three features is more than enough to build a good model, with a value of 99.5 percent. The use of a fourth feature is almost useless for this purpose, and it can be eliminated, thus simplifying the calculation.

In this case, you can reduce the system from four to three dimensions and plot the results in a 3D scatterplot.

The `scikit-learn` function that allows you to do the dimensional reduction is the `fit_transform()` function. It belongs to the PCA object. Apply the functions and methods you just used again.

```
from sklearn.decomposition import PCA
x_reduced = PCA(n_components=3).fit_transform(iris.data)
```

In addition, in order to visualize the new values, you will use a 3D scatterplot using the `mpl_toolkits.mplot3d` module of `matplotlib`. If you don't remember how to do this, see the section called "Scatter Plots in 3D" in Chapter 7.

```
import matplotlib.pyplot as plt
from mpl_toolkits.mplot3d import Axes3D
from sklearn import datasets
from sklearn.decomposition import PCA
iris = datasets.load_iris()
species = iris.target      #Species
x_reduced = PCA(n_components=3).fit_transform(iris.data)
#SCATTERPLOT 3D
fig = plt.figure()
ax = Axes3D(fig)
ax.set_title('Iris Dataset by PCA', size=14)
ax.scatter(x_reduced[:,0],x_reduced[:,1],x_reduced[:,2], c=species)
ax.set_xlabel('First eigenvector')
ax.set_ylabel('Second eigenvector')
```

```
ax.set_zlabel('Third eigenvector')
ax.w_xaxis.set_ticklabels(())
ax.w_yaxis.set_ticklabels(())
ax.w_zaxis.set_ticklabels(())
```

The result will be the scatterplot shown in Figure 8-6. The three species of iris are well characterized with respect to each other to form a cluster.

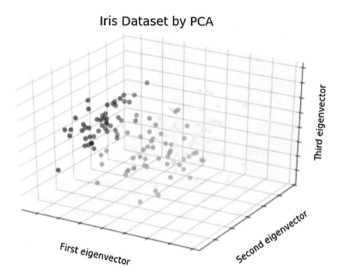

Figure 8-6. *3D scatterplot with three clusters representative of each species of iris*

K-Nearest Neighbors Classifier

Now, you will perform a *classification*, and to do this operation with the `scikit-learn` library, you need a *classifier*.

Given a new measurement of an iris flower, the task of the classifier is to determine to which of the three species it belongs. The simplest possible classifier is the *nearest neighbor*. This algorithm searches within the training set for the observation that most closely approaches the new test sample.

In fact, the mechanism behind the KNN algorithm is one of the simplest and most direct in machine learning. Despite this, KNN remains one of the most used techniques in this discipline and not only for classification, but also for other problems such as regression and outlier detection.

The algorithm first calculates the distance between the new data point to be classified in the (multidimensional) feature space and all the other data points whose class it belongs to (or has been previously evaluated). To calculate the distance between two points you can use different techniques such as Euclidean (the one we all study at school), Minkowski, Manhattan, and so on. The Euclidean method would appear to be the best, but in reality for a large number of dimensions, this is no longer so valid.

Once the distance has been calculated, the KNN algorithm selects, among all the points, the K points closest to the one to be classified. This K number is what gives the "K-Nearest Neighbors" algorithm its name. In a classification problem, the new point will be awarded to the class that has the most points among the K-nearest neighbors. See Figure 8-7.

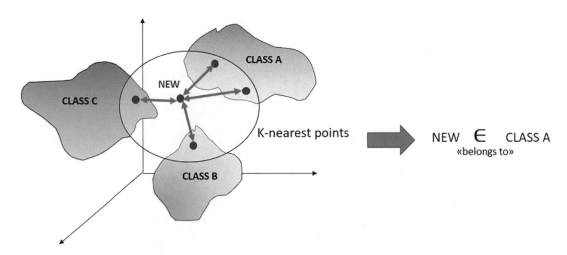

Figure 8-7. *The KNN mechanism of classification*

A very important thing to consider at this point are the concepts of *training set* and *testing set* (already seen in Chapter 1). Indeed, if you have only a single dataset of data, it is important not to use the same data both for the test and for the training. In this regard, the elements of the dataset are divided into two parts, one dedicated to train the algorithm and the other to perform its validation.

Thus, before proceeding, you have to divide your Iris Dataset into two parts. However, it is wise to randomly mix the array elements and then make the division. In fact, often the data may have been collected in a particular order, and in your case the Iris Dataset contains items sorted by species. To blend elements of the dataset, you use a NumPy function called `random.permutation()`. The mixed dataset consists of 150 different observations; the first 140 are used as the training set, the remaining 10 as the test set.

```
import numpy as np
from sklearn import datasets
np.random.seed(0)
iris = datasets.load_iris()
x = iris.data
y = iris.target
i = np.random.permutation(len(iris.data))
x_train = x[i[:-10]]
y_train = y[i[:-10]]
x_test = x[i[-10:]]
y_test = y[i[-10:]]
```

Now you can apply the K-Nearest Neighbor algorithm. Import the `KNeighborsClassifier`, call the constructor of the classifier, and then train it with the `fit()` function.

```
from sklearn.neighbors import KNeighborsClassifier
knn = KNeighborsClassifier()
knn.fit(x_train,y_train)
KNeighborsClassifier(algorithm='auto',
                     leaf_size=30,
                     metric='minkowski',
                     metric_params=None,
```

```
                    n_neighbors=5,
                    p=2,
                    weights='uniform')
```

Now that you have a predictive model that consists of the knn classifier, trained by 140 observations, you will find out how it is valid. The classifier should correctly predict the species of iris of the ten observations of the test set. In order to obtain the prediction, you have to use the predict() function, which will be applied directly to the predictive model, knn. Finally, you compare the results predicted with the actual observed results contained in y_test.

```
knn.predict(x_test)
Out[ ]: array([1, 2, 1, 0, 0, 0, 2, 1, 2, 0])
y_test
Out[ ]: array([1, 1, 1, 0, 0, 0, 2, 1, 2, 0])
```

You can see that you obtained a 10 percent error. You can visualize all this using decision boundaries in a space represented by the 2D scatterplot of sepals.

```
import numpy as np
import matplotlib.pyplot as plt
from matplotlib.colors import ListedColormap
from sklearn import datasets
from sklearn.neighbors import KNeighborsClassifier
iris = datasets.load_iris()
x = iris.data[:,:2]       #X-Axis - sepal length-width
y = iris.target           #Y-Axis - species
x_min, x_max = x[:,0].min() - .5,x[:,0].max() + .5
y_min, y_max = x[:,1].min() - .5,x[:,1].max() + .5
#MESH
cmap_light = ListedColormap(['#AAAAFF','#AAFFAA','#FFAAAA'])
h = .02
xx, yy = np.meshgrid(np.arange(x_min, x_max, h), np.arange(y_min, y_max, h))
knn = KNeighborsClassifier()
knn.fit(x,y)
Z = knn.predict(np.c_[xx.ravel(),yy.ravel()])
Z = Z.reshape(xx.shape)
plt.figure()
plt.pcolormesh(xx,yy,Z,cmap=cmap_light, shading='auto')
#Plot the training points
plt.scatter(x[:,0],x[:,1],c=y)
plt.xlim(xx.min(),xx.max())
plt.ylim(yy.min(),yy.max())
Out[ ]: (1.5, 4.900000000000003)
```

You get a subdivision of the scatterplot in decision boundaries, as shown in Figure 8-8.

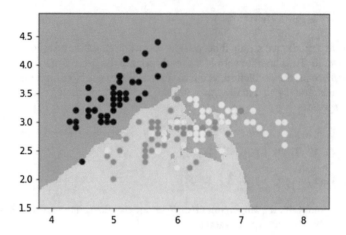

Figure 8-8. *The three decision boundaries are represented by three different colors*

You can do the same thing considering the size of the petals.

```
import numpy as np
import matplotlib.pyplot as plt
from matplotlib.colors import ListedColormap
from sklearn import datasets
from sklearn.neighbors import KNeighborsClassifier
iris = datasets.load_iris()
x = iris.data[:,2:4]       #X-Axis - petals length-width
y = iris.target            #Y-Axis - species
x_min, x_max = x[:,0].min() - .5,x[:,0].max() + .5
y_min, y_max = x[:,1].min() - .5,x[:,1].max() + .5
#MESH
cmap_light = ListedColormap(['#AAAAFF','#AAFFAA','#FFAAAA'])
h = .02
xx, yy = np.meshgrid(np.arange(x_min, x_max, h), np.arange(y_min, y_max, h))
knn = KNeighborsClassifier()
knn.fit(x,y)
Z = knn.predict(np.c_[xx.ravel(),yy.ravel()])
Z = Z.reshape(xx.shape)
plt.figure()
plt.pcolormesh(xx,yy,Z,cmap=cmap_light, shading='auto')
#Plot the training points
plt.scatter(x[:,0],x[:,1],c=y)
plt.xlim(xx.min(),xx.max())
plt.ylim(yy.min(),yy.max())
Out[ ]: (-0.40000000000000002, 2.9800000000000031)
```

As shown in Figure 8-9, you have the corresponding decision boundaries regarding the characterization of iris flowers taking into account the size of the petals.

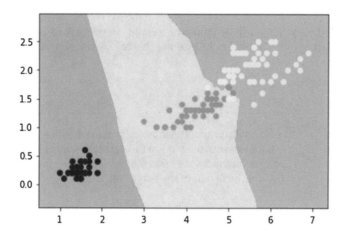

Figure 8-9. *The decision boundaries on a 2D scatterplot describing the petal sizes*

Diabetes Dataset

Among the various datasets available in the `scikit-learn` library, there is the diabetes dataset. This dataset was used for the first time in 2004 (*Annals of Statistics*, by Efron, Hastie, Johnstone, and Tibshirani). Since then it has become an example widely used to study various predictive models and their effectiveness.

To upload the data contained in this dataset, you must first import the `datasets` module of the `scikit-learn` library and then call the `load_diabetes()` function to load the dataset into a variable called `diabetes`.

```
from sklearn import datasets
 diabetes = datasets.load_diabetes()
```

This dataset contains physiological data collected on 442 patients and, as a corresponding target, an indicator of the disease progression after a year. The physiological data occupy the first ten columns with values that indicate the following, respectively:

- Age

- Sex

- Body mass index

- Blood pressure

- S1, S2, S3, S4, S5, and S6 (six blood serum measurements)

These measurements can be obtained by calling the `data` attribute. If you check the values in the dataset, you will find values very different from what you would have expected. For example, look at the ten values for the first patient.

```
diabetes.data[0]
Out[ ]:
array([ 0.03807591,  0.05068012,  0.06169621,  0.02187235, -0.0442235 ,
       -0.03482076, -0.04340085, -0.00259226,  0.01990842, -0.01764613])
```

These values are in fact the result of processing. Each of the ten values was mean-centered and subsequently scaled by the standard deviation times the number of samples. Checking will reveal that the sum of the squares of each column is equal to 1. Try doing this calculation with the age measurements; you will obtain a value very close to 1.

```
np.sum(diabetes.data[:,0]**2)
Out[ ]: 1.0000000000000746
```

Even though these values are normalized and therefore difficult to read, they continue to express the ten physiological characteristics and therefore have not lost their value or statistical information.

As for the indicators of the progress of the disease, that is, the values that must correspond to the results of your predictions, these are obtainable by means of the `target` attribute.

```
diabetes.target
Out[ ]:
array([ 151.,   75.,  141.,  206.,  135.,   97.,  138.,   63.,  110.,
        310.,  101.,   69.,  179.,  185.,  118.,  171.,  166.,  144.,
         97.,  168.,   68.,   49.,   68.,  245.,  184.,  202.,  137
       . . .
```

You obtain a series of 442 integer values between 25 and 346.

Linear Regression: The Least Square Regression

Linear regression is a procedure that uses data contained in the training set to build a linear model. The simplest is based on the equation of a `rect` with the two parameters a and b to characterize it. These parameters will be calculated so as to make the sum of the squared residuals as small as possible.

$$y = a^* x + c$$

In this expression, x is the training set, y is the target, b is the slope, and c is the intercept of the `rect` represented by the model. In `scikit-learn`, to use the predictive model for the linear regression, you must import the `linear_model` module and then use the manufacturer `LinearRegression()` constructor to create the predictive model, which you call `linreg`.

```
from sklearn import linear_model
linreg = linear_model.LinearRegression()
```

To practice with an example of linear regression, you can use the diabetes dataset described earlier. First you need to break the 442 patients into a training set (composed of the first 422 patients) and a test set (the last 20 patients).

```
from sklearn import datasets
diabetes = datasets.load_diabetes()
x_train = diabetes.data[:-20]
y_train = diabetes.target[:-20]
x_test = diabetes.data[-20:]
y_test = diabetes.target[-20:]
```

Now apply the training set to the predictive model through the use of the `fit()` function.

```
linreg.fit(x_train,y_train)
Out[ ]: LinearRegression()
```

Once the model is trained, you can get the ten b coefficients calculated for each physiological variable, using the `coef_` attribute of the predictive model.

```
linreg.coef_
Out[ ]:
array([  3.03499549e-01,  -2.37639315e+02,   5.10530605e+02,
         3.27736980e+02,  -8.14131709e+02,   4.92814588e+02,
         1.02848452e+02,   1.84606489e+02,   7.43519617e+02,
         7.60951722e+01])
```

If you apply the test set to the `linreg` prediction model, you will get a series of targets to be compared with the values actually observed.

```
linreg.predict(x_test)
Out[ ]:
array([ 197.61846908,  155.43979328,  172.88665147,  111.53537279,
        164.80054784,  131.06954875,  259.12237761,  100.47935157,
        117.0601052 ,  124.30503555,  218.36632793,   61.19831284,
        132.25046751,  120.3332925 ,   52.54458691,  194.03798088,
        102.57139702,  123.56604987,  211.0346317 ,   52.60335674])
y_test
Out[ ]:
array([ 233.,   91.,  111.,  152.,  120.,   67.,  310.,   94.,  183.,
         66.,  173.,   72.,   49.,   64.,   48.,  178.,  104.,  132.,
        220.,   57.])
```

However, a good indicator of which prediction should be perfect is the *variance*. The closer the variance is to 1, the more perfect the prediction.

```
linreg.score(x_test, y_test)
Out[ ]: 0.58507530226905713
```

Now you will start with the linear regression, taking into account a single physiological factor, for example, you can start from age.

```
import numpy as np
import matplotlib.pyplot as plt
from sklearn import linear_model
from sklearn import datasets
diabetes = datasets.load_diabetes()
x_train = diabetes.data[:-20]
y_train = diabetes.target[:-20]
x_test = diabetes.data[-20:]
y_test = diabetes.target[-20:]
x0_test = x_test[:,0]
x0_train = x_train[:,0]
x0_test = x0_test[:,np.newaxis]
```

```
x0_train = x0_train[:,np.newaxis]
linreg = linear_model.LinearRegression()
linreg.fit(x0_train,y_train)
y = linreg.predict(x0_test)
plt.scatter(x0_test,y_test,color='k')
plt.plot(x0_test,y,color='b',linewidth=3)
```

Figure 8-10 shows the line representing the linear correlation between the ages of the patients and the disease progression.

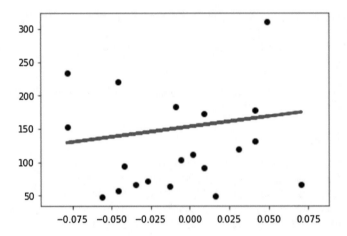

Figure 8-10. *A linear regression represents a linear correlation between a feature and the targets*

Actually, you have ten physiological factors in the diabetes dataset. Therefore, to have a more complete picture of all the training set, you can create a linear regression for every physiological feature, creating ten models and seeing the result for each of them through a linear chart.

```
import numpy as np
import matplotlib.pyplot as plt
from sklearn import linear_model
from sklearn import datasets
diabetes = datasets.load_diabetes()
x_train = diabetes.data[:-20]
y_train = diabetes.target[:-20]
x_test = diabetes.data[-20:]
y_test = diabetes.target[-20:]
plt.figure(figsize=(8,12))
for f in range(0,10):
    xi_test = x_test[:,f]
    xi_train = x_train[:,f]
    xi_test = xi_test[:,np.newaxis]
    xi_train = xi_train[:,np.newaxis]
    linreg.fit(xi_train,y_train)
    y = linreg.predict(xi_test)
    plt.subplot(5,2,f+1)
    plt.scatter(xi_test,y_test,color='k')
    plt.plot(xi_test,y,color='b',linewidth=3)
```

Figure 8-11 shows ten linear charts, each of which represents the correlation between a physiological factor and the progression of diabetes.

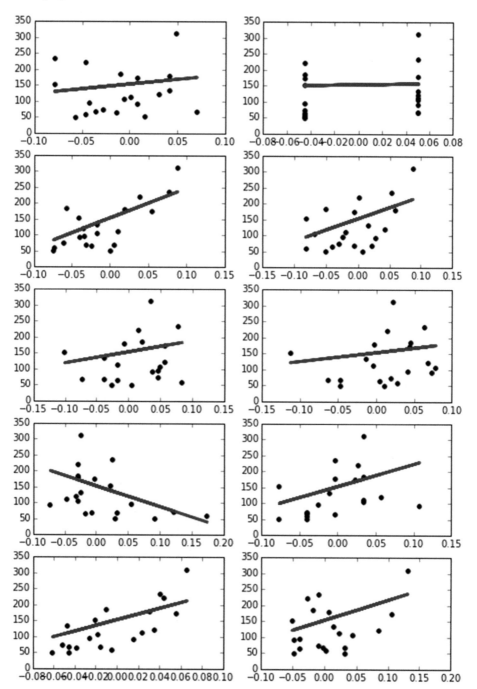

Figure 8-11. *Ten linear charts showing the correlations between physiological factors and the progression of diabetes*

Support Vector Machines (SVMs)

Support Vector Machines are a number of machine learning techniques that were first developed in the AT&T laboratories by Vapnik and colleagues in the early 90s. The basis of this class of procedures is in fact an algorithm called *Support Vector*, which is a generalization of a previous algorithm called Generalized Portrait, developed in Russia in 1963 by Vapnik as well.

In simple words, the SVM classifiers are binary or discriminating models, working on two classes of differentiation. Their main task is basically to discriminate against new observations between two classes. During the learning phase, these classifiers project the observations in a multidimensional space called *decision space* and build a separation surface called the *decision boundary* that divides this space into two areas of belonging. In the simplest case, that is, the linear case, the decision boundary will be represented by a plane (in 3D) or by a straight line (in 2D). In more complex cases, the separation surfaces are curved shapes with increasingly articulated shapes.

Also for SVM, the data points are distributed in a multidimensional space, where the values of a specific feature are distributed on each axis. The purpose of SVM is to find a hyperplane separating the space between the two classes (see Figure 8-12).

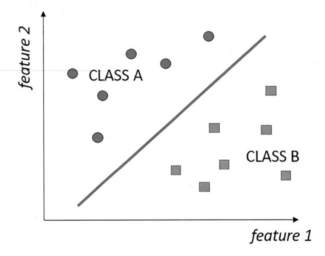

Figure 8-12. *SVM divides the decision space into two areas, each for each class, defining a straight line (decision boundary)*

The straight line that must be defined must not limit itself to separating the points of the two classes from each other, but must respond to particular requirements. SVM finds the points that are closest to the dividing line. These are called "support vectors" and they give the algorithm its name (see Figure 8-13).

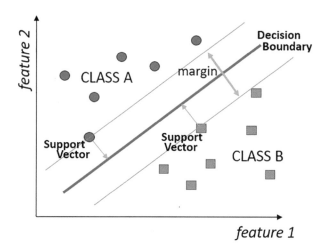

Figure 8-13. *SVM optimizes the decision boundary, maximizing the margin between support vector points*

The distance between the support vectors and the decision boundary (the straight line) is called the margin. SVM's task is to maximize this margin. When the margin reaches its maximum value, then the decision boundary will be the optimal one.

SVM is a very flexible technique that lends itself well to many applications. It can be used both in regression with the *SVR (Support Vector Regression)* and in classification with the *SVC (Support Vector Classification)*.

Support Vector Classification (SVC)

If you want to better understand how this algorithm works, you can start by referring to the simplest case, that is the linear 2D case, where the decision boundary is a straight line separating the decision area into two parts. Take for example a simple training set where some points are assigned to two different classes. The training set will consist of 11 points (observations) with two different attributes that have values between 0 and 4. These values will be contained within a NumPy array called x. Their belonging to one of two classes is defined by 0 or 1 values contained in another array, called y.

Visualize distribution of these points in space with a scatterplot, which will then be defined as a decision space (see Figure 8-14).

```
import numpy as np
import matplotlib.pyplot as plt
from sklearn import svm
x = np.array([[1,3],[1,2],[1,1.5],[1.5,2],[2,3],[2.5,1.5],
    [2,1],[3,1],[3,2],[3.5,1],[3.5,3]])
y = [0]*6 + [1]*5
plt.scatter(x[:,0],x[:,1],c=y,s=50,alpha=0.9)
```

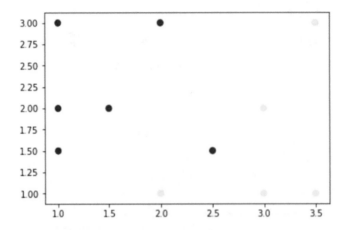

Figure 8-14. *The scatterplot of the training set displays the decision space*

Now that you have defined the training set, you can apply the SVC (Support Vector Classification) algorithm. This algorithm will create a line (decision boundary) in order to divide the decision area into two parts (see Figure 8-15), and this straight line will be placed to maximize the distance of closest observations contained in the training set. This condition should produce two different portions in which all points of a same class should be contained.

Then you apply the SVC algorithm to the training set and to do so, first you define the model with the SVC() constructor defining the kernel as linear. (A *kernel* is a class of algorithms for pattern analysis.) Then you use the fit() function with the training set as an argument. Once the model is trained, you can plot the decision boundary with the decision_function() function. Then you draw the scatterplot and provide a different color to the two portions of the decision space.

```
import numpy as np
import matplotlib.pyplot as plt
from sklearn import svm
x = np.array([[1,3],[1,2],[1,1.5],[1.5,2],[2,3],[2.5,1.5],
    [2,1],[3,1],[3,2],[3.5,1],[3.5,3]])
y = [0]*6 + [1]*5
svc = svm.SVC(kernel='linear').fit(x,y)
X,Y = np.mgrid[0:4:200j,0:4:200j]
Z = svc.decision_function(np.c_[X.ravel(),Y.ravel()])
Z = Z.reshape(X.shape)
plt.contourf(X,Y,Z > 0,alpha=0.1)
plt.contour(X,Y,Z,colors=['k'], linestyles=['-'],levels=[0])
plt.scatter(x[:,0],x[:,1],c=y,s=50,alpha=0.9)
```

In Figure 8-15, you can see the two portions containing the two classes. It can be said that the division is successful except for a dark dot in the lighter portion.

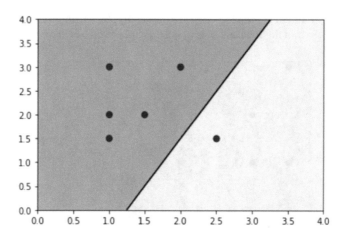

Figure 8-15. *The decision area is split into two portions*

Once the model has been trained, it is simple to understand how the predictions operate. Graphically, depending on the position occupied by the new observation, you will know its corresponding membership in one of the two classes.

Instead, from a more programmatic point of view, the predict() function will return the number of the corresponding class of belonging (0 for class in blue, 1 for the class in red).

```
svc.predict([[1.5,2.5]])
Out[ ]: array([0])
svc.predict([[2.5,1]])
Out[ ]: array([1])
```

A related concept in the SVC algorithm is *regularization*. It is set by the C parameter and a small value of C means that the margin is calculated using many or all of the observations around the line of separation (greater regularization), while a large value of C means that the margin is calculated on the observations near to the line separation (lower regularization). Unless otherwise specified, the default value of C is equal to 1. You can highlight points that participated in the margin calculation, identifying them through the support_vectors array.

```
import numpy as np
import matplotlib.pyplot as plt
from sklearn import svm
x = np.array([[1,3],[1,2],[1,1.5],[1.5,2],[2,3],[2.5,1.5],
    [2,1],[3,1],[3,2],[3.5,1],[3.5,3]])
y = [0]*6 + [1]*5
svc = svm.SVC(kernel='linear',C=1).fit(x,y)
X,Y = np.mgrid[0:4:200j,0:4:200j]
Z = svc.decision_function(np.c_[X.ravel(),Y.ravel()])
Z = Z.reshape(X.shape)
plt.contourf(X,Y,Z > 0,alpha=0.1)
plt.contour(X,Y,Z,colors=['k','k','k'], linestyles=['--','-','--'],levels=[-1,0,1])
plt.scatter(svc.support_vectors_[:,0],svc.support_vectors_[:,1],s=120,facecolors='r')
plt.scatter(x[:,0],x[:,1],c=y,s=50,alpha=0.9)
```

These points are represented by rimmed circles in the scatterplot. Furthermore, they will be within an evaluation area in the vicinity of the separation line (see the dashed lines in Figure 8-16).

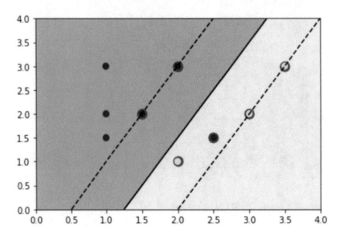

Figure 8-16. *The number of points involved in the calculation depends on the C parameter*

To see the effect on the decision boundary, you can restrict the value to C = 0.1. Take a look at how many points will be taken into consideration.

```
import numpy as np
import matplotlib.pyplot as plt
from sklearn import svm
x = np.array([[1,3],[1,2],[1,1.5],[1.5,2],[2,3],[2.5,1.5],
    [2,1],[3,1],[3,2],[3.5,1],[3.5,3]])
y = [0]*6 + [1]*5
svc = svm.SVC(kernel='linear',C=0.1).fit(x,y)
X,Y = np.mgrid[0:4:200j,0:4:200j]
Z = svc.decision_function(np.c_[X.ravel(),Y.ravel()])
Z = Z.reshape(X.shape)
plt.contourf(X,Y,Z > 0,alpha=0.1)
plt.contour(X,Y,Z,colors=['k','k','k'], linestyles=['--','-','--'],levels=[-1,0,1])
plt.scatter(svc.support_vectors_[:,0],svc.support_vectors_[:,1],s=120,facecolors='w')
plt.scatter(x[:,0],x[:,1],c=y,s=50,alpha=0.9)
```

The points taken into consideration are increased and consequently the separation line (decision boundary) has changed orientation. But now there are two points that are in the wrong decision areas (see Figure 8-17).

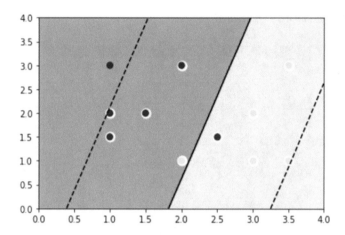

Figure 8-17. *The number of points involved in the calculation grows when C decreases*

Nonlinear SVC

So far you have seen the SVC linear algorithm defining a line of separation that was intended to split the two classes. There are also more complex SVC algorithms that can establish curves (2D) or curved surfaces (3D) based on the same principles of maximizing the distances between the points closest to the surface. This section looks at the system using a polynomial kernel.

As the name implies, you can define a polynomial curve that separates the area decision into two portions. The degree of the polynomial can be defined by the degree option. Even in this case, C is the coefficient of regularization. Try to apply an SVC algorithm with a polynomial kernel of third degree and with a C coefficient equal to 1.

```
import numpy as np
import matplotlib.pyplot as plt
from sklearn import svm
x = np.array([[1,3],[1,2],[1,1.5],[1.5,2],[2,3],[2.5,1.5],
    [2,1],[3,1],[3,2],[3.5,1],[3.5,3]])
y = [0]*6 + [1]*5
svc = svm.SVC(kernel='poly',C=1, degree=3).fit(x,y)
X,Y = np.mgrid[0:4:200j,0:4:200j]
Z = svc.decision_function(np.c_[X.ravel(),Y.ravel()])
Z = Z.reshape(X.shape)
plt.contourf(X,Y,Z > 0,alpha=0.1)
plt.contour(X,Y,Z,colors=['k','k','k'], linestyles=['--','-','--'],levels=[-1,0,1])
plt.scatter(svc.support_vectors_[:,0],svc.support_vectors_[:,1],s=120,facecolors='w')
plt.scatter(x[:,0],x[:,1],c=y,s=50,alpha=0.9)
```

You get the situation shown in Figure 8-18.

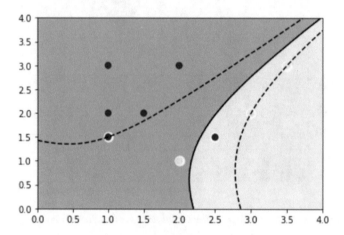

Figure 8-18. *The decision space using an SVC with a polynomial kernel*

Another type of nonlinear kernel is the *Radial Basis Function (RBF)*. In this case, the separation curves tend to define the zones radially with respect to the observation points of the training set.

```
import numpy as np
import matplotlib.pyplot as plt
from sklearn import svm
x = np.array([[1,3],[1,2],[1,1.5],[1.5,2],[2,3],[2.5,1.5],
    [2,1],[3,1],[3,2],[3.5,1],[3.5,3]])
y = [0]*6 + [1]*5
svc = svm.SVC(kernel='rbf', C=1, gamma=3).fit(x,y)
X,Y = np.mgrid[0:4:200j,0:4:200j]
Z = svc.decision_function(np.c_[X.ravel(),Y.ravel()])
Z = Z.reshape(X.shape)
plt.contourf(X,Y,Z > 0,alpha=0.1)
plt.contour(X,Y,Z,colors=['k','k','k'], linestyles=['--','-','--'],levels=[-1,0,1])
plt.scatter(svc.support_vectors_[:,0],svc.support_vectors_[:,1],s=120,facecolors='w')
plt.scatter(x[:,0],x[:,1],c=y,s=50,alpha=0.9)
```

In Figure 8-19, you can see the two portions of the decision with all points of the training set correctly positioned.

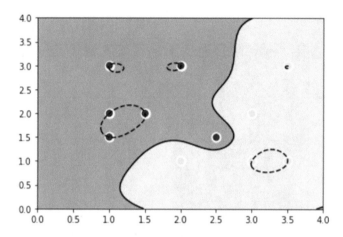

Figure 8-19. *The decision area using an SVC with the RBF kernel*

Plotting Different SVM Classifiers Using the Iris Dataset

The SVM example that you just saw is based on a very simple dataset. This section uses more complex datasets for a classification problem with SVC. In fact, it uses the previously used dataset: the Iris Dataset.

The SVC algorithm used earlier learned from a training set containing only two classes. In this case, you will extend the case to three classifications, as the Iris Dataset is split into three classes, corresponding to the three different species of flowers.

In this case, the decision boundaries intersect each other, subdividing the decision area (2D) or the decision volume (3D) into several portions.

Both linear models have linear decision boundaries (intersecting hyperplanes), while models with nonlinear kernels (polynomial or Gaussian RBF) have nonlinear decision boundaries. These boundaries are more flexible, with figures that are dependent on the type of kernel and its parameters.

```python
import numpy as np
import matplotlib.pyplot as plt
from sklearn import svm, datasets
iris = datasets.load_iris()
x = iris.data[:,:2]
y = iris.target
h = .05
svc = svm.SVC(kernel='linear',C=1.0).fit(x,y)
x_min,x_max = x[:,0].min() - .5, x[:,0].max() + .5
y_min,y_max = x[:,1].min() - .5, x[:,1].max() + .5
h = .02
X, Y = np.meshgrid(np.arange(x_min, x_max, h), np.arange(y_min,y_max,h))
Z = svc.predict(np.c_[X.ravel(),Y.ravel()])
Z = Z.reshape(X.shape)
plt.contourf(X,Y,Z,alpha=0.1)
plt.contour(X,Y,Z,colors='k')
plt.scatter(x[:,0],x[:,1],c=y)
```

In Figure 8-20, the decision space is divided into three portions separated by decision boundaries.

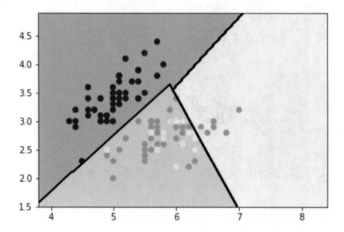

Figure 8-20. *The decision boundaries split the decision area into three different portions*

Now it's time to apply a nonlinear kernel to generate nonlinear decision boundaries, such as the polynomial kernel.

```
import numpy as np
import matplotlib.pyplot as plt
from sklearn import svm, datasets
iris = datasets.load_iris()
x = iris.data[:,:2]
y = iris.target
h = .05
svc = svm.SVC(kernel='poly',C=1.0,degree=3).fit(x,y)
x_min,x_max = x[:,0].min() - .5, x[:,0].max() + .5
y_min,y_max = x[:,1].min() - .5, x[:,1].max() + .5
h = .02
X, Y = np.meshgrid(np.arange(x_min, x_max, h), np.arange(y_min,y_max,h))
Z = svc.predict(np.c_[X.ravel(),Y.ravel()])
Z = Z.reshape(X.shape)
plt.contourf(X,Y,Z,alpha=0.1)
plt.contour(X,Y,Z,colors='k')
plt.scatter(x[:,0],x[:,1],c=y)
```

Figure 8-21 shows how the polynomial decision boundaries split the area in a very different way compared to the linear case.

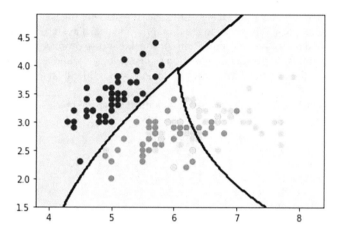

Figure 8-21. *Even in the polynomial case, the three portions remain almost unchanged*

Now you can apply the RBF kernel to see the difference in the distribution of areas. It is sufficient to update the kernel value with rbf, inside the svm.SVC() function.

```
svc = svm.SVC(kernel='rbf', gamma=3, C=1.0).fit(x,y)
```

Figure 8-22 shows how the RBF kernel generates radial areas.

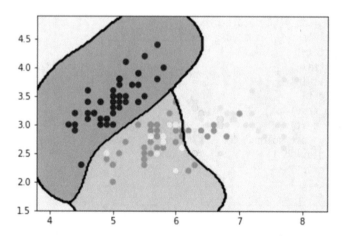

Figure 8-22. *The kernel RBF defines radial decision areas*

Support Vector Regression (SVR)

The SVC method can be extended to solve regression problems. This method is called *Support Vector Regression*.

The model produced by SVC actually does not depend on the complete training set, but uses only a subset of elements, that is, those closest to the decision boundary. In a similar way, the model produced by SVR also depends only on a subset of the training set.

This section demonstrates how the SVR algorithm uses the diabetes dataset, which you have seen in this chapter. By way of example, it refers only to the third physiological data. It performs three different regressions, one linear and two nonlinear (polynomial). The linear case produces a straight line, as the linear predictive model is very similar to the linear regression seen previously, whereas polynomial regressions are built from the second and third degrees. The SVR() function is almost identical to the SVC() function seen previously.

The only aspect to consider is that the test set of data must be sorted in ascending order.

```python
import numpy as np
import matplotlib.pyplot as plt
from sklearn import svm
from sklearn import datasets
diabetes = datasets.load_diabetes()
x_train = diabetes.data[:-20]
y_train = diabetes.target[:-20]
x_test = diabetes.data[-20:]
y_test = diabetes.target[-20:]
x0_test = x_test[:,2]
x0_train = x_train[:,2]
x0_test = x0_test[:,np.newaxis]
x0_train = x0_train[:,np.newaxis]
x0_test.sort(axis=0)
x0_test = x0_test*100
x0_train = x0_train*100
svr = svm.SVR(kernel='linear',C=1000)
svr2 = svm.SVR(kernel='poly',C=1000,degree=2)
svr3 = svm.SVR(kernel='poly',C=1000,degree=3)
svr.fit(x0_train,y_train)
svr2.fit(x0_train,y_train)
svr3.fit(x0_train,y_train)
y = svr.predict(x0_test)
y2 = svr2.predict(x0_test)
y3 = svr3.predict(x0_test)
plt.scatter(x0_test,y_test,color='k')
plt.plot(x0_test,y,color='b')
plt.plot(x0_test,y2,c='r')
plt.plot(x0_test,y3,c='g')
```

The three regression curves are represented with three colors. The linear regression will be blue (the straight line); the polynomial of the second degree (the concave line upwards, i.e. the parabolic line) will be red; and the polynomial of the third degree will be green (see Figure 8-23).

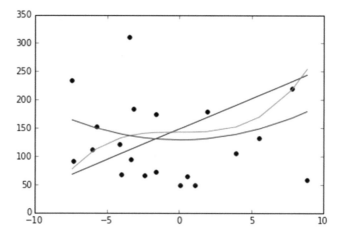

Figure 8-23. *The three regression curves produce very different trends starting from the training set*

Conclusions

In this chapter you saw simple cases of regression and classification problems solved using the `scikit-learn` library. Many concepts of the validation phase for a predictive model were presented in a practical way through some practical examples.

In the next chapter, you see a complete case in which all steps of data analysis are discussed by way of a single practical example. Everything is implemented in IPython Notebook, an interactive and innovative environment well suited for sharing every step of the data analysis. It includes interactive documentation that's useful as a report or as a web presentation.

CHAPTER 9

Deep Learning with TensorFlow

This chapter introduces an overview of the world of deep learning and the artificial neural networks on which its techniques are based. Furthermore, among the Python frameworks for deep learning, you will use *TensorFlow,* which is an excellent tool for research and development of deep learning analysis techniques. With this library, you will see how to develop different models of neural networks that are the basis of deep learning. In particular, in this third edition, the explanations and example codes are based on the new TensorFlow 2.x version, which has seen the incorporation of Keras and a complete upheaval in the modules and implementation paradigms.

Artificial Intelligence, Machine Learning, and Deep Learning

For anyone dealing with the world of data analysis, these three terms are ultimately very common on the web, in text, and on seminars related to the subject. But what is the relationship between them? And what do they really consist of?

In this section you read detailed definitions of these three terms. You discover how in recent decades, the need to create more and more elaborate algorithms, and to be able to make predictions and classify data more and more efficiently, has led to machine learning. Then you discover how, thanks to new technological innovations, and in particular to the computing power achieved by the GPU, deep learning techniques have been developed based on neural networks.

Artificial Intelligence

The term *artificial intelligence* was first used by John McCarthy in 1956, at a time full of great hope and enthusiasm for the technology world. They were at the dawn of electronics and computers as large as whole rooms that could do a few simple calculations, but they did so efficiently and quickly compared to humans. They had glimpsed possible future developments of electronic intelligence.

But without going into the world of science fiction, the current definition best suited to artificial intelligence, often referred to as AI, can be summarized briefly with the following sentence:

> *Automatic processing on a computer capable of performing operations that would seem to be exclusively relevant to human intelligence.*

Hence the concept of artificial intelligence is a variable concept that varies with the progress of the machines themselves and with the concept of "exclusive human relevance." While in the 60s and 70s, we saw artificial intelligence as the ability of computers to perform calculations and find mathematical solutions of complex problems "of exclusive relevance of great scientists," in the 80s and 90s, AI matured in its ability to assess risks, resources, and make decisions. In the year 2000, with the continuous growth of computer computing potential, the possibility of these systems to learn with machine learning was added to the definition.

Finally, in the last few years, the concept of artificial intelligence has focused on visual and auditory recognition operations, which until recently were thought of as "exclusive human relevance."

These operations include:

- Image recognition

- Object detection

- Object segmentation

- Language translation

- Natural language understanding

- Speech recognition

These problems are still under study, thanks to deep learning techniques.

Machine Learning Is a Branch of Artificial Intelligence

In the previous chapter you saw machine learning in detail, with many examples of the different techniques for classifying or predicting data.

Machine learning (ML), with all its techniques and algorithms, is a large branch of artificial intelligence. In fact, you refer to it, while remaining within the ambit of artificial intelligence, when you use systems that are able to learn (learning systems) to solve various problems that shortly before had been "considered exclusive to humans."

Deep Learning Is a Branch of Machine Learning

Within the machine learning techniques, a further subclass can be defined, called *deep learning*. You saw in Chapter 8 that machine learning uses systems that can learn, and this can be done through features inside the system (often parameters of a fixed model) that are modified in response to input data intended for learning (the training set).

Deep learning techniques take a step forward. In fact, deep learning systems are structured so as not to have these intrinsic characteristics in the model, but these characteristics are extracted and detected by the system automatically as a result of learning itself. Among these systems that can do this, this chapter refers in particular to *artificial neural networks.*

The Relationship Between Artificial Intelligence, Machine Learning, and Deep Learning

To sum up, in this section you have seen that machine learning and deep learning are actually subclasses of artificial intelligence. Figure 9-1 shows a schematization of classes in this relationship.

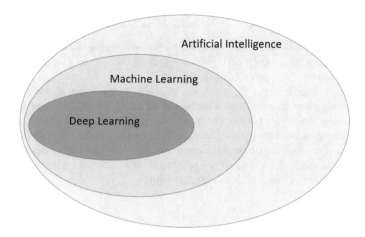

Figure 9-1. *Schematization of the relationship between artificial intelligence, machine learning, and deep learning*

Deep Learning

In this section, you learn about some significant factors that led to the development of deep learning and see how, in the last few years, there have been many steps forward.

Neural Networks and GPUs

In the previous section, you learned that in the field of artificial intelligence, deep learning has become popular only in the last few years precisely to solve problems of visual and auditory recognition.

In the context of deep learning, a lot of calculation techniques and algorithms have been developed in recent years, making the most of the potential of the Python language. But the theory behind deep learning actually dates back many years. In fact, the concept of the neural network was introduced in 1943, and the first theoretical studies on artificial neural networks and their applications were developed in the 60s.

The fact is that only in recent years the neural networks, with the related deep learning techniques that use them, have proved useful to solve many problems of artificial intelligence. This is due to the fact that only now are there technologies that can be implemented in a useful and efficient way.

In fact, at the application level, deep learning requires very complex mathematical operations that require millions or even billions of parameters. The CPUs of the 90s, even if they were powerful, were not able to perform these kinds of operations efficiently. Even today, the calculation with the CPUs, although considerably improved, requires long processing times. This inefficiency is due to the particular architecture of the CPUs, which have been designed to efficiently perform mathematical operations not required by neural networks.

A new kind of hardware has developed in recent decades, the *Graphics Processing Unit (GPU),* thanks to the enormous commercial drive of the video game market. In fact, this type of processor has been designed to manage more and more efficient vector calculations, such as multiplications between matrices, which is necessary for 3D reality simulations and rendering.

Thanks to this technological innovation, many deep learning techniques have been realized. In fact, to realize the neural networks and their learning, tensors (multidimensional matrices) are used, carrying out many mathematical operations. It is precisely this kind of work that GPUs can do more efficiently. Thanks to their contribution, the processing speed of deep learning is increased by several orders of magnitude (days instead of months).

Data Availability: Open Data Source, Internet of Things, and Big Data

Another very important factor affecting the development of deep learning is the huge amount of data that can be accessed. In fact, the data are the fundamental ingredient for the functioning of neural networks, both for the learning phase and for the verification phase.

Thanks to the spread of the Internet all over the world, now everyone can access and produce data. While a few years ago only a few organizations were providing data for analysis, today, thanks to the IoT (Internet of Things), many sensors and devices acquire data and make them available on networks. Not only that, even social networks and search engines (like Facebook, Google, and so on) can collect huge amounts of data, analyzing in real time millions of users connected to their services (called *Big Data*).

Today a lot of data related to the problems you want to solve with the deep learning techniques are easily available, many of them in free form (as open data source).

Python

Another factor that contributed to the great success and diffusion of deep learning techniques was the Python programming language.

In the past, planning neural network systems was very complex. The only language able to carry out this task was C ++, a very complex language, difficult to use and known only to a few specialists. Moreover, in order to work with the GPU (necessary for this type of calculation), it was necessary to know CUDA (Compute Unified Device Architecture), the hardware development architecture of NVIDIA graphics cards with all their technical specifications.

Today, thanks to Python, the programming of neural networks and deep learning techniques has become high level. In fact, programmers no longer have to think about the architecture and the technical specifications of the graphics card (GPU), but can focus exclusively on the part related to deep learning. Moreover, the characteristics of the Python language enable programmers to develop simple and intuitive code. You have already tried this with machine learning in the previous chapter, and the same applies to deep learning.

Deep Learning Python Frameworks

Over the past two years, many developer organizations and communities have been developing Python frameworks that are greatly simplifying the calculation and application of deep learning techniques. There is a lot of excitement about it, and many of these libraries perform the same operations almost competitively, but each of them is based on different internal mechanisms.

Among these frameworks available today for free, it is worth mentioning some that are gaining some success.

- *TensorFlow* is an open source library for numerical calculation that bases its use on data flow graphs. These are graphs where the nodes represent the mathematical operations and the edges represent tensors (multidimensional data arrays). Its architecture is very flexible and can distribute the calculations on multiple CPUs and on multiple GPUs.

- *Caffe2* is a framework developed to provide an easy and simple way to work on deep learning. It allows you to test model and algorithm calculations using the power of GPUs in the cloud.

- *PyTorch* is a scientific framework completely based on the use of GPUs. It works in a highly efficient way and was recently developed and is still not well consolidated. It is still proving a powerful tool for scientific research.

- *Theano* is the most used Python library in the scientific field for the development, definition, and evaluation of mathematical expressions and physical models. Unfortunately, the development team announced that new versions will no longer be released. However, it remains a reference framework thanks to the number of programs developed with this library, both in literature and on the web.

Artificial Neural Networks

Artificial neural networks are a fundamental element for deep learning and their use is the basis of many, if not almost all, deep learning techniques. In fact, these systems can learn, thanks to their particular structure that refers to the biological neural circuits.

In this section, you see in more detail what artificial neural networks are and how they are structured.

How Artificial Neural Networks Are Structured

Artificial neural networks are complex structures created by connecting simple basic components that are repeated in the structure. Depending on the number of these basic components and the type of connections, more and more complex networks will be formed, with different architectures, each of which will present peculiar characteristics regarding the ability to learn and solve different problems of deep learning.

Figure 9-2 shows an example of how a generic artificial neural network is structured.

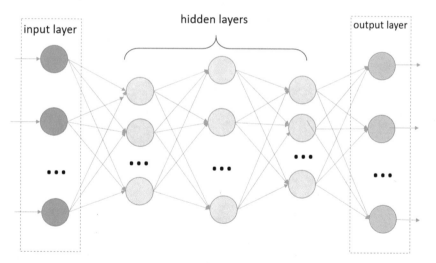

Figure 9-2. A schematization of how a generic artificial neural network is structured

The basic units are called *nodes* (the darker circles shown in Figure 9-2), which in the biological model simulate the functioning of a neuron within a neural network. These artificial neurons perform very simple operations, similar to their biological counterparts. They are activated when the total sum of the input signals they receive exceeds an activation threshold.

These nodes can transmit signals between them by means of connections, called *edges*, which simulate the functioning of biological synapses (the arrows shown in Figure 9-2). Through these edges, the signals sent by a neuron pass to the next one, behaving as a filter. That is, an edge converts the output message from a neuron, into an inhibitory or excitant signal, decreasing or increasing its intensity, according to preestablished rules (a different *weight* is generally applied for each edge).

The neural network has a certain number of nodes used to receive the input signal from the outside (see Figure 9-2). This first group of nodes is usually represented in a column at the far left end of the neural network schema. This group of nodes represents the first layer of the neural network (*input layer*). Depending on the input signals received, some (or all) of these neurons will be activated by processing the received signal and transmitting the result as output to another group of neurons, through edges.

This second group is in an intermediate position in the neural network, and it is called the *hidden layer*. This is because the neurons of this group do not communicate with the outside, neither in input nor in output, and are therefore hidden. As you can see in Figure 9-2, each of these neurons has lots of incoming edges, often with all the neurons of the previous layer. Even these hidden neurons will be activated whether the total incoming signal will exceed a certain threshold. If affirmative, they will process the signal and transmit it to another group of neurons (in the right direction of the scheme shown in Figure 9-2). This group can be another hidden layer or the *output layer,* that is, the last layer that will send the results directly to the outside.

In general, you have a flow of data that will enter the neural network (from left to right), will be processed in a more or less complex way depending on the structure, and will produce an output result.

The behavior, capabilities, and efficiency of a neural network will depend exclusively on how the nodes are connected and the total number of layers and neurons assigned to each of them. All these factors define the *neural network architecture.*

Single Layer Perceptron (SLP)

The *Single Layer Perceptron (SLP)* is the simplest neural network model and was designed by Frank Rosenblatt in 1958. Its architecture is represented in Figure 9-3.

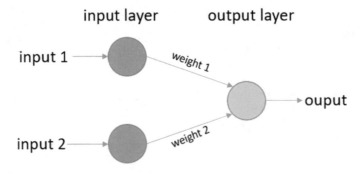

Figure 9-3. *The Single Layer Perceptron (SLP) architecture*

The Single Layer Perceptron (SLP) structure is very simple; it is a two-layer neural network, without hidden layers, in which a number of input neurons send signals to an output neuron through different connections, each with its own weight. Figure 9-4 shows in more detail the inner workings of this type of neural network.

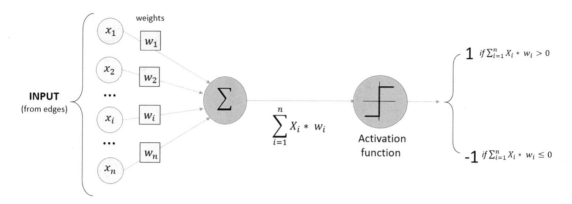

Figure 9-4. *A more detailed Single Layer Perceptron (SLP) representation with the internal operation expressed mathematically*

The edges of this structure are represented in this mathematic model by means of a weight vector consisting of the local memory of the neuron.

$$W = (w1, w2, \ldots\ldots, wn)$$

The output neuron receives an input vector signals, xi, each coming from a different neuron.

$$X = (x1, x2, \ldots\ldots, xn)$$

Then it processes the input signals via a weighed sum.

$$\sum_{i=0}^{n} w_i x_i = w_1 x_1 + w_2 x_2 + \ldots + w_n x_n = s$$

The total signal s is perceived by the output neuron. If the signal exceeds the activation threshold of the neuron, it will activate, sending 1 as a value; otherwise, it will remain inactive, sending -1.

$$\text{Output} = \begin{cases} 1, & \textit{if } s > 0 \\ -1 & \textit{otherwise} \end{cases}$$

This is the simplest *activation function* (see function A in Figure 9-5), but you can also use other more complex ones, such as the sigmoid (see function D in Figure 9-5).

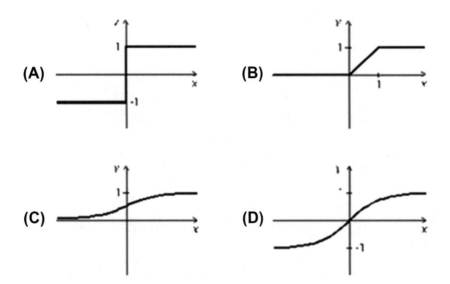

Figure 9-5. *Some examples of activation functions*

Now that you've seen the structure of the SLP neural network, you can now see how they can learn.

The learning procedure of a neural network, called the *learning phase,* works iteratively. That is, a predetermined number of cycles of operation of the neural network are carried out. The weights of the wi synapses are slightly modified in each cycle. Each learning cycle is called an *epoch*. In order to carry out the learning, you have to use appropriate input data, called the *training sets* (you have already used them in depth in Chapter 8).

In the training sets, for each input value, the expected output value is obtained. By comparing the output values produced by the neural network with the expected ones, you can analyze the differences and modify the weight values, and you can also reduce them. In practice this is done by minimizing a *cost function (loss)* that is specific of the problem of deep learning. In fact, the weights of the different connections are modified for each epoch in order to minimize the cost (*loss*).

In conclusion, supervised learning is applied to neural networks.

At the end of the learning phase, you pass to the *evaluation phase*, in which the learned SLP perceptron must analyze another set of inputs (test set) whose results are also known here. By evaluating the differences between the obtained and expected values, the degree of ability of the neural network to solve the problem of deep learning will be known. Often the percentage of cases guessed compared to the wrong ones is used to indicate this value, and it is called *accuracy*.

Multilayer Perceptron (MLP)

A more complex and efficient architecture is *Multilayer Perceptron (MLP)*. In this structure, there are one or more hidden layers interposed between the input layer and the output layer. The architecture is represented in Figure 9-6.

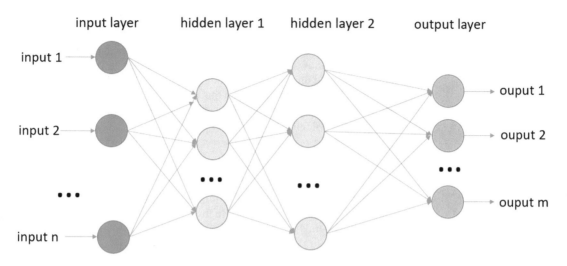

Figure 9-6. *The Multilayer Perceptron (MLP) architecture*

At the end of the learning phase, you pass to the *evaluation phase,* in which the learned SLP perceptron must analyze another set of inputs (test set) whose results are also known here. By evaluating the differences between the obtained and expected values, the degree of ability of the neural network to solve the problem of deep learning will be known. Often, the percentage of cases guessed compared to the wrong ones is used to indicate this value, and it is called accuracy.

Although more complex, the models of MLP neural networks are based primarily on the same concepts as the models of the SLP neural networks. Even in MLPs, weights are assigned to each connection. These weights must be minimized based on the evaluation of a training set, much like the SLPs. Here, too, each node must process all incoming signals through an activation function, even if this time the presence of several hidden layers makes the neural network capable of learning more, *adapting more effectively* to the type of problem deep learning is trying to solve.

On the other hand, from a practical point of view, the greater complexity of this system requires more complex algorithms both for the learning phase and for the evaluation phase. One of these is the *back propagation algorithm,* which is used to effectively modify the weights of the various connections to minimize the cost function, in order to converge the output values quickly and progressively with the expected ones.

Other algorithms are used specifically for the minimization phase of the cost (or error) function and are generally referred to as *gradient descent* techniques.

The study and detailed analysis of these algorithms is outside the scope of this text, which has only an introductory function of the argument, with the goal of trying to keep the topic of deep learning as simple and clear as possible. If you are so inclined, I suggest you go deeper into the subject, both in various books and on the Internet.

Correspondence Between Artificial and Biological Neural Networks

So far you have seen how deep learning uses basic structures, called artificial neural networks, to simulate the functioning of the human brain, particularly in the way it processes information.

There is also a real correspondence between the two systems at the highest reading level. In fact, you've just seen that neural networks have structures based on layers of neurons. The first layer processes the incoming signal, then passes it to the next layer, which in turn processes it and so on, until it reaches a final result. For each layer of neurons, incoming information is processed in a certain way, generating *different levels of representation* of the same information.

In fact, the whole operation of elaboration of an artificial neural network is nothing more than the transformation of information to ever more abstract levels.

This functioning is identical to what happens in the cerebral cortex. For example, when the eye receives an image, the image signal passes through various processing stages (such as the layers of the neural network), in which, for example, the contours of the figures are first detected (edge detection), then the geometric shape (form perception), and then to the recognition of the nature of the object with its name. Therefore, there has been a transformation at different levels of conceptuality of an incoming information, passing from an image, to lines, to geometrical figures, to arrive at a word.

TensorFlow

In a previous section of this chapter, you saw that there are several frameworks in Python that allow you to develop projects for deep learning. One of these is TensorFlow. In this section, you learn know in detail about this framework, including how it works and how it is used to realize neural networks for deep learning.

TensorFlow: Google's Framework

TensorFlow (`www.tensorflow.org`) is a library developed by the Google Brain Team, a group of Machine Learning Intelligence, a research organization headed by Google.

The purpose of this library is to have an excellent tool in the field of research for machine learning and deep learning.

The first version of TensorFlow was released by Google in February 2017, and in a year and a half, many updates have been released, in which the potential, stability, and usability of this library are greatly increased. This is mainly thanks to the large number of users among professionals and researchers who are fully using this framework. At the present time, TensorFlow is already a consolidated deep learning framework, rich in documentation, tutorials, and projects available on the Internet.

In addition to the main package, there are many other libraries that have been released over time, including:

- *TensorBoard*—A kit that allows the visualization of internal graphs of TensorFlow (`https://github.com/tensorflow/tensorboard`).

- *TensorFlow Fold*—Produces beautiful dynamic calculation charts (`https://github.com/tensorflow/fold`)

- *TensorFlow Transform*—Creates and manages input data pipelines (`https://github.com/tensorflow/transform`)

TensorFlow: Data Flow Graph

TensorFlow (`www.tensorflow.org`) is a library developed by the Google Brain Team, a group of Machine Learning Intelligence, a research organization headed by Google.

TensorFlow is based entirely on the structuring and use of graphs and on the flow of data through them, exploiting them in such a way as to make mathematical calculations.

The graph created internally in the TensorFlow runtime system is called *Data Flow Graph* and it is structured in runtime according to the mathematical model that is the basis of the calculation you want to perform. In fact, Tensor Flow allows you to define any mathematical model through a series of instructions implemented in the code. TensorFlow will take care of translating that model into the Data Flow Graph internally.

When you go to model your deep learning neural network, it will be translated into a Data Flow Graph. Given the great similarity between the structure of neural networks and the mathematical representation of graphs, it is easy to understand why this library is excellent for developing deep learning projects.

TensorFlow is not limited to deep learning and can be used to represent artificial neural networks. Many other methods of calculation and analysis can be implemented with this library, since any physical system can be represented with a mathematical model. In fact, this library can also be used to implement other machine learning techniques, for the study of complex physical systems through the calculation of partial differentials, and so on.

The nodes of the Data Flow Graph represent mathematical operations, while the edges of the graph represent tensors (multidimensional data arrays). The name TensorFlow derives from the fact that these tensors represent the flow of data through graphs, which can be used to model artificial neural networks.

Start Programming with TensorFlow

Now that you have seen in general what the TensorFlow framework consists of, you can start working with this library. In this section, you see how to install this framework, understand the differences between the old 1.x version and the new one, and the key features of the latter.

TensorFlow 2.x vs TensorFlow 1.x

As anticipated at the beginning of the chapter, in this third edition, the text and example code related to the use of TensorFlow for deep learning have been completely rewritten. This is because the new version of TensorFlow 2.x has been introduced since 2019. There are many changes that have been made. Many of the modules present in the TensorFlow 1.x release have been removed or moved. Keras has been completely incorporated as a neural network management module, and therefore most of the previous programming mechanisms and paradigms (for example, those present in the second edition with TensorFlow 1.x) are no longer compatible.

Given the large amount of programs developed in recent years with TensorFlow 1.x, it's important that these programs continued to compatible, and therefore usable. Rushing to address this issue, Google enabled a way to continue using the old code without having to rewrite much.

At the beginning of the code, you simply replace the classic import line:

```
import tensorflow as tf
```

with the following line:

```
import tensorflow.compat.v1 as tf
```

This replacement should guarantee in most cases the perfect execution of code developed with TensorFlow 1.x, even if your current version is TensorFlow 2.x.

As for present projects, if you are about to develop a new deep learning project with TensorFlow, the only reasonable path is to follow the TensorFlow 2.x paradigms. That's why this chapter only mentions this latest version, without transcribing the old code.

Installing TensorFlow

Before starting work, you need to install this library on your computer.

If the TensorFlow library is already installed, it's useful to determine which version is present, especially regarding the differences between TensorFlow 1.x and TensorFlow 2.x. You can easily find that out by opening a Python session and inserting the following lines of code:

```
import tensorflow as tf
tf.__version__
```

If the 1.x version is present, the best thing to do is create a new virtual environment on which to install the 2.x version without compromising the configuration of modules installed on your system. If the library is not present on your system, you can easily install TensorFlow 2.x. As in the previous chapters, the optimal solution is to have the Anaconda platform and graphically install the TensorFlow package from Navigator. If you prefer to use the command line, still on Anaconda, you can enter the following:

```
conda install tensorflow
```

If, on the other hand, you don't have (or don't want) the Anaconda platform, you can safely install TensorFlow via PyPI.

```
pip install tensorflow
```

■ **Note** At the time of this writing, I found some incompatibility issues in Anaconda between TensorFlow and other libraries for virtual environments based on Python 3.10 and 3.11. I then created a virtual environment with Python 3.9 and didn't encounter any problems.

Programming with the Jupyter Notebook

Once TensorFlow is installed, you can start programming with this library. The examples in this chapter use Jupyter Notebook, but you can do the same things by opening a normal Python session.

With the latest versions of TensorFlow, the demand for resources becomes more and more preponderant. Deep learning with applications like IPython and Jupyter Notebook may not be possible without proper precautions. So in this case it is necessary to add the following lines of code in the first cell of the Notebook before starting to work (also in IPython):

```
import os
os.environ["KMP_DUPLICATE_LIB_OK"]="TRUE"
```

In particular, by varying this environment variable, the OpenMP API active in the system on which you are operating is informed not to create problems if another instance of it is created (a case that leads to the crash of Jupyter Notebook).

Tensors

The basic element of the TensorFlow library is the *tensor*. In fact, the data that follow the flow within the Data Flow Graph are tensors (see Figure 9-7).

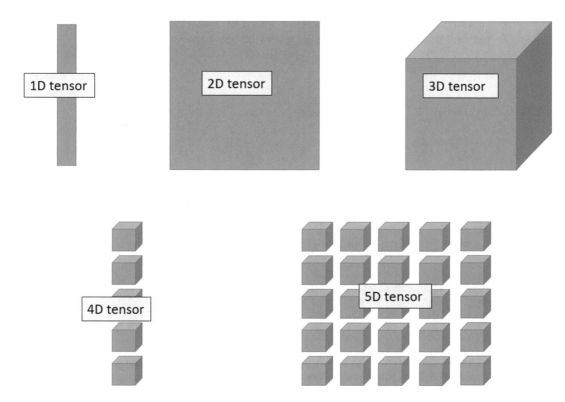

Figure 9-7. *Some representations of the tensors according to the different dimensions*

A tensor is identified by three parameters:

- **rank**—Dimension of the tensor (a matrix has rank 2, a vector has rank 1)
- **shape**—Number of rows and columns (e.g., (3.3) is a 3x3 matrix)
- **type**—The type of tensor elements

type of tensor elements and columns (eg (3.3) is a 3x3 matrix)has rank 2, a vector has rank 1)s ic

Tensors are nothing more than multidimensional arrays. In previous chapters, you saw how easy it is to get them, thanks to the NumPy library. You can start by defining one with this library.

```
import numpy as np
t = np.arange(9).reshape((3,3))
t
array([[0, 1, 2],
       [3, 4, 5],
       [6, 7, 8]])
```

You can convert this multidimensional array into a TensorFlow tensor very easily, thanks to the tf.convert_to_tensor() function, which takes as a parameter the array to convert.

```
tensor = tf.convert_to_tensor(t)
tensor
Out [ ]:
<tf.Tensor: shape=(3, 3), dtype=int64, numpy=
    array([[0, 1, 2],
           [3, 4, 5],
           [6, 7, 8]])>
```

During the conversion from array to tensor, it is also possible to change the data type. In this case, you add the second optional parameter dtyle, specifying the new data type. For example, if you wanted to convert the integer input into floating numbers, you would write:

```
tensor2 = tf.convert_to_tensor(t, dtype='float64')
tensor2
Out [ ]:
<tf.Tensor: shape=(3, 3), dtype=float64, numpy=
array([[0., 1., 2.],
          [3., 4., 5.],
          [6., 7., 8.]])>
```

As you can see, you have a tensor containing the same values and the dimensions as the multidimensional array defined with NumPy. This approach is very useful for calculating deep learning, since many input values are in the form of NumPy arrays.

But tensors can be built directly from TensorFlow, without using the NumPy library. There are a number of functions that make it possible to enhance the tensors quickly and easily.

For example, if you want to initialize a tensor with all 0 values, you can use the tf.zeros() method.

```
t0 = tf.zeros((3,3),'float64')
t0
Out [ ]:
<tf.Tensor: shape=(3, 3), dtype=float64, numpy=
array([[0., 0., 0.],
          [0., 0., 0.],
          [0., 0., 0.]])>
```

Likewise, if you want a tensor with all values of 1, you use the tf.ones() method.

```
t1 = tf.ones((3,3),'float64')
t1
Out [ ]:
<tf.Tensor: shape=(3, 3), dtype=float64, numpy=
array([[1., 1., 1.],
          [1., 1., 1.],
          [1., 1., 1.]])>
```

Finally, it is also possible to create a tensor containing random values, which follow a uniform distribution (all the values within a range are equally likely to exist), thanks to the tf.random_uniform() function.

For example, if you want a 3x3 tensor with float values between 0 and 1, you can write:

```
trand = tf.random.uniform((3, 3), minval=0, maxval=1, dtype=tf.float32)
trand
Out [ ]:
<tf.Tensor: shape=(3, 3), dtype=float32, numpy=
array([[0.99075377, 0.7289959 , 0.6183866 ],
       [0.51800334, 0.49188066, 0.01087034],
       [0.21716583, 0.29331267, 0.91550064]], dtype=float32)>
```

It can often be useful to create a tensor containing values that follow a normal distribution with a choice of mean and standard deviation. You can do this with the `tf.random_normal()` function.

For example, if you want to create a tensor of 3x3 size with mean of 0 and standard deviation of 3, you will write:

```
tnorm = tf.random.normal((3, 3), mean=0, stddev=3)
tnorm
Out [ ]:
<tf.Tensor: shape=(3, 3), dtype=float32, numpy=
array([[-1.2079163 , -0.88857937, 5.041537 ],
       [-3.7309105 , 3.157123 , -3.4958515 ],
       [ 4.219907 , 1.7997034 , -5.020906 ]], dtype=float32)>
```

Loading Data Into a Tensor from a pandas Dataframe

So far you've seen how to manually define the data within a tensor, or how to convert a NumPy array to a tensor. But a much more common operation in data analysis is having to insert the data present in a dataframe into a tensor. In fact, dataframes are one of the most commonly used formats during the data analysis process in Python. As you will see now, this operation is very simple in TensorFlow.

First import the pandas library into your Notebook.

```
import pandas as pd
```

Now define a simple dataframe as a basic example.

```
df = pd.DataFrame(np.array([[1, 2, 3],
                            [4, 5, 6],
                            [7, 8, 9]]),
                  columns=['a', 'b', 'c'])
df
```

For the output, you will get a dataframe like the one shown in Figure 9-8.

	a	b	c
0	1	2	3
1	4	5	6
2	7	8	9

Figure 9-8. *A simple example of a pandas dataframe*

To convert the data inside the dataframe into a tensor, you can use the `tf.convert_to_tensor()` function that you used previously. This in fact also accepts a pandas dataframe as an argument.

```
tensor_df = tf.convert_to_tensor(df)
tensor_df
Out [ ]:
<tf.Tensor: shape=(3, 3), dtype=int64, numpy=
array([[1, 2, 3],
       [4, 5, 6],
       [7, 8, 9]])>
```

As you can see from the result, the conversion was very easy. Rarely, however, will you have to load all the data present within a dataframe, but rather the values of one or more columns. So you will use the following form more often, making a selection of the columns (or data) that interest you within the dataframe.

```
tensor_df = tf.convert_to_tensor(df[['a','b']])
tensor_df

Out[ ]:
<tf.Tensor: shape=(3, 2), dtype=int64, numpy=
array([[1, 2],
       [4, 5],
       [7, 8]])>
```

Loading Data in a Tensor from a CSV File

Another format in which the data to be analyzed is often available is within files, especially CSV files. Also in this case the data of a CSV file can be loaded into a tensor using pandas. In fact, the pandas library provides many functions for loading data contained in CSV files (and other formats) within the dataframe. These can then be converted into tensors using the procedure seen earlier with the `tf.convert_to_tensor()` function.

For example, you can load the data present in the `training_data.csv` file containing the training data that you will use in the following examples of the chapter during the development and study of some models. To do this, you use the pandas function called `read_csv()`.

```
df = pd.read_csv('training_data.csv')
df
```

By loading the data contained in the CSV, you will obtain as a result a dataframe like the one shown in Figure 9-9.

	X	Y	label
0	1.0	3.0	0
1	1.0	2.0	0
2	1.0	1.5	0
3	1.5	2.0	0
4	2.0	3.0	0
5	2.5	1.5	1
6	2.0	1.0	1
7	3.0	1.0	1
8	3.0	2.0	1
9	3.5	1.0	1
10	3.5	3.0	1

Figure 9-9. *The pandas dataframe containing the training dataset*

The training dataset contained in the dataframe is composed of three columns. The first two represent the X,Y coordinates of the points on a Cartesian plane, while the third column contains the labels, that is, the classes to which the correlated points belong. As you learned for machine learning (Chapter 8), training datasets are composed of two parts—the features and the labels. The same rule applies to deep learning, and therefore you have to extract two distinct tensors from the dataframe, one for the features and one for the labels.

```
df_features = df.copy()
df_labels = df_features.pop('label')
data_features = tf.convert_to_tensor(df_features)
data_features
Out [ ]:
<tf.Tensor: shape=(11, 2), dtype=float64, numpy=
array([[1. , 3. ],
       [1. , 2. ],
       [1. , 1.5],
       [1.5, 2. ],
       [2. , 3. ],
       [2.5, 1.5],
       [2. , 1. ],
       [3. , 1. ],
       [3. , 2. ],
       [3.5, 1. ],
       [3.5, 3. ]])>
```

Run the same operation for the tensor of the labels.

```
data_labels = tf.convert_to_tensor(df_labels)
data_labels
Out [ ]:
<tf.Tensor: shape=(11,), dtype=int32, numpy=array([0, 0, 0, 0, 0, 1, 1, 1, 1, 1, 1])>
```

Using the tf.conver_to_tensor() function, you now have two tensors: one for the features containing the X,Y coordinates of the points and one for the labels containing the classes to which the points belong.

Operation on Tensors

Once the tensors have been defined, it will be necessary to carry out operations on them. Most mathematical calculations on tensors are based on the sum and multiplication between tensors.

Define two tensors, t1 and t2, that you will use to perform the operations between tensors.

```
t1 = tf.random.uniform((3, 3), minval=0, maxval=1, dtype=tf.float32)
t1
Out [ ]:
<tf.Tensor: shape=(3, 3), dtype=float32, numpy=
array([[0.29003692, 0.92972696, 0.41073143],
       [0.46694946, 0.46367037, 0.11636639],
       [0.31574678, 0.70260215, 0.0642364 ]], dtype=float32)>
```

```
t2 = tf.random.uniform((3, 3), minval=0, maxval=1, dtype=tf.float32)
t2
Out [ ]:
<tf.Tensor: shape=(3, 3), dtype=float32, numpy=
array([[0.23392928, 0.7185135 , 0.64518535],
       [0.6719583 , 0.7983806 , 0.10201716],
       [0.92533255, 0.32889807, 0.4179113 ]], dtype=float32)>
```

To sum these two tensors, you use the tf.add() function. To perform multiplication, you use the tf.matmul() function.

```
sum = tf.add(t1,t2)
sum
Out [ ]:
<tf.Tensor: shape=(3, 3), dtype=float32, numpy=
array([[0.5239662 , 1.6482404 , 1.0559168 ],
       [1.1389078 , 1.262051  , 0.21838355],
       [1.2410793 , 1.0315002 , 0.4821477 ]], dtype=float32)>
```

```
mul = tf.matmul(t1,t2)
mul
Out [ ]:
<tf.Tensor: shape=(3, 3), dtype=float32, numpy=
array([[1.0726491 , 1.0857601 , 0.453625  ],
       [0.5284779 , 0.7439676 , 0.39720213],
       [0.6054218 , 0.8089395 , 0.3022378 ]], dtype=float32)>
```

Another very common operation with tensors is the calculation of the determinant. TensorFlow provides the `tf.linalg.det()` method for this purpose:

```
det = tf.linalg.det(t1)
det
Out [ ]:
<tf.Tensor: shape=(), dtype=float32, numpy=0.06581897>
```

The new `tf.linalg` module (introduced in TensorFlow 2.x) contains, in addition to the determinant calculation, many other algebraic operations on matrices, which are very useful during tensor calculations. These functions, along with the basic operations, allow you to implement many mathematical expressions that use tensors.

Developing a Deep Learning Model with TensorFlow

Once you have seen the tensors that are the basis of the data to be used in TensorFlow, you can proceed further, analyzing in brief the fundamental steps to create a deep learning model with neural networks and its training and testing phases with TensorFlow 2.x. Those steps are as follows:

- Definition of tensors (training and testing sets)
- Model building
- Model compiling
- Model training
- Model testing
- Predictions making

You learned about the first part, the one related to the preparation of tensors from the training and testing datasets, in the previous section. The next section covers model building.

Model Building

Regarding the construction of the model based on a neural network, you must define how the layers will be configured. As you saw in the first part of the chapter, each neural network is composed of one or more layers of neurons. With TensorFlow 2.x, it is not necessary to define each neuron and the individual connections that compose the network. But you can use one of the layers that's predefined in Keras.

For example:

```
tf.keras.layers.Dense
```

This corresponds to a layer of neurons where all the connections are made with the adjacent layer of neurons, called a "regular densely-connected NN layer."

Another widely used layer is as follows:

```
tf.keras.layers.Flatten
```

This layer is typically used at the beginning of the neural network, when the feature dataset is not one-dimensional. This allows you to flatten the data, making them single-dimensional and thus usable by the subsequent layers of the neural network.

To better understand, if you wanted to submit an image to a neural network, this would be composed of (nxn) pixels and therefore would be two-dimensional. Inserting a Flatten layer would make this training dataset one-dimensional.

Another widely used layer is as follows:

```
tf.keras.layers.Normalization
```

This layer, like the previous one, is also placed as the first layer of the neural network and is used to normalize the input data. This practice is very common in machine learning, making the data more easily actionable.

Therefore, thanks to the integration of Keras in TensorFlow 2.x, the building of the model can have many predefined layers, with a whole series of learning parameters inside that will vary during the training phase. These are already defined and therefore the model building operation is much easier.

In fact, to build a simple neural network like the ones you saw at the beginning of the chapter, it is sufficient to define tf.keras.Sequential() with the various predefined layers inside, in the order of construction.

```
model = tf.keras.Sequential([
    tf.keras.layers.Flatten(input_shape(128,128)),
    tf.keras.layers.Dense(128),
    tf.keras.layers.Dense(12)
])
```

The model just described therefore represents a two-layer neural network (without hidden layers) in which two-dimensional tensors, such as 128x128 pixel images, are made one-dimensional in the first layer, to then be passed to the first layer composed of 128 neurons. These in turn are connected to another layer of 12 neurons, which will probably correspond to 12 classes of belonging (as you will see later).

Model Compiling

Now that a model has been defined, it is necessary to compile it. This second step requires the choice of functions and training parameters:

- Loss function: Measures how accurate the model has to be during the training phase.

- Optimizer: Takes care of how and which parameters should be updated during the training phase.

- Metrics: The parameters used to monitor the progress of the training (and testing) of the model.

All these elements are already available within the Keras module, thus already providing a set of tools ready to use in a model.

Thus, compiling the model is reduced to a simple compile() function, in which these three elements are defined as arguments.

```
model.compile( optimizer = 'adam',
               loss=tf.keras.losses.SparseCategoricalCrossentropy(from_logits=True),
               metrics = ['accuracy']
)
```

As you can see, in the new version of TensorFlow 2.x, the construction of a model and its compilation are very fast and high-level operations, contrary to the previous version, which required the detailed definition of most of these elements.

Model Training and Testing

Now that the model is compiled, you can move on to training and testing it. For these two phases, a dataset is used in which there are *features* (a series of variables that describe the subject under study) and *labels* (the solutions such as, for example, the classes to which they belong). This is very similar to what you saw with machine learning in the previous chapter (Chapter 8). The dataset is divided into a training dataset and a testing dataset, with the former being much larger.

Each feature is then subdivided into two tensors, until four tensors are obtained:

- – `train_features`
- – `train_labels`
- – `test_features`
- – `test_labels`

The first two tensors are used for model learning, using the `fit()` function.

```
model.fit(train_features, train_labels, epochs=100)
```

The number of epochs is the third parameter, which is the number of times the learning phase is performed. At each of these stages, the accuracy metrics should improve, thus signaling successful model learning. Once done, you need to have an educated model ready for testing. Here, you use the other two tensors (`test_features` and `test_labels`) to evaluate the model's ability to make predictions. The difference between these and the values contained in `test_labels` will provide the accuracy of these predictions.

For the testing phase, the `evaluate()` function is used.

```
test_loss, test_acc = model.evaluate( test_features, test_labels, verbose=2)
```

This function returns the loss and accuracy of the model as values.

Prediction Making

The last phase submits the model to the purpose for which it was created, making predictions on input data whose solution you do not know (for example, the class it belongs to).

During this phase, an additional layer is often added to the model, called Softmax. This is placed at the end of the neural network and is used to convert the output of the last layer (logits) into probability values. In this case, it is not necessary to recompile the newly trained model again, but this can be extended by adding this layer at the end. You have thus defined a new extended model.

```
probability_model = tf.keras.Sequential([
      model,
      tf.keras.layers.Softmax()
])
```

To complete the deep learning steps, the model makes predictions from data whose solutions you do not know. You create a new tensor and submit it to the predict() function.

```
predictions = probability_model.predict(samples_features)
```

An array with all predictions divided by the percentage is obtained as the returned value. It can be used to obtain the class to which it belongs.

```
np.argmax(prediction[i])
```

i is the ith element of the dataset.

Practical Examples with TensorFlow 2.x

At this point, in theory, you should have acquired enough basic knowledge to be able to start working with real examples. Let's put into practice what you have seen so far with different types of neural networks.

- Single Layer Perceptron (SLP)
- Multilayer Neural Network with One Hidden Layer
- Multilayer Neural Network with Two Hidden Layers

Single Layer Perceptron with TensorFlow

To better understand how to develop neural networks with TensorFlow, you will begin to implement a Single Layer Perceptron (SLP) neural network that is as simple as possible. You will use the tools made available in the TensorFlow library. By using the concepts you learned during the chapter and gradually introducing new ones, you can implement a Single Layer Perceptron (SLP) neural network.

Before Starting

Before starting, open a new Jupyter Notebook or shut down and start the kernel again. Once the session is open it imports all the necessary modules:

```
import os
os.environ["KMP_DUPLICATE_LIB_OK"]="TRUE"

import tensorflow as tf
import numpy as np
import matplotlib.pyplot as plt
```

Data To Be Analyzed

For the examples that you consider in this chapter, you will use a series of data that you used in Chapter 8, in particular in the section entitled "Support Vector Machines (SVMs)."

The set of data that you will study is a set of 11 points distributed in a Cartesian axis, divided into two classes of membership. The first six belong to the first class, the other five to the second. The coordinates (x, y) of the points are contained within a NumPy inputX array, while the class to which they belong is indicated in inputY. This is a list of two-element arrays, with an element for each class they belong to. The value 1 in the first or second element indicates the class to which it belongs.

If the element has value [1.0], it will belong to the first class. If it has value [0,1], it belongs to the second class. The fact that they are floating point values is due to the optimization calculation of deep learning. You will see later that the test results of the neural networks are floating numbers, indicating the probability that an element belongs to the first or second class.

Suppose, for example, that the neural network will give you the result of an element that will have the following values:

```
[0.910, 0.090]
```

This result will mean that the neural network considers that the element under analysis belongs 91 percent to the first class and 9 percent to the second class. You will see this in practice at the end of the section, but it is important to explain the concept to better understand the purpose of some values.

Based on the values taken from the example of SVMs in Chapter 8, you can define the following values.

```
#Training set
inputX = np.array([[1.,3.],[1.,2.],[1.,1.5],
                   [1.5,2.],[2.,3.],[2.5,1.5],
                   [2.,1.],[3.,1.],[3.,2.],
                   [3.5,1.],[3.5,3.]])
inputY = [[1.,0.]]*6+ [[0.,1.]]*5
print(inputX)
print(inputY)
Out [ ]:
[[1.  3. ]
 [1.  2. ]
 [1.  1.5]
 [1.5 2. ]
 [2.  3. ]
 [2.5 1.5]
 [2.  1. ]
 [3.  1. ]
 [3.  2. ]
 [3.5 1. ]
 [3.5 3. ]]
[[1.0, 0.0], [1.0, 0.0], [1.0, 0.0], [1.0, 0.0], [1.0, 0.0], [1.0, 0.0], [0.0, 1.0], [0.0,
1.0], [0.0, 1.0], [0.0, 1.0], [0.0, 1.0]]
```

In reality, you have already seen these values previously, when we talked about how to import data from a CSV file by converting it to a pandas dataframe. Here, you use the NumPy array version (the same one used for machine learning, in particular in the section about the Support Vector Machines technique) to give an additional version of the starting data to be converted into tensors. Feel free to choose the format you prefer for the initial datasets.

To better see how these points are arranged spatially and which classes they belong to, there is no better approach than to plot everything with matplotlib.

```
yc = [0]*5 + [1]*6
print(yc)
plt.scatter(inputX[:,0],inputX[:,1],c=yc, s=50, alpha=0.9)
plt.show()
```

```
Out [ ]:
[0, 0, 0, 0, 0, 1, 1, 1, 1, 1, 1]
```

You will get the graph in Figure 9-8 as a result.

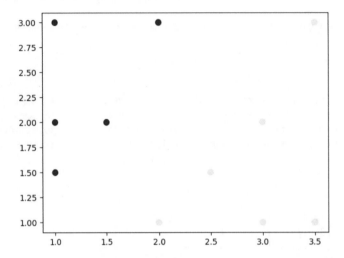

Figure 9-10. *The training set is a set of Cartesian points divided into two classes of membership (light and dark)*

To help in the graphic representation (as shown in Figure 9-8) of the color assignment, the inputY array has been replaced with the yc array.

As you can see, the two classes are easily identifiable in two opposite regions. The first region covers the upper-left part, and the second region covers the lower-right part. All this would seem to be simply subdivided by an imaginary diagonal line, but to make the system more complex, there is an exception with the point 6 that is internal to the other points.

It will be interesting to see how and if the neural networks that you implement can correctly assign the class to points of this kind.

You then convert the values of the training arrays into tensors, as you have seen done previously.

```
train_features = tf.convert_to_tensor(inputX)
train_labels = tf.convert_to_tensor(inputY)
train_features
Out [ ]:
<tf.Tensor: shape=(11, 2), dtype=float64, numpy=
array([[1. , 3. ],
       [1. , 2. ],
       [1. , 1.5],
       [1.5, 2. ],
       [2. , 3. ],
       [2.5, 1.5],
       [2. , 1. ],
       [3. , 1. ],
       [3. , 2. ],
       [3.5, 1. ],
       [3.5, 3. ]])>
```

```
train_labels
Out [ ]:
<tf.Tensor: shape=(11, 2), dtype=float32, numpy=
array([[1., 0.],
       [1., 0.],
       [1., 0.],
       [1., 0.],
       [1., 0.],
       [1., 0.],
       [0., 1.],
       [0., 1.],
       [0., 1.],
       [0., 1.],
       [0., 1.]], dtype=float32)>
```

Now you can build the Single Layer Perceptron model based on the neural network in Figure 9-11, adding the various Keras layers to the model definition.

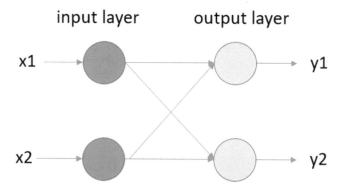

Figure 9-11. *The Single Layer Perceptron model used in this example*

```
model = tf.keras.Sequential([
    tf.keras.layers.Dense(2),
    tf.keras.layers.Dense(2)
])
```

As you can see, it is a very simple model, but more than sufficient for making predictions in simple cases such as the one in question.

Because this is a binary classification problem, you can choose `BinaryCrossentropy` as the function loss for the compilation.

```
model.compile(optimizer='SGD',
              loss=tf.keras.losses.BinaryCrossentropy(from_logits=True),
              metrics=['accuracy'])
```

Once the model is compiled, you can move on to training it. Choose 200 epochs, assuming more than enough time for the model to learn.

```
h = model.fit(train_features, train_labels, epochs=2000)
Out [ ]:
Epoch 1/2000
1/1 [==============================] - 1s 637ms/step - loss: 2.0271 - accuracy: 0.5455
Epoch 2/2000
1/1 [==============================] - 0s 4ms/step - loss: 1.9807 - accuracy: 0.5455
Epoch 3/2000
1/1 [==============================] - 0s 5ms/step - loss: 1.9356 - accuracy: 0.5455
Epoch 4/2000
1/1 [==============================] - 0s 9ms/step - loss: 1.8918 - accuracy: 0.5455
Epoch 5/2000
1/1 [==============================] - 0s 8ms/step - loss: 1.8493 - accuracy: 0.5455
Epoch 6/2000
1/1 [==============================] - 0s 6ms/step - loss: 1.8079 - accuracy: 0.5455
Epoch 7/2000
...
```

In the output, you will have all the learning situations epoch by epoch, through a scroll bar that shows the completion for each of them. The loss and accuracy value will then be shown next to each line of output.

However, it is clear that this output is not the easiest way to understand how this neural network model behaved during the learning phase. For this purpose, I have saved the output in the return value h (for history), which you can use for graphical visualizations that can help you.

Extract from the history variable the loss values corresponding to the various periods.

```
acc_set = h.history['loss']
epoch_set = h.epoch
```

Arrange these values in a plotting chart to see the learning progress graphically, thanks to the matplotlib library.

```
# return list of every 100th item in a larger list
plt.plot(epoch_set[0::100],acc_set[0::100], 'o', label='Training phase')
plt.ylabel('loss')
plt.xlabel('epoch')
plt.legend()
```

As a result, you will obtain a graph similar to the one shown in Figure 9-12.

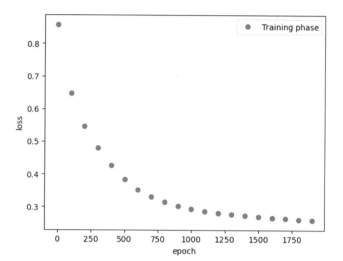

Figure 9-12. *The less value decreases during the learning phase (less optimization)*

Now move on to the testing phase, defining the other two tensors.

```
testX = np.array([[1.,2.25],[1.25,3.],
                  [2,2.5],[2.25,2.75],
                  [2.5,3.],[2.,0.9],
                  [2.5,1.2],[3.,1.25],
                  [3.,1.5],[3.5,2.],
                  [3.5,2.5]])
testY = [[1.,0.]]*5 + [[0.,1.]]*6

test_features = tf.convert_to_tensor(testX)
test_labels = tf.convert_to_tensor(testY)
```

Evaluate the accuracy of the newly educated SLP model.

```
test_loss, test_acc = model.evaluate(test_features, test_labels, verbose=2)
Out [ ]:
1/1 - 0s - loss: 0.1812 - accuracy: 1.0000 - 233ms/epoch - 233ms/step
```

As you can see, the accuracy is at best, given the simplicity of the classification. So you can expect a very good level of prediction of the newly educated model.

Now move on to the proper classification, passing to the neural network a very large amount of data (points on the Cartesian plane) without knowing to which class they belong. This is, in fact, the moment that the neural network informs you about the possible classes.

To this end, the program simulates experimental data, creating points on the Cartesian plane that are completely random. For example, you can generate an array containing 1,000 random points.

```
exp_features = 3*np.random.random((1000,2))
```

Now you extend the model with Softmax to obtain an output of the probability of the different points belonging to the two classes.

315

```
probability_model = tf.keras.Sequential([
    model,
    tf.keras.layers.Softmax()
])
```

Now make the predictions of the experimental data with the model you just extended.

```
predictions = probability_model.predict(exp_features)
```

Let's determine the probability of a single point belonging to the two classes. For example, the first:

```
predictions[0]
Out [ ]:
array([0.073105 , 0.9268949], dtype=float32)
```

If, on the other hand, we want to know directly which of the two classes it belongs to, we write the following code obtaining the class it belongs to (0 for the first and 1 for the second):

```
np.argmax(predictions[0])
Out [ ]:
1
```

Instead of analyzing point by point in a textual way, there is a graphical way to visualize the result of these predictions. In the previous scatterplots, you classified the points on the Cartesian plane with two colors (yellow and purple). Now that you considered the probability of a point belonging to these two classes, for all intermediate probabilities, an intermediate color of the gradient will be displayed. It will fade from yellow to purple depending on how close it is to one of the two classes.

```
yc = predictions[:,1]
plt.scatter(exp_features[:,0],exp_features[:,1],c=yc, s=50, alpha=1)
plt.show()
```

Running the code, you will get a scatterplot like the one shown in Figure 9-13.

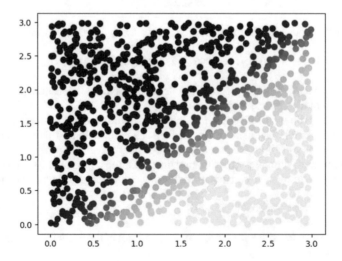

Figure 9-13. *A scatterplot with all the experimental points and the estimate of the classes to which they belong*

As you can see according to the shades, two areas of classification are delimited on the plane, with a color gradient in the central part (green color) indicating the zones of uncertainty.

The classification results can be made more comprehensible and clearer by deciding to establish based them on the probability of the point belonging to one or the other class. If the probability of a point belonging to a class is greater than 0.5, it will belong to it.

You can modify the previous scatterplot by requiring that each point belong to one or another class, according to the most probable option.

```
yc = np.round(predictions[:,1])
plt.scatter(exp_features[:,0],exp_features[:,1],c=yc, s=50, alpha=1)
plt.show()
```

Running the code, you will get a scatterplot similar to the one shown in Figure 9-14.

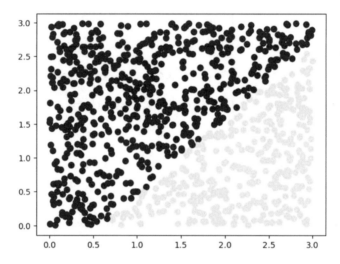

Figure 9-14. *The points delimit the two regions corresponding to the two classes*

In the scatterplot shown in Figure 9-14, you can clearly see the two regions of the Cartesian plane that characterize the two classes of belonging.

Multilayer Perceptron (with One Hidden Layer) with TensorFlow

In this section, you deal with the same problem as in the previous section, but using an MLP (Multilayer Perceptron) neural network.

Start a new Jupyter Notebook, or continue with the same one, but reset the kernel.

As you saw earlier in the chapter, an MLP neural network differs from an SLP neural network in that it can have one or more hidden layers.

To build this model, which compared to the previous one has a hidden layer of two neurons, you define a model similar to the previous one, but with an intermediate Dense layer.

```
model = tf.keras.Sequential([
    tf.keras.layers.Dense(2),
    tf.keras.layers.Dense(2),
    tf.keras.layers.Dense(2)
])
```

The next step is to compile the model, choosing an optimization method. For MLP neural networks, a good choice is the Adam optimization method. Instead, keep the loss function and the metrics unchanged.

```
model.compile(optimizer='Adam',
              loss=tf.keras.losses.BinaryCrossentropy(from_logits=True),
              metrics=['accuracy'])
```

As with the previous SLP model, train it using the same training dataset.

```
h = model.fit(train_features, train_labels, epochs=2000)
Out [ ]:
Epoch 1/2000
1/1 [==============================] - 0s 426ms/step - loss: 1.8568 - accuracy: 0.4545
Epoch 2/2000
1/1 [==============================] - 0s 3ms/step - loss: 1.8473 - accuracy: 0.4545
Epoch 3/2000
1/1 [==============================] - 0s 3ms/step - loss: 1.8378 - accuracy: 0.4545
Epoch 4/2000
1/1 [==============================] - 0s 646us/step - loss: 1.8284 - accuracy: 0.4545
Epoch 5/2000
1/1 [==============================] - 0s 0s/step - loss: 1.8190 - accuracy: 0.4545
Epoch 6/2000
1/1 [==============================] - 0s 0s/step - loss: 1.8097 - accuracy: 0.4545
Epoch 7/2000
1/1 [==============================] - 0s 14ms/step - loss: 1.8004 - accuracy: 0.4545
Epoch 8/2000
...
```

Extract the training history and create the same graph showing the behavior of the model during the training process.

```
acc_set = h.history['loss']
epoch_set = h.epoch# return list of every 50th item in a larger list
plt.plot(epoch_set[0::50],acc_set[0::50], 'o', label='Training phase')
plt.ylabel('loss')
plt.xlabel('epoch')
plt.legend()
```

Running the previous code, you will get a chart like the one shown in Figure 9-15.

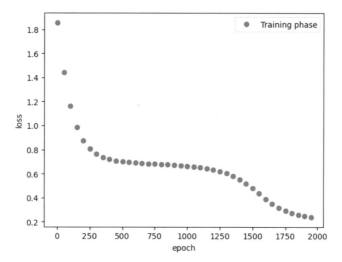

Figure 9-15. *The learning curve of the MLP model shows two distinct optimization phases*

As you can see, even a more complex neural network shows different stages of learning. In this case, the choice of 2,000 epochs was fundamental to obtain an increase in the forecasting capacity of the model. If you had stopped at 1,000 epochs, the loss would have been 0.7 instead of 0.2.

As the result of testing this last model, you get an accuracy of 100 percent and a loss value of 01.

```
test_loss, test_acc = model.evaluate(test_features, test_labels, verbose=2)
Out [ ]:
1/1 - 0s - loss: 0.1602 - accuracy: 1.0000 - 105ms/epoch - 105ms/step
```

Multilayer Perceptron (with Two Hidden Layers) with TensorFlow

In this section, you extend the previous structure by adding two neurons to the first hidden layer (four in all) and adding a second hidden layer with two neurons.

As you did previously, start a new Jupyter Notebook and write the necessary code of the previous examples, or restart the kernel and execute the cells with the necessary code.

```
).
model = tf.keras.Sequential([
    tf.keras.layers.Dense(2),
    tf.keras.layers.Dense(4),
    tf.keras.layers.Dense(2),
    tf.keras.layers.Dense(2)
])
model.compile(optimizer='Adam',
              loss=tf.keras.losses.BinaryCrossentropy(from_logits=True),
              metrics=['accuracy'])
```

Then train the new two hidden layer MLP model with the same training dataset you used previously.

```
h = model.fit(train_features, train_labels, epochs=2000)
Out [ ]:
Epoch 1/2000
1/1 [==============================] - 1s 826ms/step - loss: 0.6980 - accuracy: 0.4545
Epoch 2/2000
1/1 [==============================] - 0s 3ms/step - loss: 0.6970 - accuracy: 0.4545
Epoch 3/2000
1/1 [==============================] - 0s 3ms/step - loss: 0.6961 - accuracy: 0.4545
Epoch 4/2000
1/1 [==============================] - 0s 0s/step - loss: 0.6953 - accuracy: 0.4545
Epoch 5/2000
1/1 [==============================] - 0s 0s/step - loss: 0.6945 - accuracy: 0.4545
Epoch 6/2000
1/1 [==============================] - 0s 0s/step - loss: 0.6938 - accuracy: 0.4545
Epoch 7/2000
1/1 [==============================] - 0s 3ms/step - loss: 0.6931 - accuracy: 0.4545
Epoch 8/2000
1/1 [==============================] - 0s 0s/step - loss: 0.6925 - accuracy: 0.4545
...
```

Also for this model, you can analyze the learning phase of the neural network by displaying the loss values as the epochs increase in a chart.

```
acc_set = h.history['loss']
epoch_set = h.epoch
# return list of every 50th item in a larger list
plt.plot(epoch_set[0::50],acc_set[0::50], 'o', label='Training phase')
plt.ylabel('loss')
plt.xlabel('epoch')
plt.legend()
```

Running the code, you will get a plot like the one in Figure 9-16.

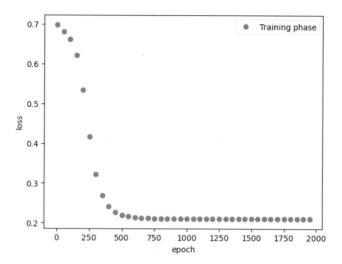

Figure 9-16. *The trend of the loss during the learning phase for an MLP with two hidden layers*

From what you can see in Figure 9-16, learning in this case is much faster than the previous case (at 1,000 epochs, you would be fine).

```
test_loss, test_acc = model.evaluate(test_features, test_labels, verbose=2)
Out [ ]:
1/1 - 0s - loss: 0.0951 - accuracy: 1.0000 - 124ms/epoch - 124ms/step
```

The optimized loss is the best of those obtained so far, and only at 700 epochs (0.0951 versus 0.16 in the previous case, at 2,000 epochs). It is clear that adding the hidden layer of four neurons has made the model faster and more efficient.

Conclusions

In this chapter, you learned about the branch of machine learning that uses neural networks as a computing structure, called deep learning. You read an overview of the basic concepts of deep learning, which involves neural networks and their structure. Finally, thanks to the TensorFlow library, you implemented different types of neural networks, such as Perceptron Single Layer and Perceptron Multilayer.

Deep learning, with all its techniques and algorithms, is a very complex subject, and it is practically impossible to treat it properly in one chapter. However, you have now become familiar with deep learning and can begin implementing more complex neural networks.

CHAPTER 10

■ ■ ■

An Example—Meteorological Data

One type of data that's easier to find on the Internet is meteorological data. Many sites provide historical data on many meteorological parameters, such as pressure, temperature, humidity, rain, and so on. You only need to specify the location and the date to get a file with datasets of measurements collected by weather stations. These data are a source of a wide range of information. As you read in the first chapter of this book, the purpose of data analysis is to transform the raw data into information and then convert it into knowledge.

In this chapter, you will see a simple example of how to use meteorological data. This example is useful for getting a general idea of how to apply many of the techniques seen in the previous chapters.

A Hypothesis to Be Tested: The Influence of the Proximity of the Sea

At the time of writing of this chapter, I find myself at the beginning of summer and temperatures rising. On the weekend, many inland people travel to mountain villages or cities close to the sea, in order to enjoy a little refreshment and get away from the sultry weather of the inland cities. This has always made me wonder what effect the proximity of the sea has on the climate.

This simple question can be a good starting point for data analysis. I don't want to pass this chapter off as something scientific; it's just a way for someone passionate about data analysis to put knowledge into practice in order to answer this question—what influence, if any, does the proximity of the sea have on local climate?

The System in the Study: The Adriatic Sea and the Po Valley

Now that the problem has been defined, it is necessary to look for a system that is well suited to the study of the data and to provide a setting suitable for this question.

First you need a sea. Well, I'm in Italy and I have many seas to choose from, since Italy is a peninsula surrounded by seas. Why limit myself to Italy? Well, the problem involves a behavior typical of the Italians, that is, they take refuge in places close to the sea during the summer to avoid the heat of the hinterland. Not knowing if this behavior is the same for people of other nations, I will only consider Italy as a system of study.

But what areas of Italy might we consider studying? Can we assess the effects of the sea at various distances? This creates a lot of problems. In fact, Italy is rich in mountainous areas and doesn't have a lot territory that uniformly extends for many kilometers inland. So, to assess the effects of the sea, I exclude the mountains, as they may introduce many other factors that also affect climate, such as altitude, for example.

© Fabio Nelli 2023
F. Nelli, *Python Data Analytics*, https://doi.org/10.1007/978-1-4842-9532-8_10

A part of Italy that is well suited to this assessment is the Po Valley. This plain starts from the Adriatic Sea and spreads inland for hundreds of kilometers (see Figure 10-1). It is surrounded by mountains, but the width of the valley mitigates any mountain effects. It also has many towns and so it is easy to choose a set of cities increasingly distant from the sea, to cover a distance of almost 400 km in this evaluation.

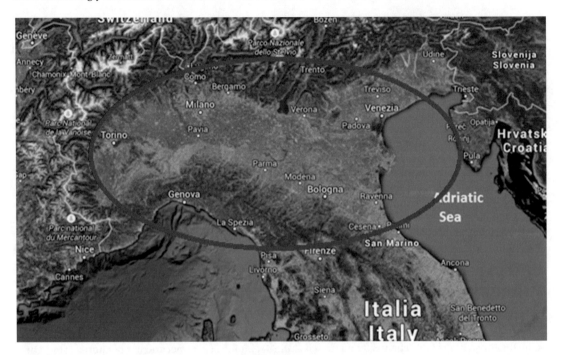

Figure 10-1. *An image of the Po Valley and the Adriatic Sea (Google Maps)*

The first step is to choose a set of ten cities that will serve as reference standards. These cities are selected in order to cover the entire range of the plain (see Figure 10-2).

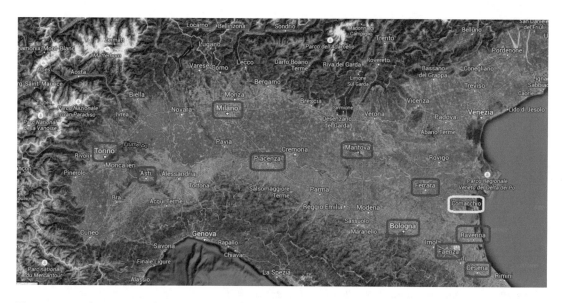

Figure 10-2. *The ten cities chosen as samples (there is another one used as a reference for distances from the sea)*

In Figure 10-2, you can see the ten cities that were chosen to analyze weather data: five cities within the first 100 km and the other five distributed in the remaining 300 km.

Here are the chosen cities:

- Ferrara

- Torino

- Mantova

- Milano

- Ravenna

- Asti

- Bologna

- Piacenza

- Cesena

- Faenza

Now you have to determine the distances of these cities from the sea. You can follow many procedures to obtain these values. In this case, you can use the service provided by the site TheDistanceNow (`www. thedistancenow.com/`), which is available in many languages (see Figure 10-3).

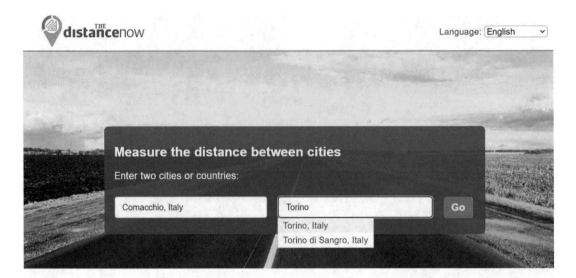

Figure 10-3. *TheDistanceNow website allows you to calculate distances between two cities*

Thanks to this service, it is possible to calculate the approximate distances of the cities from the sea. You can do this by selecting a city on the coast as the destination. For many of them, you can choose the city of Comacchio as a reference to calculate the distance from the sea (see Figure 10-2). Once you have determined the distances from the ten cities, you will get the values shown in Table 10-1.

Table 10-1. *The Distances from the Sea of the Ten Cities*

City	Distance (km)	Note
Ravenna	8	Measured with Google Earth
Cesena	14	Measured with Google Earth
Faenza	37	Distance Faenza-Ravenna +8km
Ferrara	47	Distance Ferrara-Comacchio
Bologna	71	Distance Bologna-Comacchio
Mantova	121	Distance Mantova-Comacchio
Piacenza	200	Distance Piacenza-Comacchio
Milano	250	Distance Milano-Comacchio
Asti	315	Distance Asti-Comacchio
Torino	357	Distance Torino-Comacchio

Finding the Data Source

Once you have defined the system under study, you need to establish a data source from which to obtain the needed data. By browsing the Internet, you can discover many sites that provide meteorological data measured from various locations around the world. One such site is OpenWeather, available at openweathermap.org (see Figure 10-4).

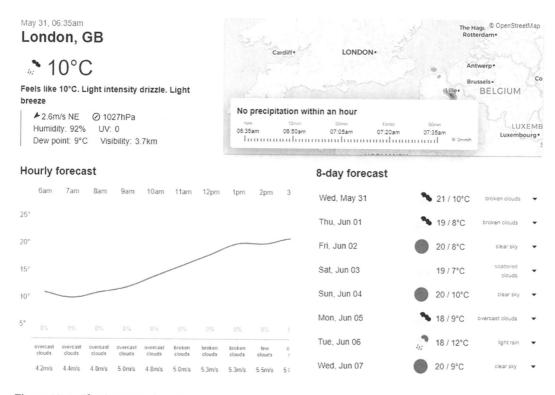

Figure 10-4. *The OpenWeather site*

After you've signed up for an account and received an app ID code, this site enables you to capture data by specifying the city through a request via an URL.

```
https://api.openweathermap.org/data/2.5/weather?q=Atlanta,US&appid=5807ad2a45eb6bf4e81d13
7dafe74e15
```

This request will return a JSON file containing all the information about the current weather situation in the city in question (see Figure 10-5). This JSON file will be submitted for data analysis using the Python pandas library.

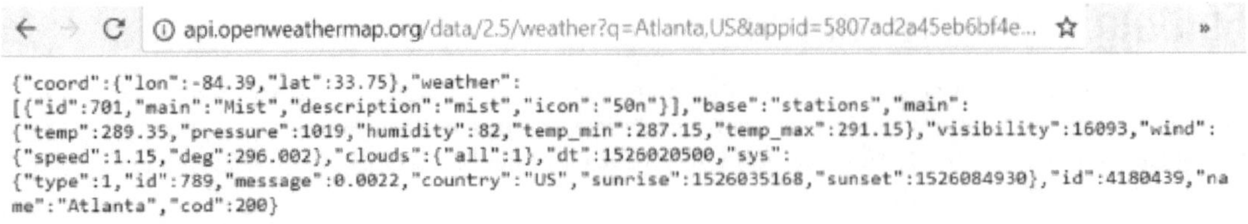

Figure 10-5. *The JSON file containing the meteorological data on the city requested*

Data Analysis on Jupyter Notebook

This chapter addresses data analysis using Jupyter Notebook. It allows you to enter and study portions of code gradually.

After the service has started, create a new Notebook.

Start by importing the necessary libraries:

```
import numpy as np
import pandas as pd
import datetime
```

The first step is to study the structure of the data received from the site through a specific request.

Choose a city from those chosen for the study, for example Ferrara, and make a request for its current meteorological data, using the URL specified. Without a browser, you can get the text content of a page by using the request.get() text function. Because the content obtained is in JSON format, you can directly read the text received following this format with the json.load() function.

```
import json
import requests
ferrara = json.loads(requests.get('https://api.openweathermap.org/data/2.5/weather?q=Ferrara,
IT&appid=5807ad2a45eb6bf4e81d137dafe74e15').text)
```

Now you can see the contents of the JSON file with the meteorological data related to the city of Ferrara.

```
ferrara
Out [ ]:
{'coord': {'lon': 11.8333, 'lat': 44.8},
 'weather': [{'id': 804,
   'main': 'Clouds',
   'description': 'overcast clouds',
   'icon': '04d'}],
 'base': 'stations',
 'main': {'temp': 292.51,
  'feels_like': 292.22,
  'temp_min': 292.04,
  'temp_max': 293.74,
  'pressure': 1017,
  'humidity': 66,
  'sea_level': 1017,
  'grnd_level': 1017},
 'visibility': 10000,
```

```
'wind': {'speed': 2.68, 'deg': 50, 'gust': 7.76},
'clouds': {'all': 100},
'dt': 1685511558,
'sys': {'type': 2,
 'id': 2007888,
 'country': 'IT',
 'sunrise': 1685503859,
 'sunset': 1685558996},
'timezone': 7200,
'id': 3177088,
'name': 'Provincia di Ferrara',
'cod': 200}
```

When you want to analyze the structure of a JSON file, the following command is useful:

```
list(ferrara.keys())
Out [ ]:
['coord',
 'weather',
 'base',
 'main',
 'visibility',
 'wind',
 'clouds',
 'dt',
 'sys',
 'id',
 'name',
 'cod']
```

This way, you can have a list of all the keys that make up the internal structure of the JSON file. Once you know the name of these keys, you can easily access internal data.

```
print('Coordinates = ', ferrara['coord'])
print('Weather = ', ferrara['weather'])
print('base = ', ferrara['base'])
print('main = ', ferrara['main'])
print('visibility = ', ferrara['visibility'])
print('wind = ', ferrara['wind'])
print('clouds = ', ferrara['clouds'])
print('dt = ', ferrara['dt'])
print('sys = ', ferrara['sys'])
print('id = ', ferrara['id'])
print('name = ', ferrara['name'])
print('cod = ', ferrara['cod'])
Out [ ]:
Coordinates =  {'lon': 11.8333, 'lat': 44.8}
Weather =  [{'id': 804, 'main': 'Clouds', 'description': 'overcast clouds', 'icon': '04d'}]
base =  stations
main =  {'temp': 292.51, 'feels_like': 292.22, 'temp_min': 292.04, 'temp_max': 293.74,
'pressure': 1017, 'humidity': 66, 'sea_level': 1017, 'grnd_level': 1017}
visibility =  10000
```

```
wind =  {'speed': 2.68, 'deg': 50, 'gust': 7.76}
clouds =  {'all': 100}
dt =  1685511558
sys =  {'type': 2, 'id': 2007888, 'country': 'IT', 'sunrise': 1685503859, 'sunset':
1685558996}
id =  3177088
name =  Provincia di Ferrara
cod =  200
```

Now choose the values that you consider most interesting or useful for this type of analysis. For example, an important value is temperature:

```
ferrara['main']['temp']
Out [ ]:
292.51
```

The purpose of this analysis of the initial structure is to identify the data that could be most important in the JSON structure. These data must be processed for analysis. That is, the data must be extracted from the structure, cleaned or modified according to your needs, and ordered in a dataframe. This way, you can apply all the data analysis techniques presented in this book.

A convenient way to avoid repeating the same code is to insert some extraction procedures into a function, such as the following:

```
def prepare(city,city_name):
    temp = [ ]
    humidity = [ ]
    pressure = [ ]
    description = [ ]
    dt = [ ]
    wind_speed = [ ]
    wind_deg = [ ]
    temp.append(city['main']['temp']-273.15)
    humidity.append(city['main']['humidity'])
    pressure.append(city['main']['pressure'])
    description.append(city['weather'][0]['description'])
    dt.append(city['dt'])
    wind_speed.append(city['wind']['speed'])
    wind_deg.append(city['wind']['deg'])
    headings = ['temp','humidity','pressure','description','dt','wind_speed','wind_deg']
    data = [temp,humidity,pressure,description,dt,wind_speed,wind_deg]
    df = pd.DataFrame(data,index=headings)
    city = df.T
    city['city'] = city_name
    city['day'] = city['dt'].apply(datetime.datetime.fromtimestamp)
    return city
```

This function does nothing more than take the meteorological data you are interested in from the JSON structure, and once they are cleaned or modified (for example, dates and times), that data are collected in a row of a dataframe (as shown in Figure 10-6).

```
t1 = prepare(ferrara,'ferrara')
t1
```

330

	temp	humidity	pressure	description	dt	wind_speed	wind_deg	city	day
0	19.36	66	1017	overcast clouds	1685511558	2.68	50	ferrara	2023-05-31 07:39:18

Figure 10-6. The dataframe obtained with the data processed from JSON extraction

Among all the parameters described in the JSON structure in the list column, these are the most appropriate for the study:

- Temperature
- Humidity
- Pressure
- Description
- Wind speed
- Wind degree

All these properties will be related to the time of acquisition expressed from the dt column, which contains a timestamp as the type of data. This value is difficult to read, so you can convert it into a datetime format that allows you to express the date and time in a manner more familiar to you. The new column will be called day.

```
city['day'] = city['dt'].apply(datetime.datetime.fromtimestamp)
```

Temperature is expressed in degrees Kelvin, and you can convert these values to Celsius by subtracting 273.15 from each value.

Finally, add the name of the city passed as a second argument of the prepare() function.

Data is collected at regular intervals, during different times of the day. For example, you could use a program that executes these requests every hour. Each acquisition will have a row of the dataframe structure that will be added to a general dataframe related to the city, called for example, df_ferrara (as shown in Figure 10-7).

```
df_ferrara = t1
t2 = prepare(ferrara,'ferrara')
df_ferrara = pd.concat([df_ferrara, t2])
df_ferrara
```

	temp	humidity	pressure	description	dt	wind_speed	wind_deg	city	day
0	19.36	66	1017	overcast clouds	1685511558	2.68	50	ferrara	2023-05-31 07:39:18
0	19.36	66	1017	overcast clouds	1685511558	2.68	50	ferrara	2023-05-31 07:39:18

Figure 10-7. The dataframe structure corresponding to a city

It often happens that data that's useful to this analysis is not present in the JSON source. In that case, you have to resort to other data sources and import the missing data into the structure. In this example, the distances of the cities to the sea are indispensable. You repeat the procedure just described for all the cities in the list that you want to analyze. Then you add the distance values to the dataframe you obtain.

```
.
df_ravenna['dist'] = 8
df_cesena['dist'] = 14
df_faenza['dist'] = 37
df_ferrara['dist'] = 47
df_bologna['dist'] = 71
df_mantova['dist'] = 121
df_piacenza['dist'] = 200
df_milano['dist'] = 250
df_asti['dist'] = 315
df_torino['dist'] = 357
.
```

Analysis of Processed Meteorological Data

For practical purposes, I have already collected data from all the cities involved in the analysis. I have already processed and collected them in a dataframe, which I saved as a CSV file.

If you want to refer to the data used in this chapter, you have to load the ten CSV files that I saved at the time of writing. These files contain data already processed to be used for this analysis.

```
df_ferrara=pd.read_csv('ferrara_270615.csv')
df_milano=pd.read_csv('milano_270615.csv')
df_mantova=pd.read_csv('mantova_270615.csv')
df_ravenna=pd.read_csv('ravenna_270615.csv')
df_torino=pd.read_csv('torino_270615.csv')
df_asti=pd.read_csv('asti_270615.csv')
df_bologna=pd.read_csv('bologna_270615.csv')
df_piacenza=pd.read_csv('piacenza_270615.csv')
df_cesena=pd.read_csv('cesena_270615.csv')
df_faenza=pd.read_csv('faenza_270615.csv')
```

Thanks to the read_csv() function of pandas, you can convert CSV files to the dataframe in just one step.

Once you have uploaded data for each city as a dataframe, you can easily see the content.

```
df_cesena
```

As you can see in Figure 10-8, Jupyter Notebook makes it much easier to read dataframes with the generation of graphical tables. Furthermore, you can see that each row shows the measured values for each hour of the day, covering a timeline of about 20 hours in the past.

	Unnamed: 0	temp	humidity	pressure	description	dt	wind_speed	wind_deg	city	day	dist
0	0	23.34	82	1017	very heavy rain	1435387623	1.91	175.511	Cesena	2015-06-27 08:47:03	14
1	1	24.95	69	1018	very heavy rain	1435390801	2.01	159.500	Cesena	2015-06-27 09:40:01	14
2	2	25.67	73	1017	very heavy rain	1435394204	2.10	100.000	Cesena	2015-06-27 10:36:44	14
3	3	26.17	69	1017	very heavy rain	1435398652	3.10	120.000	Cesena	2015-06-27 11:50:52	14
4	4	27.07	61	1016	very heavy rain	1435402083	3.10	110.000	Cesena	2015-06-27 12:48:03	14
5	5	27.41	69	1016	very heavy rain	1435405721	3.60	110.000	Cesena	2015-06-27 13:48:41	14
6	6	27.38	65	1015	very heavy rain	1435409381	5.70	110.000	Cesena	2015-06-27 14:49:41	14
7	7	26.59	65	1014	very heavy rain	1435416585	5.10	110.000	Cesena	2015-06-27 16:49:45	14
8	8	27.16	65	1014	very heavy rain	1435420195	6.20	120.000	Cesena	2015-06-27 17:49:55	14
9	9	27.10	65	1014	very heavy rain	1435423927	6.70	120.000	Cesena	2015-06-27 18:52:07	14
10	10	26.01	73	1013	very heavy rain	1435427556	6.20	120.000	Cesena	2015-06-27 19:52:36	14
11	11	23.37	94	1015	very heavy rain	1435438070	2.60	90.000	Cesena	2015-06-27 22:47:50	14
12	12	22.48	83	1016	very heavy rain	1435441857	5.70	90.000	Cesena	2015-06-27 23:50:57	14
13	13	21.94	83	1016	very heavy rain	1435445495	2.10	210.000	Cesena	2015-06-28 00:51:35	14
14	14	20.26	94	1016	very heavy rain	1435452847	2.01	107.004	Cesena	2015-06-28 02:54:07	14
15	15	19.65	93	1016	very heavy rain	1435456185	2.10	330.000	Cesena	2015-06-28 03:49:45	14
16	16	19.29	93	1016	very heavy rain	1435459689	1.50	320.000	Cesena	2015-06-28 04:48:09	14
17	17	18.41	93	1016	very heavy rain	1435463462	0.50	300.000	Cesena	2015-06-28 05:51:02	14
18	18	19.48	88	1016	very heavy rain	1435466850	0.50	270.000	Cesena	2015-06-28 06:47:30	14
19	19	22.00	88	1016	very heavy rain	1435470541	2.10	260.000	Cesena	2015-06-28 07:49:01	14

Figure 10-8. *The dataframe structure corresponding to a city*

In the case shown in Figure 10-8, note that there are only 19 rows. In fact, observing other cities, it looks like the meteorological measurement systems sometimes failed during the measuring process, leaving holes in the acquisition. If the data collected make up 19 rows, as in this case, they are sufficient to describe the trend of the meteorological properties during the day. However, it is good practice to check the size of all ten dataframes. If a city provides insufficient data to describe the daily trend, you need to replace it with another city.

There is an easy way to check the size, without having to put one table after another. Thanks to the shape() function, you can determine the number of data acquired (lines) for each city.

```
print(df_ferrara.shape)
print(df_milano.shape)
print(df_mantova.shape)
print(df_ravenna.shape)
print(df_torino.shape)
print(df_asti.shape)
print(df_bologna.shape)
print(df_piacenza.shape)
print(df_cesena.shape)
print(df_faenza.shape)
```

This will give the following result:

```
(20, 9)
(18, 9)
(20, 9)
```

```
(18, 9)
(20, 9)
(20, 9)
(20, 9)
(20, 9)
(20, 9)
(19, 9)
```

As you can see, the choice of ten cities is optimal, since the control units have provided enough data to continue with the data analysis.

A normal way to analyze the data you just collected is to use data visualization. You saw that the `matplotlib` library includes a set of tools to generate charts on which to display data. In fact, data visualization helps you during data analysis to discover features of the system you are studying.

Next, activate the necessary libraries:

```
%matplotlib inline
import matplotlib.pyplot as plt
import matplotlib.dates as mdates
```

For example, there is a simple way to analyze the trend of the temperature during the day. Consider the city of Milan.

```
y1 = df_milano['temp']
df_milano['day'] = pd.to_datetime(df_milano['day'])
x1 = df_milano['day']
fig, ax = plt.subplots()
plt.xticks(rotation=70)
hours = mdates.DateFormatter('%H:%M')
ax.xaxis.set_major_formatter(hours)
ax.plot(x1,y1,'r')
```

Executing this code, you get the graph shown in Figure 10-9. As you can see, the temperature trend follows a nearly sinusoidal pattern characterized by a temperature that rises in the morning, to reach the maximum value during the heat of the afternoon (between 2:00 and 6:00 pm). Then the temperature decreases to a minimum value corresponding to just before dawn, that is, at 6:00 am.

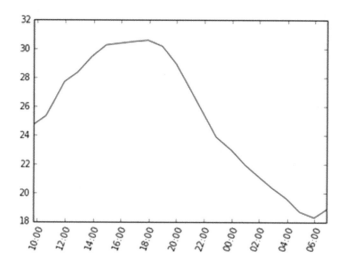

Figure 10-9. *Temperature trend of Milan during the day*

Because the purpose of your analysis is to try to interpret the weather, it is possible to assess how and if the sea influences this trend. This time, you evaluate the trends of different cities simultaneously. This is the only way to see if the analysis is going in the right direction. Thus, choose the three cities closest to the sea and the three cities farthest from it.

```
y1 = df_ravenna['temp']
df_ravenna['day'] = pd.to_datetime(df_ravenna['day'])
x1 = df_ravenna['day']
y2 = df_faenza['temp']
df_faenza['day'] = pd.to_datetime(df_faenza['day'])
x2 = df_faenza['day']
y3 = df_cesena['temp']
df_cesena['day'] = pd.to_datetime(df_cesena['day'])
x3 = df_cesena['day']
y4 = df_milano['temp']
df_milano['day'] = pd.to_datetime(df_milano['day'])
x4 = df_milano['day']
y5 = df_asti['temp']
df_asti['day'] = pd.to_datetime(df_asti['day'])
x5 = df_asti['day']
y6 = df_torino['temp']
df_torino['day'] = pd.to_datetime(df_torino['day'])
x6 = df_torino['day']
fig, ax = plt.subplots()
plt.xticks(rotation=70)
hours = mdates.DateFormatter('%H:%M')
ax.xaxis.set_major_formatter(hours)
plt.plot(x1,y1,'r',x2,y2,'r',x3,y3,'r')
plt.plot(x4,y4,'g',x5,y5,'g',x6,y6,'g')
```

This code will produce the chart shown in Figure 10-10. The temperatures of the three cities closest to the sea are shown in red, while the temperatures of the three cities farthest away are in green.

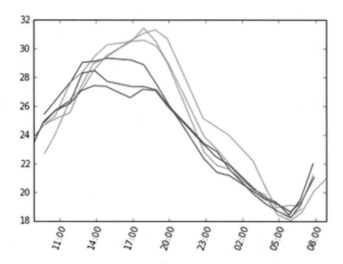

Figure 10-10. *The trend of the temperatures of six different cities (red is the closest to the sea; green is the farthest)*

Looking at Figure 10-10, the results seem promising. In fact, the three closest cities have maximum temperatures much lower than those farthest away, whereas there seems to be little difference in the minimum temperatures.

In order to go deep into this aspect, you can collect the maximum and minimum temperatures of all ten cities and display a line chart that charts these temperatures compared to their distances from the sea.

```
dist = [df_ravenna['dist'][0],
     df_cesena['dist'][0],
     df_faenza['dist'][0],
     df_ferrara['dist'][0],
     df_bologna['dist'][0],
     df_mantova['dist'][0],
     df_piacenza['dist'][0],
     df_milano['dist'][0],
     df_asti['dist'][0],
     df_torino['dist'][0]
]temp_max = [df_ravenna['temp'].max(),
     df_cesena['temp'].max(),
     df_faenza['temp'].max(),
     df_ferrara['temp'].max(),
     df_bologna['temp'].max(),
     df_mantova['temp'].max(),
     df_piacenza['temp'].max(),
     df_milano['temp'].max(),
     df_asti['temp'].max(),
     df_torino['temp'].max()
]
temp_min = [df_ravenna['temp'].min(),
     df_cesena['temp'].min(),
     df_faenza['temp'].min(),
     df_ferrara['temp'].min(),
```

```
    df_bologna['temp'].min(),
    df_mantova['temp'].min(),
    df_piacenza['temp'].min(),
    df_milano['temp'].min(),
    df_asti['temp'].min(),
    df_torino['temp'].min()
]
```

Start by representing the maximum temperatures.

```
plt.plot(dist,temp_max,'ro')
```

The result is shown in Figure 10-11.

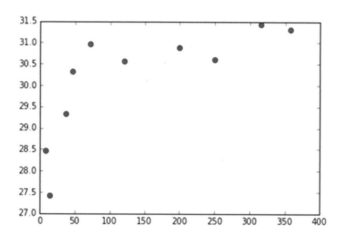

Figure 10-11. *Trend of maximum temperature in relation to distance from the sea*

As shown in Figure 10-11, you can affirm the hypothesis that the presence of the sea somehow influences meteorological parameters is true (at least in the day today ☺).

Furthermore, you can see that the effect of the sea decreases rapidly, and after about 60-70 km, the maximum temperatures reach a plateau.

An interesting idea would be to represent the two different trends with two straight lines obtained by linear regression. To do this, you can use the SVR method provided by the scikit-learn library.

If you haven't installed the scikit-learn library yet, do so now. If you are working with Anaconda, you can install it via Anaconda Navigator by selecting it from the packages available in your virtual environment (see Figure 10-12), or from the CMD.exe Prompt console, by entering this command:

```
conda install scikit-learn
```

Figure 10-12. *Installation of the scikit-learn library with Anaconda Navigator*

At this point, you can continue with the example by inserting the following code into a cell in the Notebook:

```
x = np.array(dist)
y = np.array(temp_max)
x1 = x[x<100]
x1 = x1.reshape((x1.size,1))
y1 = y[x<100]
x2 = x[x>50]
x2 = x2.reshape((x2.size,1))
y2 = y[x>50]
from sklearn.svm import SVR
svr_lin1 = SVR(kernel='linear', C=1e3)
svr_lin2 = SVR(kernel='linear', C=1e3)
svr_lin1.fit(x1, y1)
svr_lin2.fit(x2, y2)
xp1 = np.arange(10,100,10).reshape((9,1))
xp2 = np.arange(50,400,50).reshape((7,1))
yp1 = svr_lin1.predict(xp1)
yp2 = svr_lin2.predict(xp2)
plt.plot(xp1, yp1, c='r', label='Strong sea effect')
plt.plot(xp2, yp2, c='b', label='Light sea effect')
plt.axis((0,400,27,32))
plt.scatter(x, y, c='k', label='data')
```

This code will produce the chart shown in Figure 10-13.

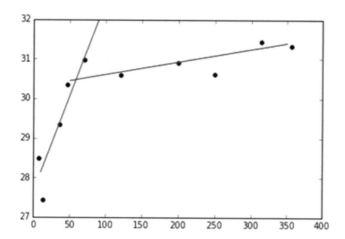

Figure 10-13. *The two trends described by the maximum temperatures in relation to distance*

As you can see, temperature increase in the first 60 km is very rapid, rising from 28 to 31 degrees. It then increases very mildly (if at all) over longer distances. The two trends are described by two straight lines that have the following expression

$$x = ax + b$$

where *a* is the slope and the *b* is the intercept.

```
print( svr_lin1.coef_)
print( svr_lin1.intercept_)
print( svr_lin2.coef_)
print( svr_lin2.intercept_)
Out [ ]:
[[-0.04794118]]
[ 27.65617647]
[[-0.00317797]]
[ 30.2854661]
```

You might consider the intersection point of the two lines as the point between the area where the sea exerts its influence and the area where it doesn't, or at least not as strongly.

```
from scipy.optimize import fsolve
def line1(x):
    a1 = svr_lin1.coef_[0][0]
    b1 = svr_lin1.intercept_[0]
    return a1*x + b1
def line2(x):
    a2 = svr_lin2.coef_[0][0]
    b2 = svr_lin2.intercept_[0]
    return a2*x + b2
def findIntersection(fun1,fun2,x0):
 return fsolve(lambda x : fun1(x) - fun2(x),x0)
result = findIntersection(line1,line2,0.0)
print("[x,y] = [ %d , %d ]" % (result,line1(result)))
```

```
x = np.linspace(0,300,31)
plt.plot(x,line1(x),x,line2(x),result,line1(result),'ro')
```

Executing the code, you can find the point of intersection as follows:

```
Out [ ]:
[x,y] = [ 58, 30 ]
```

This point is represented in the chart shown in Figure 10-14.

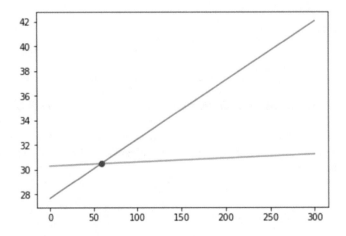

Figure 10-14. *The point of intersection between two straight lines obtained by linear regression*

You can say that the average distance in which the effects of the sea vanish is 58 km.
Now you can analyze the minimum temperatures.

```
plt.axis((0,400,15,25))
plt.plot(dist,temp_min,'bo')
```

Doing this, you'll obtain the chart shown in Figure 10-15.

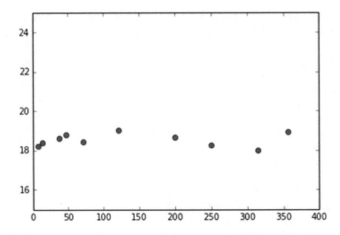

Figure 10-15. *The minimum temperatures appear to be independent of the distance from the sea*

In this case, it appears very clear that the sea has no effect on minimum temperatures recorded during the night, or rather, around six in the morning. If I remember well, when I was a child I was taught that the sea mitigated the cold temperatures, or that the sea released the heat absorbed during the day. This does not seem to be the case. This case tracks summer in Italy; it would be interesting to see if this hypothesis is true in the winter or somewhere else.

Another meteorological measure contained in the ten dataframes is the humidity. Even for this measure, you can see the trend of the humidity during the day for the three cities closest to the sea and for the three farthest away.

```
y1 = df_ravenna['humidity']
x1 = df_ravenna['day']
y2 = df_faenza['humidity']
x2 = df_faenza['day']
y3 = df_cesena['humidity']
x3 = df_cesena['day']
y4 = df_milano['humidity']
x4 = df_milano['day']
y5 = df_asti['humidity']
x5 = df_asti['day']
y6 = df_torino['humidity']
x6 = df_torino['day']
fig, ax = plt.subplots()
plt.xticks(rotation=70)
hours = mdates.DateFormatter('%H:%M')
ax.xaxis.set_major_formatter(hours)
plt.plot(x1,y1,'r',x2,y2,'r',x3,y3,'r')
plt.plot(x4,y4,'g',x5,y5,'g',x6,y6,'g')
```

This code will create the chart shown in Figure 10-16.

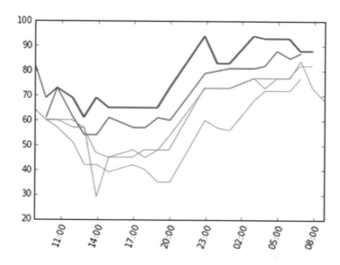

Figure 10-16. *The trend of the humidity during the day for three cities nearest the sea (shown in red) and three cities farthest away (indicated in green)*

At first glance, it would seem that the cities closest to the sea experience more humidity than those farthest away and that this difference in moisture (about 20 percent) extends throughout the day. You can see if this remains true when you report the maximum and minimum humidity with respect to the distances from the sea.

```
hum_max = [df_ravenna['humidity'].max(),
    df_cesena['humidity'].max(),
    df_faenza['humidity'].max(),
    df_ferrara['humidity'].max(),
    df_bologna['humidity'].max(),
    df_mantova['humidity'].max(),
    df_piacenza['humidity'].max(),
    df_milano['humidity'].max(),
    df_asti['humidity'].max(),
    df_torino['humidity'].max()
]
plt.plot(dist,hum_max,'bo')
```

The maximum humidity of ten cities according to their distance from the sea is represented in the chart in Figure 10-17.

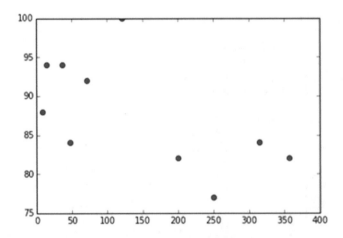

Figure 10-17. *The trend of the maximum humidity function with respect to the distance from the sea*

```
hum_min = [df_ravenna['humidity'].min(),
    df_cesena['humidity'].min(),
    df_faenza['humidity'].min(),
    df_ferrara['humidity'].min(),
    df_bologna['humidity'].min(),
    df_mantova['humidity'].min(),
    df_piacenza['humidity'].min(),
    df_milano['humidity'].min(),
    df_asti['humidity'].min(),
    df_torino['humidity'].min()
]
plt.plot(dist,hum_min,'bo')
```

The minimum humidity of ten cities according to their distance from the sea is represented in the chart in Figure 10-18.

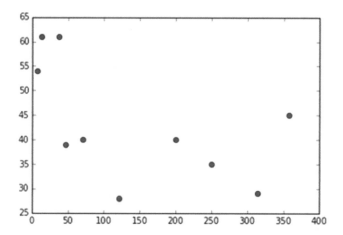

Figure 10-18. *The trend of the minimum humidity as a function of distance from the sea*

Looking at Figures 10-17 and 10-18, you can certainly see that the humidity, both the minimum and maximum, is greater in the cities closest to the sea. However, in my opinion, it is not possible to say that there is a linear relationship or some other kind of relationship to draw a curve. The collected points (ten) are too few to highlight a trend in this case.

The RoseWind

Among the various meteorological data that were collected for each city are those related to the wind:

- Wind degree (direction)
- Wind speed

If you analyze the dataframe, you will notice that the wind speed is relative to the direction it blows and the time of day. For instance, each measurement shows the direction in which the wind blows (see Figure 10-19).

	wind_deg	wind_speed	day
0	159.5000	2.01	2015-06-27 09:42:05
1	100.0000	2.10	2015-06-27 10:37:24
2	80.0000	4.60	2015-06-27 11:57:01
3	90.0000	4.60	2015-06-27 12:53:43
4	80.0000	6.20	2015-06-27 13:54:20
5	80.0000	6.70	2015-06-27 14:55:06
6	90.0000	6.70	2015-06-27 16:55:00
7	90.0000	5.70	2015-06-27 17:55:43
8	90.0000	4.60	2015-06-27 18:58:17
9	97.0000	2.06	2015-06-27 19:58:58
10	89.0000	2.06	2015-06-27 22:52:39
11	88.0147	2.86	2015-06-27 23:57:25
12	107.0040	2.01	2015-06-28 00:57:46
13	107.0040	2.01	2015-06-28 03:00:34
14	132.5030	1.06	2015-06-28 03:54:49
15	132.5030	1.06	2015-06-28 04:54:04
16	132.5030	1.06	2015-06-28 05:58:15
17	251.0000	1.54	2015-06-28 06:52:59

Figure 10-19. *The wind data contained in the dataframe*

To better analyze this kind of data, it is necessary to visualize them. In this case, a linear chart in Cartesian coordinates is not the most optimal approach.

If you use the classic scatterplot with the points contained in a single dataframe:

```
plt.plot(df_ravenna['wind_deg'],df_ravenna['wind_speed'],'ro')
```

You get a chart like the one shown in Figure 10-20, which certainly is not very educational.

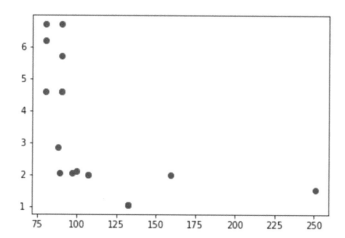

Figure 10-20. *A scatterplot representing a distribution of 360 degrees*

To represent a distribution of points in 360 degrees, it's best to use another type of visualization: the *polar chart*. You have already seen this kind of chart in Chapter 8.

First you need to create a histogram, whereby the data are distributed over the interval of 360 degrees divided into eight bins, each of which is 45 degrees.

```
hist, bins = np.histogram(df_ravenna['wind_deg'],8,[0,360])
print(hist)
print(bins)
```

The values returned are occurrences within each bin expressed by an array called `hist`:

```
Out [ ]:
[ 0  5 11  1  0  1  0  0]
```

and an array called `bins`, which defines the edges of each bin within the range of 360 degrees.

```
Out [ ]:
[   0.   45.   90.  135.  180.  225.  270.  315.  360.]
```

These arrays will be useful to correctly define the polar chart to be drawn. For this purpose, you have to create a function in part by using the code contained in Chapter 8. This function is called `showRoseWind()`, and it will need three different arguments: `values` is the array containing the values to be displayed, which in this case is the `hist` array; `city_name` is a string containing the name of the city to be shown as the chart title; and `max_value` is an integer that establishes the maximum value for presenting the blue color.

Defining a function of this kind helps you avoid rewriting the same code many times, and it produces more modular code, which allows you to focus on the concepts related to a particular operation within a function.

```
def showRoseWind(values,city_name,max_value):
    N = 8
    theta = np.arange(0.,2 * np.pi, 2 * np.pi / N)
    radii = np.array(values)
    plt.axes([0.025, 0.025, 0.95, 0.95], polar=True)
```

345

```
colors = [(1-x/max_value, 1-x/max_value, 0.75) for x in radii]
plt.bar(theta +np.pi/8, radii, width=(2*np.pi/N), bottom=0.0, color=colors)
plt.title(city_name,x=0.2, fontsize=20)
```

One thing that changed is the color map. In this case, the closer to blue the slice is, the greater the value it represents.

Once you define a function, you can use it:

```
showRoseWind(hist,'Ravenna',max(hist))
```

Executing this code, you will obtain a polar chart like the one shown in Figure 10-21.

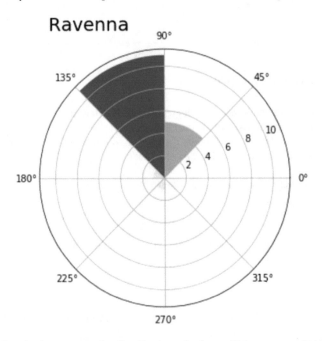

Figure 10-21. *The polar chart represents the distribution of values within a range of 360 degrees*

As you can see in Figure 10-20, you have a range of 360 degrees divided into eight areas of 45 degrees each (bin), in which a scale of values is represented radially. In each of the eight areas, a slice is represented with a variable length that corresponds precisely to the corresponding value. The more radially extended the slice is, the greater the value represented. In order to increase the readability of the chart, a color scale has been entered that corresponds to the extension of its slice. The wider the slice is, the more the color tends to a deep blue.

This polar chart provides you with information about how the wind direction will be distributed radially. In this case, the wind has blown purely toward the southwest/west most of the day.

Once you have defined the showRoseWind function, it is very easy to observe the winds with respect to any of the ten sample cities.

```
hist, bin = np.histogram(df_ferrara['wind_deg'],8,[0,360])
print(hist)
showRoseWind(hist,'Ferrara', 15.0)
Out [ ]:
[7 2 3 3 3 2 0 0]
```

Figure 10-22 shows the polar charts of the ten cities.

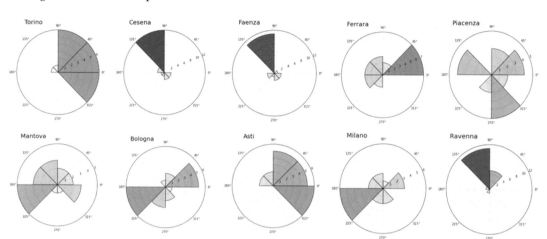

Figure 10-22. *The polar charts display the distribution of the wind direction*

Calculating the Mean Distribution of the Wind Speed

Even the other quantity that relates the speed of the winds can be represented as a distribution on 360 degrees.

Now define a feature called RoseWind_Speed that will allow you to calculate the mean wind speeds for each of the eight bins into which 360 degrees are divided.

```
def RoseWind_Speed(df_city):
    degs = np.arange(45,361,45)
    tmp = []
    for deg in degs:
        tmp.append(df_city[(df_city['wind_deg']>(deg-46)) & (df_city['wind_deg']<deg)]['wind_
        speed'].mean())
    return np.nan_to_num(tmp)
```

This function returns a NumPy array containing the eight mean wind speeds. This array will be used as the first argument of the ShowRoseWind_Speed() function, which is an improved version of the previous ShowRoseWind() function used to represent the polar chart.

```
def showRoseWind_Speed(speeds,city_name):
    N = 8
    theta = np.arange(0,2 * np.pi, 2 * np.pi / N)
    radii = np.array(speeds)
    plt.axes([0.025, 0.025, 0.95, 0.95], polar=True)
    colors = [(1-x/10.0, 1-x/10.0, 0.75) for x in radii]
    bars = plt.bar(theta+np.pi/8, radii, width=(2*np.pi/N), bottom=0.0, color=colors)
    plt.title(city_name,x=0.2, fontsize=20)
showRoseWind_Speed(RoseWind_Speed(df_ravenna),'Ravenna')
```

Figure 10-23 represents the RoseWind corresponding to the wind speeds distributed around 360 degrees.

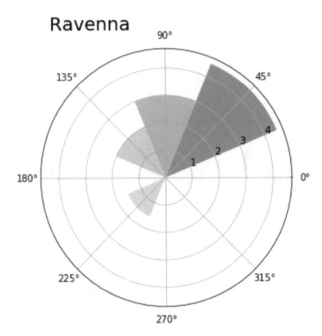

Figure 10-23. *This polar chart represents the distribution of wind speeds within 360 degrees*

At the end of all this work, you can save the dataframe as a CSV file, thanks to the to_csv () function of the pandas library.

```
df_ferrara.to_csv('ferrara.csv')
df_milano.to_csv('milano.csv')
df_mantova.to_csv('mantova.csv')
df_ravenna.to_csv('ravenna.csv')
df_torino.to_csv('torino.csv')
df_asti.to_csv('asti.csv')
df_bologna.to_csv('bologna.csv')
df_piacenza.to_csv('piacenza.csv')
df_cesena.to_csv('cesena.csv')
df_faenza.to_csv('faenza.csv')
```

Conclusions

The purpose of this chapter was mainly to show how you can get information from raw data. Some of this information will not lead to important conclusions, while other information will lead to the confirmation of a hypothesis, thus increasing your state of knowledge. These are the cases in which data analysis has led to a success.

In the next chapter, you see another case related to real data obtained from an open data source. You also see how you can further enhance the graphical representation of the data using the D3 JavaScript library. This library, although not Python, can be easily integrated into Python.

CHAPTER 11

■ ■ ■

Embedding the JavaScript D3 Library in the IPython Notebook

In this chapter, you will learn how to extend the capabilities of the graphical representation including the JavaScript D3 library in your Jupyter Notebook. This library has enormous potential graphics and allows you to build graphical representations that even the `matplotlib` library cannot represent.

In the course of the various examples, you will see how to implement JavaScript code in a Python environment, using the large capacity of the integrative Jupyter Notebook. You'll also see different ways to use the data contained in pandas dataframes and representations based on JavaScript code.

The Open Data Source for Demographics

In this chapter, you use demographic data as the dataset on which to perform the analysis. This chapter uses the United States Census Bureau site (`www.census.gov`) as the data source for the demographics (see Figure 11-1).

Figure 11-1. This is the home page of the United States Census Bureau

© Fabio Nelli 2023
F. Nelli, *Python Data Analytics*, https://doi.org/10.1007/978-1-4842-9532-8_11

The United States Census Bureau is part of the United States Department of Commerce, and it is officially in charge of collecting demographic data on the U.S. population and reporting statistics about it. Its site provides a large amount of data as CSV files, which, as you have seen in previous chapters, are easily imported in the form of pandas dataframes.

For the purposes of this chapter, you want the data that estimates the population of the states and counties in the United States. On the site there is a series of datasets made available for studies at the link www2.census.gov/programs-surveys/popest/datasets/. Among the available datasets, look for the most recent one and download it to your computer. This example uses the CSV file called co-est2022-alldata.csv.

Now, open a Jupyter Notebook and import all the necessary libraries for this kind of analysis in the first cell.

```
import numpy as np
import pandas as pd
import matplotlib.pyplot as plt
```

You can start by importing data from Census.gov in your Notebook. You need to upload the co-est2022--alldata.csv file directly in the form of a pandas dataframe. The pd.read_csv() function will convert tabular data contained in a CSV file to a pandas dataframe, which you should name pop2022. Using the dtype option, you can force some fields that could be interpreted as numbers to be interpreted as strings instead.

```
pop2022 =pd.read_csv('co-est2022-alldata.csv' ,encoding='latin-1',dtype={'STATE': 'str',
'COUNTY': 'str'})
```

Once you have acquired and collected data in the pop2022 dataframe, you can see how the data are structured by simply writing:

```
pop2022
```

You will obtain an image like the one shown in Figure 11-2.

	SUMLEV	REGION	DIVISION	STATE	COUNTY	STNAME	CTYNAME	ESTIMATESBASE2020	POPESTIMATE2020	POPESTIMATE2021	...	RDEATH2021	RDEATH2022	RNATURALCHG2021
0	40	3	6	01	000	Alabama	Alabama	5024356	5031362	5049846	...	13.699945	13.210008	-2.369557
1	50	3	6	01	001	Alabama	Autauga County	58802	58902	59210	...	11.548361	11.313872	0.203197
2	50	3	6	01	003	Alabama	Baldwin County	231761	233219	239361	...	12.882475	11.976220	-2.865123
3	50	3	6	01	005	Alabama	Barbour County	25224	24960	24539	...	15.475060	15.108133	-4.323320
4	50	3	6	01	007	Alabama	Bibb County	22300	22183	22370	...	14.275133	14.467606	-3.321886
...	...	...	...	...	...	...	...	...	...	...	...	...	...	...
3190	50	4	8	56	037	Wyoming	Sweetwater County	42267	42190	41582	...	10.122714	10.515272	0.716230
3191	50	4	8	56	039	Wyoming	Teton County	23323	23377	23622	...	3.957531	4.476753	5.404370
3192	50	4	8	56	041	Wyoming	Uinta County	20446	20457	20655	...	8.999805	9.621196	2.043199
3193	50	4	8	56	043	Wyoming	Washakie County	7682	7658	7712	...	14.313598	14.905061	-4.814574
3194	50	4	8	56	045	Wyoming	Weston County	6840	6818	6766	...	12.956419	14.677822	-4.564193

3195 rows × 51 columns

Figure 11-2. *The pop2022 dataframe contains all demographics for the years 2020, 2021, and 2022*

Carefully analyzing the nature of the data, you can see how they are organized within the dataframe. The SUMLEV column contains the geographic level of the data; for example, 40 indicates a state and 50 indicates data covering a single county.

The REGION, DIVISION, STATE, and COUNTY columns contain hierarchical subdivisions of all areas in which the U.S. territory has been divided. STNAME and CTYNAME indicate the name of the state and the county, respectively. The following columns contain the data on population. POPESTIMATE2020 is the column that contains the population estimate for 2020, followed by those for 2021 and 2022.

You will use these values of population estimates as data to be represented in the examples discussed in this chapter.

The pop2022 dataframe contains a large number of columns and rows that you are not interested in, so it is smart to eliminate unnecessary information. First, you are interested in the values of the people who relate to entire states, and so you can extract only the rows with SUMLEV equal to 40. Collect these data within the pop2022_by_state dataframe.

```
pop2022_by_state = pop2022[pop2022.SUMLEV == 40]
pop2022_by_state
```

You get a dataframe like the one shown in Figure 11-3.

	SUMLEV	REGION	DIVISION	STATE	COUNTY	STNAME	CTYNAME	ESTIMATESBASE2020	POPESTIMATE2020	POPESTIMATE2021	...	RDEATH2021	RDEATH2022	RNATURALCH(
0	40	3	6	01	000	Alabama	Alabama	5024356	5031362	5049846	...	13.699945	13.210008	-2.3
68	40	4	9	02	000	Alaska	Alaska	733378	732923	734182	...	7.312360	8.715292	5.5
99	40	4	8	04	000	Arizona	Arizona	7151507	7179943	7264877	...	11.136172	10.849097	-0.6
115	40	3	7	05	000	Arkansas	Arkansas	3011555	3014195	3028122	...	13.297548	13.232991	-1.6
191	40	4	9	06	000	California	California	39538245	39501653	39142991	...	8.787630	8.148586	1.6
250	40	4	8	08	000	Colorado	Colorado	5773733	5784865	5811297	...	8.044731	8.476535	2.6
315	40	1	1	09	000	Connecticut	Connecticut	3605942	3597362	3623355	...	9.597108	9.628722	-0.2
325	40	3	5	10	000	Delaware	Delaware	989957	992114	1004807	...	11.247315	11.330549	-1.0
329	40	3	5	11	000	District of Columbia	District of Columbia	689546	670868	668791	...	8.706693	8.288863	3.6

Figure 11-3. *The pop2022_by_state dataframe contains all demographics related to the states*

The dataframe just obtained still contains too many columns with unnecessary information. Given the large number of columns, instead of removing them with the drop() function, it is more convenient to perform an extraction.

```
states = pop2022_by_state[['STNAME','POPESTIMATE2020', 'POPESTIMATE2021',
'POPESTIMATE2022']]
```

Now that you have the essential information, you can start to make graphical representations. For example, you could determine the five most populated states in the country.

```
states.sort_values(['POPESTIMATE2022'], ascending=False)[:5]
```

Listing them in descending order, you will receive the dataframe shown in Figure 11-4.

	STNAME	POPESTIMATE2020	POPESTIMATE2021	POPESTIMATE2022
191	California	39501653	39142991	39029342
2568	Texas	29232474	29558864	30029572
331	Florida	21589602	21828069	22244823
1862	New York	20108296	19857492	19677151
2284	Pennsylvania	12994440	13012059	12972008

Figure 11-4. *The five most populous states in the United States*

For example, you could use a bar chart to represent the five most populous states in descending order. This work is easily achieved using `matplotlib`, but in this chapter, you take advantage of this simple representation to see how you can use the JavaScript D3 library to create the same representation.

The JavaScript D3 Library

D3 is a JavaScript library that allows direct inspection and manipulation of the DOM object (HTML5), but it is intended solely for data visualization and it does its job excellently. In fact, the name D3 is derived from the three Ds contained in "data-driven documents." D3 was entirely developed by Mike Bostock.

This library is proving to be very versatile and powerful, thanks to the technologies upon which it is based: JavaScript, SVG, and CSS. D3 combines powerful visualization components with a data-driven approach to the DOM manipulation. In so doing, D3 takes full advantage of the capabilities of the modern browser.

Given that even Jupyter Notebooks are web objects and use the same technologies that are the basis of the current browser, the idea of using this library in a notebook is not as preposterous as it may seem at first, even though it's a JavaScript library.

For those not familiar with the JavaScript D3 library and want to know more about this topic, I recommend reading another book, entitled *Create Web Charts with D3,* by F. Nelli (Apress, 2014).

Indeed, Jupyter Notebook has the magic function called `%% javascript` that integrates JavaScript code into Python code.

But the JavaScript code, in a manner similar to Python, requires you to import some libraries. The libraries are available online and must be loaded each time you launch the execution. In HTML, the process of importing a library has a particular construct:

```
<script src="https://cdnjs.cloudflare.com/ajax/libs/d3/3.5.5/d3.min.js"></script>
```

This is an HTML tag. To make the import within an Jupyter Notebook, you should use this different construct:

```
%%javascript
require.config({
    paths: {
        d3: '//cdnjs.cloudflare.com/ajax/libs/d3/3.5.5/d3.min'
    }
});
```

Using `require.config()`, you can import all the necessary JavaScript libraries.

In addition, if you are familiar with HTML code, you will know for sure that you need to define CSS styles if you want to strengthen the capacity of visualization of an HTML page. In parallel, also in the Jupyter Notebook, you can define a set of CSS styles. To do this, you can write HTML code, thanks to the `HTML()` function belonging to the `IPython.core.display` module. Therefore, make the appropriate CSS definitions as follows:

```
from IPython.display import display, Javascript, HTML
display(HTML("""
<style>
.bar {
    fill: steelblue;
}
.bar:hover{
    fill: brown;
}
.axis {
    font: 10px sans-serif;
}
.axis path,
.axis line {
    fill: none;
    stroke: #000;
}
.x.axis path {
    display: none;
}
</style>
<div id="chart_d3" />
"""))
```

At the bottom of the previous code, note that the `<div>` HTML tag is identified as `chart_d3`. This tag identifies the location where it will be represented.

Now you have to write the JavaScript code by using the functions provided by the D3 library. Using the `Template` object provided by the `Jinja2` library, you can define dynamic JavaScript code, where you can replace the text depending on the values contained in a pandas dataframe.

If there is still not a `Jinja2` library installed on your system, you can always install it with Anaconda.

```
conda install jinja2
```

Or by using this

```
pip install jinja2
```

After you have installed this library, you can define the template.

```
import jinja2
myTemplate = jinja2.Template("""
require(["d3"], function(d3){
    var data = []
    {% for row in data %}
```

```
        data.push({ 'state': '{{ row[1] }}', 'population': '{{ row[4] }}'  });
            {% endfor %}
    d3.select("#chart_d3 svg").remove()
        var margin = {top: 20, right: 20, bottom: 30, left: 40},
            width = 800 - margin.left - margin.right,
            height = 400 - margin.top - margin.bottom;
        var x = d3.scale.ordinal()
            .rangeRoundBands([0, width], .25);
        var y = d3.scale.linear()
            .range([height, 0]);
        var xAxis = d3.svg.axis()
            .scale(x)
            .orient("bottom");
        var yAxis = d3.svg.axis()
            .scale(y)
            .orient("left")
            .ticks(10)
            .tickFormat(d3.format('.1s'));
        var svg = d3.select("#chart_d3").append("svg")
            .attr("width", width + margin.left + margin.right)
            .attr("height", height + margin.top + margin.bottom)
            .append("g")
            .attr("transform", "translate(" + margin.left + "," + margin.top + ")");
        x.domain(data.map(function(d) { return d.state; }));
        y.domain([0, d3.max(data, function(d) { return d.population; })]);
        svg.append("g")
            .attr("class", "x axis")
            .attr("transform", "translate(0," + height + ")")
            .call(xAxis);
        svg.append("g")
            .attr("class", "y axis")
            .call(yAxis)
            .append("text")
            .attr("transform", "rotate(-90)")
            .attr("y", 6)
            .attr("dy", ".71em")
            .style("text-anchor", "end")
            .text("Population");
        svg.selectAll(".bar")
            .data(data)
            .enter().append("rect")
            .attr("class", "bar")
            .attr("x", function(d) { return x(d.state); })
            .attr("width", x.rangeBand())
            .attr("y", function(d) { return y(d.population); })
            .attr("height", function(d) { return height - y(d.population); });
    });
    """);
```

You aren't finished. Now is the time to launch the representation of the D3 chart you just defined. You also need to write the commands needed to pass data contained in the pandas dataframe to the template, so they can be directly integrated into the JavaScript code written previously. The representation of JavaScript code, or rather the template just defined, is executed by launching the render() function.

```
display(Javascript(myTemplate.render(
    data=states.sort_values(['POPESTIMATE2022'], ascending=False)[:10].itertuples()
)))
```

The bar chart will appear in the previous frame in which the <div> was placed, as shown in Figure 11-5, which shows all the population estimates for the year 2022.

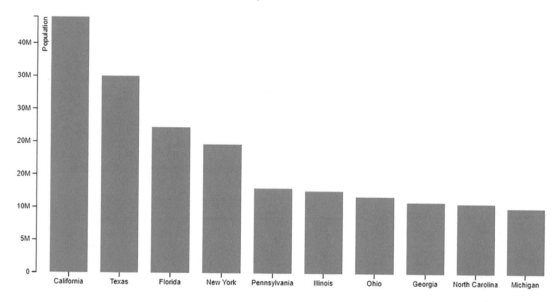

Figure 11-5. *The five most populous states of the United States represented by a bar chart relative to 2022*

Drawing a Clustered Bar Chart

So far you have relied broadly on what had been described in the fantastic article written by Barto. However, the type of data that you extracted has given you the trend of population estimates in the last four years for the United States. A more useful chart for visualizing data would be to show the trend of the population of each state over time.

To do that, a good choice is to use a clustered bar chart, where each cluster is one of the five most populous states and each cluster will have four bars that represent the population in a given year.

At this point you can modify the previous code or write new code in your Jupyter Notebook.

```
display(HTML("""
<style>
.bar2020 {
    fill: steelblue;
}
```

```
.bar2021 {
   fill: red;
}
.bar2022 {
   fill: yellow;
}
.axis {
   font: 10px sans-serif;
}
.axis path,
.axis line {
   fill: none;
   stroke: #000;
}
.x.axis path {
   display: none;
}
</style>
<div id="chart_d3" />
"""))
```

You have to modify the template as well, by adding the other three sets of data corresponding to the years 2020 and 2021. These years will be represented by a different color on the clustered bar chart.

```
import jinja2
myTemplate = jinja2.Template("""
require(["d3"], function(d3){
   var data = []
   var data2 = []
   var data3 = []

   {% for row in data %}
   data.push ({ 'state': '{{ row[1] }}', 'population': '{{ row[2] }}'  });
   data2.push({ 'state': '{{ row[1] }}', 'population': '{{ row[3] }}'  });
   data3.push({ 'state': '{{ row[1] }}', 'population': '{{ row[4] }}'  });
   {% endfor %}
d3.select("#chart_d3 svg").remove()
   var margin = {top: 20, right: 20, bottom: 30, left: 40},
       width = 800 - margin.left - margin.right,
       height = 400 - margin.top - margin.bottom;
   var x = d3.scale.ordinal()
       .rangeRoundBands([0, width], .25);
   var y = d3.scale.linear()
       .range([height, 0]);
   var xAxis = d3.svg.axis()
       .scale(x)
       .orient("bottom");
   var yAxis = d3.svg.axis()
       .scale(y)
       .orient("left")
       .ticks(10)
       .tickFormat(d3.format('.1s'));
```

```
    var svg = d3.select("#chart_d3").append("svg")
        .attr("width", width + margin.left + margin.right)
        .attr("height", height + margin.top + margin.bottom)
        .append("g")
        .attr("transform", "translate(" + margin.left + "," + margin.top + ")");
    x.domain(data.map(function(d) { return d.state; }));
    y.domain([0, d3.max(data, function(d) { return d.population; })]);
    svg.append("g")
        .attr("class", "x axis")
        .attr("transform", "translate(0," + height + ")")
        .call(xAxis);
    svg.append("g")
        .attr("class", "y axis")
        .call(yAxis)
        .append("text")
        .attr("transform", "rotate(-90)")
        .attr("y", 6)
        .attr("dy", ".71em")
        .style("text-anchor", "end")
        .text("Population");
    svg.selectAll(".bar2020")
        .data(data)
        .enter().append("rect")
        .attr("class", "bar2020")
        .attr("x", function(d) { return x(d.state); })
        .attr("width", x.rangeBand()/4)
        .attr("y", function(d) { return y(d.population); })
        .attr("height", function(d) { return height - y(d.population); });
    svg.selectAll(".bar2021")
        .data(data2)
        .enter().append("rect")
        .attr("class", "bar2021")
        .attr("x", function(d) { return (x(d.state)+x.rangeBand()/3); })
        .attr("width", x.rangeBand()/3)
        .attr("y", function(d) { return y(d.population); })
        .attr("height", function(d) { return height - y(d.population); });
    svg.selectAll(".bar2022")
        .data(data3)
        .enter().append("rect")
        .attr("class", "bar2022")
        .attr("x", function(d) { return (x(d.state)+2*x.rangeBand()/3); })
        .attr("width", x.rangeBand()/3)
        .attr("y", function(d) { return y(d.population); })
        .attr("height", function(d) { return height - y(d.population); });
});
""");
```

The series of data to be passed from the dataframe to the template are now four, so you have to refresh the data and the changes that you just made to the code. Therefore, you need to rerun the code of the render() function.

```
display(Javascript(myTemplate.render(
    data=states.sort_values(['POPESTIMATE2022'], ascending=False)[:5].itertuples()
)))
```

Once you launch the render() function again, you get a chart like the one shown in Figure 11-6.

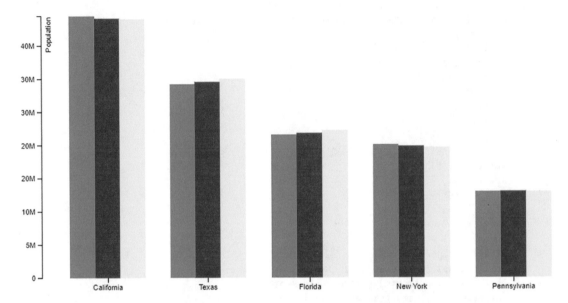

Figure 11-6. *A clustered bar chart representing the populations of the five most populous states from 2020 to 2022*

The Choropleth Maps

In the previous sections, you saw how to use JavaScript code and the D3 library to represent a bar chart. Well, these achievements would have been easy with matplotlib and perhaps implemented in an even better way. The purpose of the previous code was only for educational purposes.

Something quite different is the use of much more complex views that are unobtainable by matplotlib. This section illustrates the true potential made available by the D3 library. The *choropleth maps* are very complex types of representations.

The choropleth maps are geographical representations where the land areas are divided into portions characterized by different colors. The colors and the boundaries between a portion geographical and another are themselves representations of data.

This type of representation is very useful for representing the results of data analysis carried out on demographic or economic information, and this is also the case for data that correlates to their geographical distributions.

The representation of choropleth is based on a particular file called TopoJSON. This type of file contains all the inside information representing a choropleth map, such as the United States (see Figure 11-7).

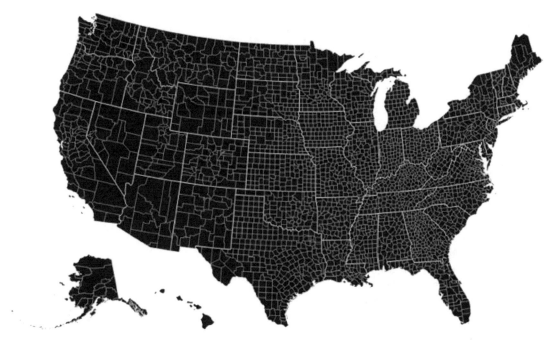

Figure 11-7. *The representation of a choropleth map of U.S. territories with no value related to each county or state*

A good link to find such material is the U.S. Atlas TopoJSON (`https://github.com/mbostock/us-atlas`), but a lot of literature about it is available online.

A representation of this kind is not only possible but is also customizable. Thanks to the D3 library, you can correlate the geographic portions based on the value of particular columns contained in a dataframe.

First, start with an example already on the Internet, in the D3 library, `http://bl.ocks.org/mbostock/4060606`, but fully developed in HTML. So now you learn how to adapt a D3 example in HTML in an IPython Notebook.

If you look at the code shown on the web page of the example, you can see that there are three necessary JavaScript libraries. This time, in addition to the D3 library, you need to import the queue and TopoJSON libraries.

```
<script src="https://cdnjs.cloudflare.com/ajax/libs/d3/3.5.5/d3.min.js"></script>
<script src="https://cdnjs.cloudflare.com/ajax/libs/queue-async/1.0.7/queue.min.js"></script>
<script src="https://cdnjs.cloudflare.com/ajax/libs/topojson/1.6.19/topojson.min.js"></script>
```

You have to use `require.config()` as you did in the previous sections.

```
%%javascript
require.config({
    paths: {
        d3: '//cdnjs.cloudflare.com/ajax/libs/d3/3.5.5/d3.min',
        queue: '//cdnjs.cloudflare.com/ajax/libs/queue-async/1.0.7/queue.min',
        topojson: '//cdnjs.cloudflare.com/ajax/libs/topojson/1.6.19/topojson.min'
    }
});
```

The pertinent part of the CSS is shown again, all within the HTML() function.

```
from IPython.display import display, Javascript, HTML
display(HTML("""
<style>
.counties {
  fill: none;
}
.states {
  fill: none;
  stroke: #fff;
  stroke-linejoin: round;
}
.q0-9 { fill:rgb(247,251,255); }
.q1-9 { fill:rgb(222,235,247); }
.q2-9 { fill:rgb(198,219,239); }
.q3-9 { fill:rgb(158,202,225); }
.q4-9 { fill:rgb(107,174,214); }
.q5-9 { fill:rgb(66,146,198); }
.q6-9 { fill:rgb(33,113,181); }
.q7-9 { fill:rgb(8,81,156); }
.q8-9 { fill:rgb(8,48,107); }
</style>
<div id="choropleth" />
"""))
```

Here is the new template that mirrors the code shown in the Bostock example, with some changes:

```
import jinja2
choropleth = jinja2.Template("""
require(["d3","queue","topojson"], function(d3,queue,topojson){
d3.select("#choropleth svg").remove()
var width = 960,
    height = 600;
var rateById = d3.map();
var quantize = d3.scale.quantize()
    .domain([0, .15])
    .range(d3.range(9).map(function(i) { return "q" + i + "-9"; }));
var projection = d3.geo.albersUsa()
    .scale(1280)
    .translate([width / 2, height / 2]);
var path = d3.geo.path()
    .projection(projection);
//row to modify
var svg = d3.select("#choropleth").append("svg")
    .attr("width", width)
    .attr("height", height);
queue()
    .defer(d3.json, "us.json")
    .defer(d3.tsv, "unemployment.tsv", function(d) { rateById.set(d.id, +d.rate); })
    .await(ready);
```

```
function ready(error, us) {
  if (error) throw error;
  svg.append("g")
      .attr("class", "counties")
    .selectAll("path")
      .data(topojson.feature(us, us.objects.counties).features)
    .enter().append("path")
      .attr("class", function(d) { return quantize(rateById.get(d.id)); })
      .attr("d", path);
  svg.append("path")
      .datum(topojson.mesh(us, us.objects.states, function(a, b) { return a !== b; }))
      .attr("class", "states")
      .attr("d", path);
}
});
""");
```

Now you launch the representation, this time without any value for the template, since all the values are contained in the us.json and unemployment.tsv files (you can find them in the source code of this book).

```
display(Javascript(choropleth.render()))
```

The results are identical to those shown in the Bostock example (see Figure 11-8).

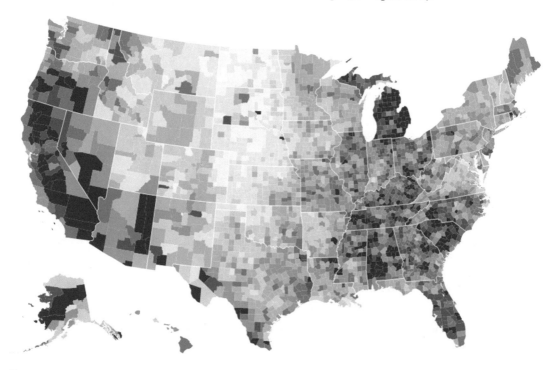

Figure 11-8. *The choropleth map of the United States with the coloring of the counties based on the values contained in the file TSV*

The Choropleth Map of the U.S. Population in 2022

Now that you have seen how to extract demographic information from the U.S. Census Bureau and you can create a choropleth map, you can unify both things to represent a choropleth map showing the population values. The more populous the county, the deeper blue it will be. In counties with very low population levels, the hue will tend toward white.

In the first section of the chapter, you extracted information about the states using the pop2022 dataframe. This was done by selecting the rows of the dataframe with SUMLEV values equal to 40. In this example, you instead need the values of the populations of each county. Therefore, you have to take out a new dataframe by taking pop2022 using only lines with a SUMLEV of 50.

You must instead select the rows to level 50.

```
pop2022_by_county = pop2022[pop2022.SUMLEV == 50]
pop2022_by_county
```

You get a dataframe that contains all U.S. counties, as shown in Figure 11-9.

	SUMLEV	REGION	DIVISION	STATE	COUNTY	STNAME	CTYNAME	ESTIMATESBASE2020	POPESTIMATE2020	POPESTIMATE2021	...	RDEATH2021	RI
1	50	3	6	01	001	Alabama	Autauga County	58802	58902	59210	...	11.548361	
2	50	3	6	01	003	Alabama	Baldwin County	231761	233219	239361	...	12.882475	
3	50	3	6	01	005	Alabama	Barbour County	25224	24960	24539	...	15.475060	
4	50	3	6	01	007	Alabama	Bibb County	22300	22183	22370	...	14.275133	
5	50	3	6	01	009	Alabama	Blount County	59130	59102	59085	...	14.637820	
...	...	...	...	...	...	...	...	...	...	...	...	...	
3190	50	4	8	56	037	Wyoming	Sweetwater County	42267	42190	41582	...	10.122714	
3191	50	4	8	56	039	Wyoming	Teton County	23323	23377	23622	...	3.957531	
3192	50	4	8	56	041	Wyoming	Uinta County	20446	20457	20655	...	8.999805	
3193	50	4	8	56	043	Wyoming	Washakie County	7682	7658	7712	...	14.313598	
3194	50	4	8	56	045	Wyoming	Weston County	6840	6818	6766	...	12.956419	

3144 rows × 51 columns

Figure 11-9. *The pop2022_by_county dataframe contains all demographics of all U.S. counties*

You must use your data instead of the TSV previously used. Inside it, there are the ID numbers corresponding to the various counties. You can use a file on the web to determine their names. You can download it and turn it into a dataframe.

```
USJSONnames = pd.read_table('us-county-names.tsv')
USJSONnames
```

Thanks to this file, you see the codes with the corresponding counties (see Figure 11-10).

	id	name
0	1000	Alabama
1	1001	Autauga
2	1003	Baldwin
3	1005	Barbour
4	1007	Bibb
...	...	...
3349	78222	Ngardmau
3350	78224	Ngatpang
3351	78226	Ngchesar
3352	78350	Peleliu
3353	78370	Sonsorol

Figure 11-10. *The codes of the counties are contained in the TSV file*

If you take for example the Baldwin county:

```
USJSONnames[USJSONnames['name'] == 'Baldwin']
```

You can see that there are actually two counties with the same name, but they are identified by two different identifiers (Figure 11-11).

	id	name
2	1003	Baldwin
399	13009	Baldwin

Figure 11-11. *There are two Baldwin counties*

You get a table and see that there are two counties and two different codes. Now you see this in your dataframe with data taken from the data source at census.gov (see Figure 11-12).

```
pop2022_by_county[pop2022_by_county['CTYNAME'] == 'Baldwin County']
```

	SUMLEV	REGION	DIVISION	STATE	COUNTY	STNAME	CTYNAME	ESTIMATESBASE2020	POPESTIMATE2020	POPESTIMATE2021	...
2	50	3	6	01	003	Alabama	Baldwin County	231761	233219	239361	...
404	50	3	5	13	009	Georgia	Baldwin County	43795	43794	43671	...

Figure 11-12. *The ID codes in the TSV files correspond to the combination of the values contained in the STATE and COUNTY columns*

You can recognize that there is a match. The ID contained in TOPOJSON matches the numbers in the STATE and COUNTY columns if combined, but removing the 0 when it is the digit at the beginning of the code. So now you can reconstruct all the data needed to replicate the TSV example of choropleth from the counties dataframe. The file will be saved as population.csv.

```
counties = pop2022_by_county[['STATE','COUNTY','POPESTIMATE2022']]
counties.is_copy = False
counties['id'] = counties['STATE'].str.lstrip('0') + "" + counties['COUNTY']
del counties['STATE']
del counties['COUNTY']
counties.columns = ['pop','id']
counties = counties[['id','pop']]
counties.to_csv('population.csv')
```

Now you rewrite the contents of the HTML() function by specifying a new <div> tag with the ID as choropleth2.

```
from IPython.display import display, Javascript, HTML
display(HTML("""
<style>
.counties {
  fill: none;
}
.states {
  fill: none;
  stroke: #fff;
  stroke-linejoin: round;
}
.q0-9 { fill:rgb(247,251,255); }
.q1-9 { fill:rgb(222,235,247); }
.q2-9 { fill:rgb(198,219,239); }
.q3-9 { fill:rgb(158,202,225); }
.q4-9 { fill:rgb(107,174,214); }
.q5-9 { fill:rgb(66,146,198); }
.q6-9 { fill:rgb(33,113,181); }
.q7-9 { fill:rgb(8,81,156); }
.q8-9 { fill:rgb(8,48,107); }
</style>
<div id="choropleth2" />
"""))
```

You also have to define a new Template object.

```
choropleth2 = jinja2.Template("""
require(["d3","queue","topojson"], function(d3,queue,topojson){
    var data = []
d3.select("#choropleth2 svg").remove()
var width = 960,
    height = 600;
var rateById = d3.map();
var quantize = d3.scale.quantize()
    .domain([0, 1000000])
    .range(d3.range(9).map(function(i) { return "q" + i + "-9"; }));
var projection = d3.geo.albersUsa()
    .scale(1280)
    .translate([width / 2, height / 2]);
var path = d3.geo.path()
    .projection(projection);
var svg = d3.select("#choropleth2").append("svg")
    .attr("width", width)
    .attr("height", height);
queue()
    .defer(d3.json, "us.json")
    .defer(d3.csv,"population.csv", function(d) { rateById.set(d.id, +d.pop); })
    .await(ready);
function ready(error, us) {
  if (error) throw error;
  svg.append("g")
      .attr("class", "counties")
    .selectAll("path")
      .data(topojson.feature(us, us.objects.counties).features)
    .enter().append("path")
      .attr("class", function(d) { return quantize(rateById.get(d.id)); })
      .attr("d", path);
  svg.append("path")
      .datum(topojson.mesh(us, us.objects.states, function(a, b) { return a !== b; }))
      .attr("class", "states")
      .attr("d", path);
}
});
""");
```

Finally, you can execute the render() function to get the chart.

```
display(Javascript(choropleth2.render()))
```

The choropleth map will be shown with the counties differently colored depending on their population, as shown in Figure 11-13.

Figure 11-13. *A choropleth map of the United States showing the density of the population of all counties*

Conclusions

In this chapter, you learned how it is possible to further extend the ability to display data using a JavaScript library called D3. Choropleth maps are just one of many examples of advanced graphics that are used to represent data. This is also a very good way to see the Jupyter Notebook in action. The world does not revolve around Python alone, but Python can provide additional capabilities for your work.

In the next chapter, you learn how to apply data analysis to images. You also see how easy it is to build a model that can recognize handwritten numbers.

CHAPTER 12

Recognizing Handwritten Digits

So far you have seen how to apply the techniques of data analysis to pandas dataframes containing numbers and strings. However, data analysis is not limited to numbers and strings, because images and sounds can also be analyzed and classified.

In this short but no-less-important chapter, you will learn about handwriting recognition.

Handwriting Recognition

Recognizing handwritten text is a problem that can be traced back to the first automatic machines that needed to recognize individual characters in handwritten documents. Think about, for example, the ZIP codes on letters at the post office and the automation needed to recognize these five digits. Perfect recognition of these codes is necessary in order to sort mail automatically and efficiently.

Included among the other applications that may come to mind is OCR (Optical Character Recognition) software. OCR software must read handwritten text, or pages of printed books, for general electronic documents in which each character is well defined.

But the problem of handwriting recognition goes farther back in time, more precisely to the early 20th century (1920s), when Emanuel Goldberg (1881–1970) began his studies regarding this issue and suggested that a statistical approach would be an optimal choice.

To address this issue in Python, the scikit-learn library provides a good example. This library can help you better understand this technique, the issues involved, and the possibility of making predictions.

Recognizing Handwritten Digits with scikit-learn

The scikit-learn library (http://scikit-learn.org/) enables you to approach this type of data analysis in a way that is slightly different from what you've used in the book so far. The data to be analyzed is closely related to numerical values or strings, but can also involve images and sounds.

The problem you face in this chapter involves predicting a numeric value, and then reading and interpreting an image that uses a handwritten font.

In this case, you have an *estimator* with the task of learning through a fit() function, and once it reaches a degree of predictive capability (the model is sufficiently valid), it will produce a prediction with the predict() function. The training and validation sets are created this time from a series of images.

This chapter uses Jupyter Notebook to run through the Python code examples, so open Jupyter and create a new Notebook.

An estimator that is useful in this case is sklearn.svm.SVC, which uses the technique of Support Vector Classification (SVC).

© Fabio Nelli 2023
F. Nelli, *Python Data Analytics*, https://doi.org/10.1007/978-1-4842-9532-8_12

Thus, you have to import the svm module of the scikit-learn library. You can create an estimator of SVC type and then choose an initial setting, assigning the values C and gamma generic values. These values can then be adjusted in a different way during the course of the analysis.

```
from sklearn import svm
svc = svm.SVC(gamma=0.001, C=100.)
```

The Digits Dataset

As you saw in Chapter 8, the scikit-learn library provides numerous datasets that are useful for testing many problems of data analysis and prediction of the results. Also in this case there is a dataset of images called *Digits*.

This dataset consists of 1,797 images that are 8x8 pixels in size. Each image is a handwritten digit in grayscale, as shown in Figure 12-1.

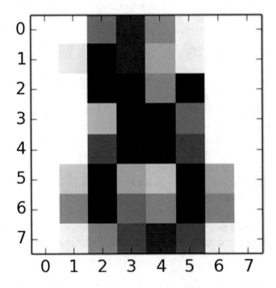

Figure 12-1. *One of 1,797 handwritten number images that make up the Digits dataset*

Thus, you can load the Digits dataset into your Notebook.

```
from sklearn import datasets
digits = datasets.load_digits()
```

After loading the dataset, you can analyze the content. First, you can read lots of information about the datasets by calling the DESCR attribute.

```
print(digits.DESCR)
```

For a textual description of the dataset, the authors who contributed to its creation and the references appear as shown in Figure 12-2.

```
print(digits.DESCR)
```

```
.. _digits_dataset:

Optical recognition of handwritten digits dataset
--------------------------------------------------

**Data Set Characteristics:**

    :Number of Instances: 1797
    :Number of Attributes: 64
    :Attribute Information: 8x8 image of integer pixels in the range 0..16.
    :Missing Attribute Values: None
    :Creator: E. Alpaydin (alpaydin '@' boun.edu.tr)
    :Date: July; 1998

This is a copy of the test set of the UCI ML hand-written digits datasets
https://archive.ics.uci.edu/ml/datasets/Optical+Recognition+of+Handwritten+Digits

The data set contains images of hand-written digits: 10 classes where
each class refers to a digit.

Preprocessing programs made available by NIST were used to extract
normalized bitmaps of handwritten digits from a preprinted form. From a
total of 43 people, 30 contributed to the training set and different 13
to the test set. 32x32 bitmaps are divided into nonoverlapping blocks of
4x4 and the number of on pixels are counted in each block. This generates
an input matrix of 8x8 where each element is an integer in the range
0..16. This reduces dimensionality and gives invariance to small
distortions.

For info on NIST preprocessing routines, see M. D. Garris, J. L. Blue, G.
T. Candela, D. L. Dimmick, J. Geist, P. J. Grother, S. A. Janet, and C.
L. Wilson, NIST Form-Based Handprint Recognition System, NISTIR 5469,
```

Figure 12-2. Each dataset in the scikit-learn library has a field containing all the information

The images of the handwritten digits are contained in a digits.images array. Each element in this array is an image that is represented by an 8x8 matrix of numerical values that correspond to grayscale array. White has a value of 0 and black has a value of 15.

```
digits.images[0]
```

You will get the following result:

```
array([[  0.,   0.,   5.,  13.,   9.,   1.,   0.,   0.],
       [  0.,   0.,  13.,  15.,  10.,  15.,   5.,   0.],
       [  0.,   3.,  15.,   2.,   0.,  11.,   8.,   0.],
       [  0.,   4.,  12.,   0.,   0.,   8.,   8.,   0.],
       [  0.,   5.,   8.,   0.,   0.,   9.,   8.,   0.],
```

```
[  0.,   4.,  11.,   0.,   1.,  12.,   7.,   0.],
[  0.,   2.,  14.,   5.,  10.,  12.,   0.,   0.],
[  0.,   0.,   6.,  13.,  10.,   0.,   0.,   0.]])
```

You can visually check the contents of this result using the `matplotlib` library.

```
import matplotlib.pyplot as plt
plt.imshow(digits.images[0], cmap=plt.cm.gray_r, interpolation='nearest')
```

When you launch this command, you obtain the grayscale image shown in Figure 12-3.

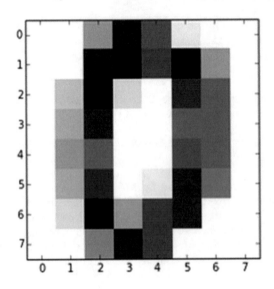

Figure 12-3. *One of the 1,797 handwritten digits*

The numerical values represented by images, that is, the targets, are contained in the `digits.target` array.

```
digits.targetOut [ ]:
array([0, 1, 2, ..., 8, 9, 8])
```

It was reported that the dataset is a training set consisting of 1,797 images. You can determine if that is true.

```
digits.target.sizeOut [ ]:
1797
```

Learning and Predicting

Now that you have loaded the Digits dataset into your Notebook and have defined an SVC estimator, you can start learning.

As you learned in Chapter 8, once you define a predictive model, you must instruct it with a training set, which is a set of data in which you already know the class. Given the large quantity of elements contained in the Digits dataset, you will certainly obtain a very effective model, that is, one that's capable of recognizing with good certainty the handwritten number.

This dataset contains 1,797 elements, so you can consider the first 1,791 as a training set and use the last 6 as a validation set.

You can see in detail these six handwritten digits by using the matplotlib library:

```
import matplotlib.pyplot as plt
plt.subplot(321)
plt.imshow(digits.images[1791], cmap=plt.cm.gray_r, interpolation='nearest')
plt.subplot(322)
plt.imshow(digits.images[1792], cmap=plt.cm.gray_r, interpolation='nearest')
plt.subplot(323)
plt.imshow(digits.images[1793], cmap=plt.cm.gray_r, interpolation='nearest')
plt.subplot(324)
plt.imshow(digits.images[1794], cmap=plt.cm.gray_r, interpolation='nearest')
plt.subplot(325)
plt.imshow(digits.images[1795], cmap=plt.cm.gray_r, interpolation='nearest')
plt.subplot(326)
plt.imshow(digits.images[1796], cmap=plt.cm.gray_r, interpolation='nearest')
```

This will produce an image with six digits, as shown in Figure 12-4.

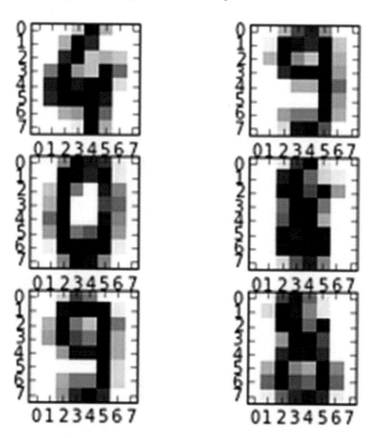

Figure 12-4. The six digits of the validation set

Now you can train the `svc` estimator that you defined earlier.

```
svc.fit(digits.data[1:1790], digits.target[1:1790])
```

This will produce an image as shown in Figure 12-5.

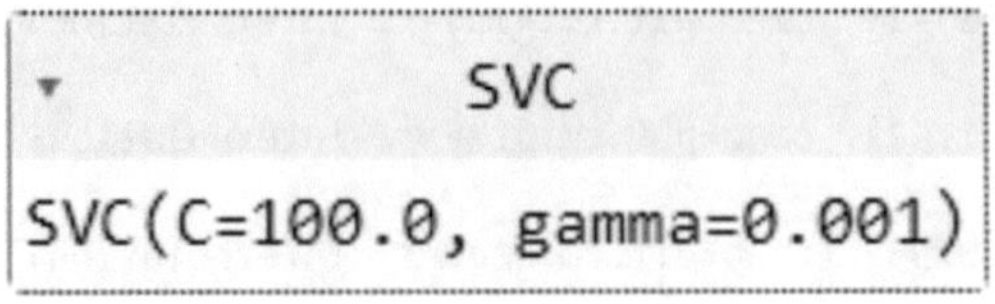

```
▼                    SVC

SVC(C=100.0, gamma=0.001)
```

Figure 12-5. *The parameters of the SVC estimator*

Now you have to test your estimator, making it interpret the six digits of the validation set.

```
svc.predict(digits.data[1791:1976])
Out [ ]: array([4, 9, 0, 8, 9, 8])
```

If you compare them with the actual digits, as follows:

```
digits.target[1791:1976]
Out [ ]:
array([4, 9, 0, 8, 9, 8])
```

You can see that the `svc` estimator has learned correctly. It recognizes the handwritten digits, interpreting correctly all six digits of the validation set.

Recognizing Handwritten Digits with TensorFlow

You have just seen an example of how machine learning techniques can recognize handwritten numbers. Now the same problem is applied to the deep learning techniques that you used in Chapter 9. As was the case in Chapter 9, the following section regarding TensorFlow has been completely rewritten from the previous edition. In fact, here too you will use the new TensorFlow 2.x version, which is completely different from TensorFlow 1.x. The code used here is therefore not present in older editions of this book.

Given the great value of the MNIST dataset, the TensorFlow library also contains a copy of it. It will therefore be very easy to perform studies and tests on neural networks with this dataset, without having to download or import them from other data sources.

In addition to TensorFlow, install the `tensorflow-dataset` package. You can do this either using Anaconda Navigator or via the command console:

```
conda install tensorflow-dataset
```

If you don't have the Anaconda platform, you can install the package through the PyPI system.

```
pip install tensorflow-dataset
```

Importing the MNIST dataset into the Jupyter Notebook (in any Python session) is very simple. Indeed, with the new version of TensorFlow 2.x, there is no need to import test datasets like MNIST from other libraries, as they are available within Keras, which is integrated within the tensorflow module you already imported. You can simply import the libraries like numpy and matplotlib along with tensorflow, which also contains the MNIST dataset.

```
import numpy as np
import matplotlib.pyplot as plt
import tensorflow as tf
ì
```

Now load the dataset directly into your Notebook by simply writing the following line of code.

```
(x_train, y_train),(x_test, y_test) = tf.keras.datasets.mnist.load_data()
```

Then you can load the dataset directly into your Notebook.

```
x_validation = x_train[55000:]
x_train = x_train[:55000]
y_validation = y_train[55000:]
y_train = y_train[:55000]
len(x_train)
Out [ ]:
55000

len(x_test)
Out [ ]:
10000

len(x_validation)
Out [ ]:
5000
```

The MNIST data is split into three parts: 55,000 data points of training data (x_train), 10,000 points of test data (x_test), and 5,000 points of validation data (x_validation).

All this data will be submitted to the model: x is the feature dataset and y is the label dataset. As you saw earlier, these are images of handwritten letters. You can look at the first image of the training dataset (features).

```
x_train[0].shape
Out [ ]:
(28, 28)
```

This is a square image of 28 pixels per side.

```
plt.imshow(x_train[0], cmap=plt.cm.gray_r, interpolation='nearest')
```

You will get the black and white image of a handwritten number, similar to the one shown in Figure 12-6.

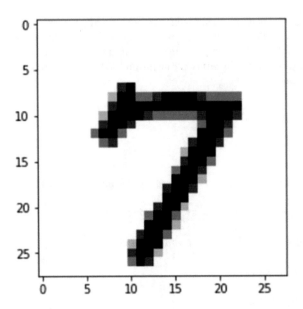

Figure 12-6. *A digit of the training set in the MNIST dataset provided by the TensorFlow library*

To give you an idea of the contents of the MNIST dataset, a better visualization is the following:

```
fig, ax = plt.subplots(10, 10)
k = 0
for i in range(10):
    for j in range(10):
        ax[i][j].imshow(x_train[k].reshape(28, 28),
                        cmap=plt.cm.gray_r,
                        interpolation='nearest',
                        aspect='auto')
        ax[i][j].set_xticks([])
        ax[i][j].set_yticks([])
        k += 1
```

Running the code, you will get a pattern of 100 numbers in the dataset, as shown in Figure 12-7.

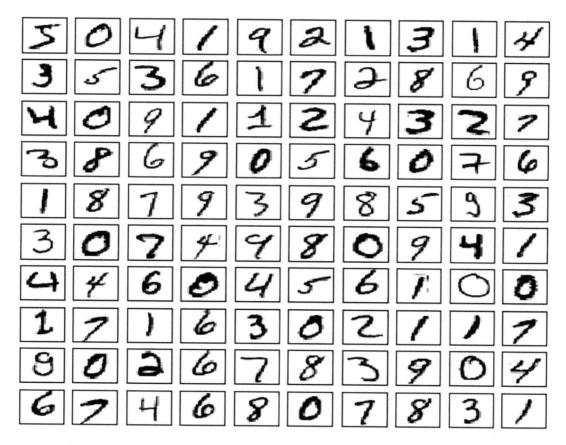

Figure 12-7. *100 digits of the training dataset from MNIST dataset provided by the TensorFlow library*

Because these are images and therefore two-dimensional arrays, as you learned in Chapter 9, you have to use a flatten layer at the beginning of the neural network to flatten the input data and make them one-dimensional.

As for the types of values to be submitted to the neural network, you have integer values ranging from 0 to 255.

```
print(np.max(x_train))
np.min(x_train)
Out [ ]:
255
0
```

In fact, these are grayscale images which, like RGB colors, are included in a range of values between 0 and 255. You therefore also have to add a normalization layer to the neural network model.

Now convert all the arrays to tensors for use in TensorFlow.

```
train_features = tf.convert_to_tensor(x_train)
train_labels = tf.convert_to_tensor(y_train)
test_features = tf.convert_to_tensor(x_test)
test_labels = tf.convert_to_tensor(y_test)
exp_features = tf.convert_to_tensor(x_validation)
```

Let's look at the characteristics of one of the tensors as an example.

```
train_features
Out [ ]:
<tf.Tensor: shape=(55000, 28, 28), dtype=uint8, numpy=
array([[[0, 0, 0, ..., 0, 0, 0],
        [0, 0, 0, ..., 0, 0, 0],
        [0, 0, 0, ..., 0, 0, 0],
        ...,
        [0, 0, 0, ..., 0, 0, 0],
        [0, 0, 0, ..., 0, 0, 0],
        [0, 0, 0, ..., 0, 0, 0]],
...
```

The results obtained are what you would expect.

Learning and Predicting with an SLP

Now that you've seen how to get the training set, the testing set, and the validation set with TensorFlow, it's time to do an analysis with a neural network, very similar to the one you used in Chapter 9. Let's start by using a Single Layer Perceptron (SLP).

First define a model with a single dense layer with ten outputs corresponding to the ten numerical digits ranging from 0 to 9, and which correspond to the ten classes of membership of the handwritten digits to be identified. To this single layer, you will add the two layers: Normalization and Flatten. The former will normalize the pixel values of the images from 0 to 255 in the range of 0 to 1, and the latter will convert the two-dimensional array of 28x28 images into a single one-dimensional array.

```
model = tf.keras.Sequential([
    tf.keras.layers.Normalization(),
    tf.keras.layers.Flatten(input_shape=(28, 28)),
    tf.keras.layers.Dense(10, activation='sigmoid')
])
```

Once the model has been defined, you can compile it, setting Adam as an optimizer and sparse_categorical_crossentropy as a function. Then you can start learning the model with 20 epochs.

```
model.compile(
    optimizer='adam',
    loss='sparse_categorical_crossentropy',
    metrics=['accuracy'])
h = model.fit(train_features, train_labels, epochs=20)
Out [ ]:
Epoch 1/20
1719/1719 [==============================] - 3s 1ms/step - loss: 10.6817 - accuracy: 0.8341
Epoch 2/20
1719/1719 [==============================] - 2s 1ms/step - loss: 6.1368 - accuracy: 0.8759
Epoch 3/20
1719/1719 [==============================] - 3s 2ms/step - loss: 5.7782 - accuracy: 0.8798
Epoch 4/20
1719/1719 [==============================] - 3s 1ms/step - loss: 5.5405 - accuracy: 0.8824
```

```
Epoch 5/20
1719/1719 [==============================] - 2s 1ms/step - loss: 5.4812 - accuracy: 0.8840
Epoch 6/20
1719/1719 [==============================] - 2s 1ms/step - loss: 5.3873 - accuracy: 0.8851
Epoch 7/20
1719/1719 [==============================] - 2s 1ms/step - loss: 5.3120 - accuracy: 0.8861
...
```

Now check the learning phase through the history by graphically monitoring the trend of the loss.

```
acc_set = h.history['loss']
epoch_set = h.epoch
plt.plot(epoch_set,acc_set, 'o', label='Training phase')
plt.ylabel('loss')
plt.xlabel('epoch')
plt.legend()
```

Running the code, you get a plot like the one shown in Figure 12-8.

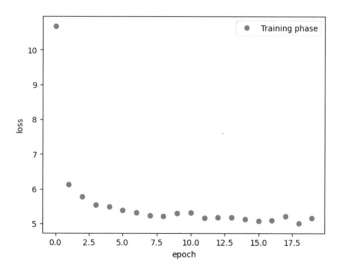

Figure 12-8. *The loss trend during the learning phase of the SLP neural network*

You can also evaluate the model numerically using the following line of code:

```
model.evaluate(test_features, test_labels)
Out [ ]:
313/313 [==============================] - 1s 1ms/step - loss: 5.8654 - accuracy: 0.8925
[5.865407466888428, 0.8924999833106995]
```

As you can see from the numerical values, an accuracy of 0.89 is not optimal and the loss value does not seem to drop too much, stabilizing at a value of 5.86.

Let's see how this model can recognize handwritten numbers that have not been used for learning or for testing. For this purpose, a third dataset has been set aside: exp_features. You extend the model with the Softmax layer to get the probabilities of belonging to the various classes as a result. You then let the newly educated SLP model make the predictions.

```
probability_model = tf.keras.Sequential([
    model,
    tf.keras.layers.Softmax()
])
predictions = probability_model.predict(exp_features)
Out [ ]:
157/157 [==============================] - 0s 981us/step
```

Now take the first image to be predicted, with the probabilities of recognition at each of the ten numerical digits.

```
predictions[0]
Out [ ]:
array([0.04717345, 0.12823072, 0.12823072, 0.12823072, 0.1282307 ,
       0.08905466, 0.04721416, 0.04717345, 0.12823072, 0.12823072],
      dtype=float32)
```

From the list of ten probabilities in the output, the situation is not so legible. If you use a graphical approach, representing the various probabilities in a barplot, you get better results.

```
p = plt.bar(np.arange(10),predictions[0])
plt.xticks(np.arange(10))
predicted_label = np.argmax(predictions[0])
p[predicted_label].set_color('red')
```

Running the previous code will give you a barplot similar to the one shown in Figure 12-9.

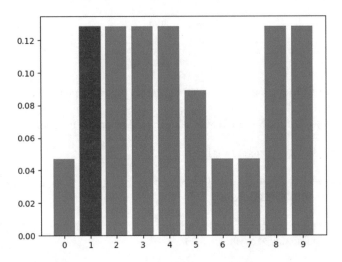

Figure 12-9. *The loss trend during the learning phase of the SLP neural network*

You can immediately see that many digits have the same probability of being the one represented in the image. Although the barplot shows the most probable figure in red, in this case there is an error since many other figures have the same probability (the graph shows only the first maximum in the case of parity of values). So the forecast was not successful. Now determine the true value of the image submitted to the model.

```
y_validation[0]
Out [ ]:
1
```

You can also look at it graphically.

```
plt.imshow(x_validation[0], cmap=plt.cm.gray_r, interpolation='nearest')
```

Executing the previous code, you obtain the number shown in Figure 12-10.

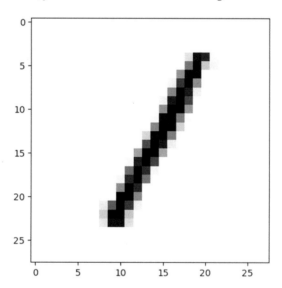

Figure 12-10. *The image shows the handwritten number 1*

As you can see, it is the number 1, which is present among the most probable results. However, there are too many probable options, so you cannot consider this a good prediction. Now take a number that is easier to recognize and see if the SLP model can recognize it correctly.

Choose the 14th number, which is easily recognizable.

```
plt.imshow(x_validation[13], cmap=plt.cm.gray_r, interpolation='nearest')
```

By running the code, you will get the image of this easily recognizable number, as shown in Figure 12-11.

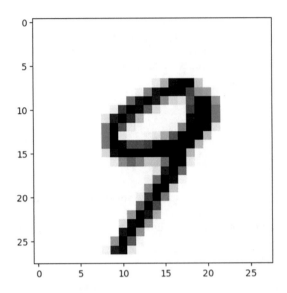

Figure 12-11. *The image shows the handwritten number 9*

The image clearly shows the number 9. Now check the corresponding label.

```
y_validation[13]
Out [ ]:
9
```

Let's see, in this very simple case, if the model recognized the number 9 clearly.

```
predictions[13]
Out [ ]:
array([0.07372559, 0.07270355, 0.07317524, 0.07424378, 0.13360192,
       0.07305884, 0.07261127, 0.13158722, 0.09798288, 0.1973097 ],
     dtype=float32)
```

You can also look at this graphically.

```
p = plt.bar(np.arange(10),predictions[13])
plt.xticks(np.arange(10))
predicted_label = np.argmax(predictions[13])
p[predicted_label].set_color('red')
```

Running the code will result in a barplot like the one shown in Figure 12-12.

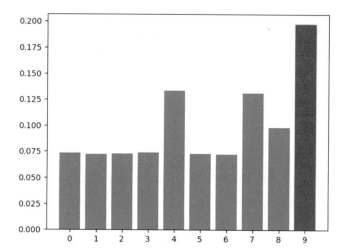

Figure 12-12. *The barplot shows the 9 digit as the most probable result (red bar)*

Although it guessed the number correctly this time, giving it a probability of around 20 percent is certainly not a good prediction. The number is easily recognizable, and it should have a much higher probability of recognition than the other digits.

Learning and Predicting with an MLP

Given the moderate success of the SLP model, this section builds a much more complex neural network, with more layers and more neurons in play. This network uses a Multiple Layer Perceptron (MLP) model with a hidden layer to predict handwritten digits. Furthermore, it brings the number of neurons of the first layer to 256, to which you will add another 128 for the hidden layer. It leaves the output layer unchanged at ten neurons (the ten digits to be classified).

```
model = tf.keras.Sequential([
    tf.keras.layers.Normalization(),
    tf.keras.layers.Flatten(input_shape=(28, 28)),
    tf.keras.layers.Dense(256, activation='sigmoid'),
    tf.keras.layers.Dense(128, activation='sigmoid'),
    tf.keras.layers.Dense(10, activation='sigmoid')
])
```

You can compile the model and train it with the same number of epochs as the previous one (20).

```
model.compile(optimizer='adam',
              loss='sparse_categorical_crossentropy',
              metrics=['accuracy'])
h = model.fit(train_features, train_labels, epochs=20)
Out [ ]:
Epoch 1/20
1719/1719 [==============================] - 4s 2ms/step - loss: 0.4885 - accuracy: 0.8658
Epoch 2/20
1719/1719 [==============================] - 4s 2ms/step - loss: 0.3243 - accuracy: 0.9013
```

```
Epoch 3/20
1719/1719 [==============================] - 3s 2ms/step - loss: 0.2935 - accuracy: 0.9093
Epoch 4/20
1719/1719 [==============================] - 3s 2ms/step - loss: 0.2630 - accuracy: 0.9171
Epoch 5/20
1719/1719 [==============================] - 3s 2ms/step - loss: 0.2423 - accuracy: 0.9250
Epoch 6/20
1719/1719 [==============================] - 3s 2ms/step - loss: 0.2370 - accuracy: 0.9268
Epoch 7/20
...
```

You can also graphically see the trend of the loss during the learning phase of the model.

```
acc_set = h.history['loss']
epoch_set = h.epoch
plt.plot(epoch_set,acc_set, 'o', label='Training phase')
plt.ylabel('loss')
plt.xlabel('epoch')
plt.legend()
```

By executing the previous code, you obtain a plot like the one shown in Figure 12-13.

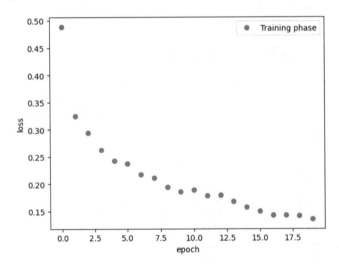

Figure 12-13. *The image shows how the value of the loss is optimized during the training phase*

You can also evaluate the model learning process numerically.

```
model.evaluate(test_features, test_labels)
Out [ ]:
313/313 [==============================] - 1s 1ms/step - loss: 0.1537 - accuracy: 0.9511
[0.1537058800458908, 0.9510999917984009]
```

As you can clearly see, this time the learning, in addition to being regular, also reaches a good accuracy value and a low loss value. You could also increase the number of epochs during the training phase to further increase the predictive power of this MLP model. However, leave things unchanged to compare the performance of this model to the previous one.

Now you can see how this model recognizes numbers using the same two images that you submitted to the SLP model.

```
probability_model = tf.keras.Sequential([
    model,
    tf.keras.layers.Softmax()
])
predictions = probability_model.predict(exp_features)
Out [ ]:
157/157 [==============================] - 0s 1ms/step
```

First you saw how the model assigns the probabilities of belonging to the ten digits of the image with the number 1.

```
predictions[0]
array([0.07896608, 0.21424502, 0.08096407, 0.07952367, 0.08464722,
       0.07919335, 0.08132026, 0.11285783, 0.10845622, 0.07982624],
     dtype=float32)
```

You can represent these graphically in a barplot.

```
p = plt.bar(np.arange(10),predictions[0])
plt.xticks(np.arange(10))
predicted_label = np.argmax(predictions[0])
p[predicted_label].set_color('red')
```

Running this code, you get the barplot shown in Figure 12-14.

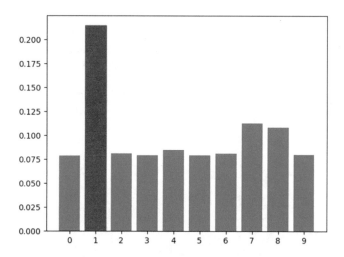

Figure 12-14. *The barplot shows the 1 digit as the most probable result (red bar)*

As you can see, this model guessed the number represented in the image, with a much higher probability than the other digits. This time, the number 1 has been clearly identified.

If you carry out the same operations with the second image (representing the number 9), you will get a similar result, as shown in the barplot in Figure 12-15.

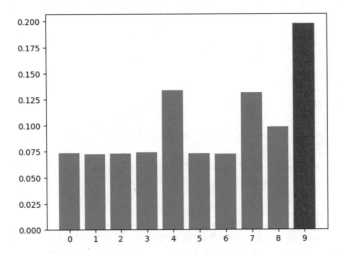

Figure 12-15. *The barplot shows the 1 digit as the most probable (red bar)*

It is therefore clear that a more complex neural network model such, as the one used in the example, is more efficient: it learns faster and is more adept at identifying the handwritten numbers in the images. But it is not always true that a more complex neural network model leads to an increase in potential. Only an adequate study of the various models, the optimizations used, the loss functions chosen, and all the other parameters used can prove the accuracy of a model. I therefore invite you to study this topic further, if you are fascinated by it.

Conclusions

In this chapter, you learned how many application possibilities this data analysis process has. It is not limited only to the analysis of numerical and textual data, but also can analyze images, such as the handwritten digits read by a camera or a scanner.

Furthermore, you have seen that predictive models can provide optimal results, thanks to machine learning and deep learning techniques, which are powerful analysis tools thanks to libraries such as TensorFlow.

CHAPTER 13

■ ■ ■

Textual Data Analysis with NLTK

In this book, you have seen various analysis techniques and numerous examples that worked on data in numerical or tabular form, which is easily processed through mathematical expressions and statistical techniques. But most of the data is composed of text, which responds to grammatical rules (or sometimes not even that :)) that differ from language to language. In text, the words and the meanings attributable to the words (as well as the emotions they transmit) can be a very useful source of information.

In this chapter, you will learn about some text analysis techniques using the NLTK (Natural Language Toolkit) library, which allows you to perform otherwise complex operations. Furthermore, the topics covered will help you understand this important part of data analysis.

Text Analysis Techniques

In recent years, with the advent of Big Data and the immense amount of textual data coming from the Internet, a lot of text analysis techniques have been developed by necessity. In fact, this form of data can be very difficult to analyze, but at the same time represents a source of a lot of useful information, also given the enormous availability of data. Just think of all the literature produced—the numerous posts published on the Internet, for example. Comments on social networks and chats can also be a great source of data, especially to understand the degree of approval or disapproval of a particular topic.

Analyzing these texts has therefore become a source of enormous interest, and there are many techniques that have been introduced for this purpose, creating a real discipline in itself. Some of the more important techniques are listed here.

For preprocessing:

- Lowercase conversion
- Word and sentence tokenization
- Punctuation mark removal
- Stopword removal
- Stemming
- Lemmatization

For text analysis:

- Analysis of the frequency distribution of words
- Pattern recognition

F. Nelli, *Python Data Analytics*, https://doi.org/10.1007/978-1-4842-9532-8_13

- Tagging
- Analysis of links and associations
- Sentiment analysis

The Natural Language Toolkit (NLTK)

If you program in Python and want to analyze data in text form, one of the most commonly used tools at the moment is the Python Natural Language Toolkit (NLTK).

NLTK is nothing more than a Python library (www.nltk.org) in which there are many tools specialized in processing and text data analysis. NLTK was created in 2001 for educational purposes, then over time it developed to such an extent that it became a real analysis tool.

Within the NLTK library, there is also a large collection of sample texts, called *corpora*. This collection of texts is taken largely from literature and is very useful as a basis for the application of the techniques developed with the NLTK library. In particular, it's used to perform tests (a role similar to the MNIST dataset present in TensorFlow, which is discussed in Chapter 9).

Installing NLTK on your computer is a very simple operation.

If you are not currently using an Anaconda platform, you can install it using the PyPI system.

```
pip install nltk
```

If, on the other hand, you have an Anaconda platform to develop your projects in Python, on the virtual environment you want to use, install the nltk package graphically via Anaconda Navigator, or via the command console:

```
conda install nltk
```

Import the NLTK Library and the NLTK Downloader Tool

In order to be more confident with NLTK, there is no better method than working directly with the Python code. This way, you can see and gradually understand the operation of this library.

The first thing you need to do is open a Jupyter Notebook. The first command imports the NLTK library.

```
import nltk
```

Then you need to import text from the corpora collection. To do this, there is a function called nltk.download_shell(), which opens a tool called NLTK Downloader. The downloader allows you to make selections through a guided choice of options.

If you enter this command on the terminal:

```
nltk.download_shell()
```

You will see in output the NLTK Downloader suggesting various options in text format, as shown in Figure 13-1.

```
NLTK Downloader
-----------------------------------------------------------------------
    d) Download   l) List    u) Update    c) Config    h) Help    q) Quit
-----------------------------------------------------------------------

Downloader> [                                                        ]
```

Figure 13-1. *The NTLK Downloader in a Jupyter Notebook*

Now the tool is waiting for an option. If you want to see a list of possible NLTK extensions, enter L for list and press Enter. You will immediately see a list of all the possible packages belonging to NLTK that you can download to extend the functionality of NLTK, including the texts of the corpora collection.

```
Packages:
  [ ] abc.................. Australian Broadcasting Commission 2006
  [ ] alpino.............. Alpino Dutch Treebank
  [ ] averaged_perceptron_tagger Averaged Perceptron Tagger
  [ ] averaged_perceptron_tagger_ru Averaged Perceptron Tagger (Russian)
  [ ] basque_grammars..... Grammars for Basque
  [ ] biocreative_ppi..... BioCreAtIvE (Critical Assessment of Information
                           Extraction Systems in Biology)
  [ ] bllip_wsj_no_aux.... BLLIP Parser: WSJ Model
  [ ] book_grammars....... Grammars from NLTK Book
  [ ] brown............... Brown Corpus
  [ ] brown_tei.......... Brown Corpus (TEI XML Version)
  [ ] cess_cat........... CESS-CAT Treebank
  [ ] cess_esp........... CESS-ESP Treebank
  [ ] chat80............. Chat-80 Data Files
  [ ] city_database...... City Database
  [ ] cmudict............ The Carnegie Mellon Pronouncing Dictionary (0.6)
  [ ] comparative_sentences Comparative Sentence Dataset
  [ ] comtrans........... ComTrans Corpus Sample
  [ ] conll2000.......... CONLL 2000 Chunking Corpus
  [ ] conll2002.......... CONLL 2002 Named Entity Recognition Corpus
Hit Enter to continue:
```

Pressing Enter again will continue displaying the list by showing other packages in alphabetical order. Press Enter until the list is finished to see all the possible packages. At the end of the list, the different initial options of the NLTK Downloader will reappear.

To create a series of examples to learn about the library, you need a series of texts to work on. An excellent source of texts suitable for this purpose is the Gutenberg corpus, present in the corpora collection. The Gutenberg corpus is a small selection of texts extracted from the electronic archive called the Project Gutenberg (www.gutenberg.org). There are over 25,000 e-books in this archive.

■ **Note**   Attention, in some countries such as Italy, this site is not accessible.

To download this package, first enter the **d** option to download it. The tool will ask you for the package name, so you then enter the name gutenberg.

```
-----------------------------------------------------------------------
    d) Download   l) List    u) Update   c) Config   h) Help   q) Quit
-----------------------------------------------------------------------
Downloader> d
Download which package (l=list; x=cancel)?
  Identifier> gutenberg
```

At this point the package will start to download.

When you already know the name of the package you want to download, just enter the command nltk.download() with the package name as an argument. This will not open the NLTK Downloader tool, but will directly download the required package. So the previous operation is equivalent to writing:

```
nltk.download ('gutenberg')
```

Once it's completed, you can see the contents of the package thanks to the fileids() function, which shows the names of the files contained in it.

```
gb = nltk.corpus.gutenberg
print ("Gutenberg files:", gb.fileids ())
```

An array will appear on the terminal with all the text files contained in the gutenberg package.

```
Out [ ]:
Gutenberg files : ['austen-emma.txt', 'austen-persuasion.txt', 'austen-sense.txt', 'bible-
kjv.txt', 'blake-poems.txt', 'bryant-stories.txt', 'burgess-busterbrown.txt', 'carroll-
alice.txt', 'chesterton-ball.txt', 'chesterton-brown.txt', 'chesterton-thursday.txt',
'edgeworth-parents.txt', 'melville-moby_dick.txt', 'milton-paradise.txt', 'shakespeare-
caesar.txt', 'shakespeare-hamlet.txt', 'shakespeare-macbeth.txt', 'whitman-leaves.txt']
```

To access the internal content of one of these files, you first select one, for example Shakespeare's Macbeth (shakespeare-macbeth.txt), and then assign it to a variable of convenience. An extraction mode is for words, that is, you want to create an array containing words as elements. In this regard, you need to use the words() function.

```
macbeth = nltk.corpus.gutenberg.words ('shakespeare-macbeth.txt')
```

If you want to see the length of this text (in words), you can use the len() function.

```
len (macbeth)
Out [ ]:
23140
```

The text used for these examples is therefore composed of 23140 words.

The macbeth variable is a long array containing the words of the text. If you want to see the first ten words of the text, you can write the following command.

```
macbeth [:10]
Out [ ]:
['[',
```

```
'The',
'Tragedie',
'of',
'Macbeth',
'by',
'William',
'Shakespeare',
'1603',
']']
```

As you can see, the first ten words contain the title of the work, but also the square brackets, which indicate the beginning and end of a sentence. If you had used the sentence extraction mode with the sents() function, you would have obtained a more structured array, with each sentence as an element. These elements, in turn, would be arrays with words for elements.

```
macbeth_sents = nltk.corpus.gutenberg.sents ('shakespeare-macbeth.txt')
macbeth_sents [: 5]
Out [ ]:
[['[',
  'The',
  'Tragedie',
  'of',
  'Macbeth',
  'by',
  'William',
  'Shakespeare',
  '1603',
  ']'],
 ['Actus', 'Primus', '.'],
 ['Scoena', 'Prima', '.'],
 ['Thunder', 'and', 'Lightning', '.'],
 ['Enter', 'three', 'Witches', '.']]
```

Search for a Word with NLTK

One of the most basic things you need to do when you have an NLTK corpus (that is, an array of words extracted from a text) is to do research inside it. The concept of research is slightly different than what you are used to.

The concordance() function looks for all occurrences of a word passed as an argument within a corpus.

The first time you run this command, the system will take several seconds to return a result. The subsequent times will be faster. In fact, the first time this command is executed on a corpus, it creates an indexing of the content to perform the search, which once created will be used in subsequent calls. This explains why the system takes longer the first time.

First, make sure that the corpus is an object nltk.Text, and then search internally for the word 'Stage'.

```
text = nltk.Text(macbeth)
text.concordance('Stage')
Out [ ]:
Displaying 3 of 3 matches:
nts with Dishes and Seruice ouer the Stage . Then enter Macbeth Macb . If it we
with mans Act , Threatens his bloody Stage : byth ' Clock ' tis Day , And yet d
 struts and frets his houre vpon the Stage , And then is heard no more . It is
```

You have obtained three different occurrences of the text.

Another form of searching for a word present in NLTK is that of context. That is, the previous word and the word next to the one you are looking for. To do this, you must use the common_contexts() function.

```
text.common_contexts(['Stage'])
Out [ ]:
the_ bloody_: the_,
```

If you look at the results of the previous research, you can see that the three results correspond to what has been said.

Once you understand how NLTK conceives the concept of the word and its context during the search, it is easy to understand the concept of a synonym. That is, it is assumed that all words that have the same context can be possible synonyms. To search for all words that have the same context as the searched one, you must use the similar() function.

```
text.similar('Stage')
Out [ ]:
fogge ayre bleeding reuolt good shew heeles skie other sea feare
consequence heart braine seruice herbenger lady round deed doore
```

These methods of research may seem rather strange for those who are not used to processing and analyzing text, but you will soon understand that these methods of research are perfectly suited to the words and their meaning in relation to the text in which they are present.

Analyze the Frequency of Words

One of the simplest and most basic examples for the analysis of a text is to calculate the frequency of the words contained in it. This operation is so common that it has been incorporated into a single nltk.FreqDist() function to which the variable containing the word array is passed as an argument.

So to get a statistical distribution of all the words in the text, you enter a simple command.

```
fd = nltk.FreqDist(macbeth)
```

If you want to see the first ten most common words in the text, you can use the most_common() function.

```
fd.most_common(10)
Out [ ]:
[(',', 1962),
 ('.', 1235),
 ("'", 637),
 ('the', 531),
 (':', 477),
 ('and', 376),
 ('I', 333),
 ('of', 315),
 ('to', 311),
 ('?', 241)]
```

From the result obtained, you can see that the most common elements are punctuation, prepositions, and articles, and this applies to many languages, including English. Because these have little meaning during text analysis, it is often necessary to eliminate them. These are called *stopwords*.

Stopwords are words that have little meaning in the analysis and must be filtered. There is no general rule to determine whether a word is a stopword (to be deleted) or not. However, the NLTK library comes to the rescue by providing you with an array of pre-selected stopwords. To download these stopwords, you can use the nltk.download() command.

```
nltk.download('stopwords')
Out [ ]:
[nltk_data] Downloading package stopwords to
[nltk_data]     C:\Users\nelli\AppData\Roaming\nltk_data...
[nltk_data]   Package stopwords is already up-to-date!
True
```

Once you have downloaded all the stopwords, you can select only those related to English, saving them in a variable sw.

```
sw = set(nltk.corpus.stopwords.words ('english'))
print(len(sw))
list(sw) [:10]
Out [ ]:
179
['through',
 'are',
 'than',
 'nor',
 'ain',
 "didn't",
 'didn',
 "shan't",
 'down',
 'our']
```

There are 179 stopwords in the English vocabulary according to NLTK. Now you can use these stopwords to filter the macbeth variable.

```
macbeth_filtered = [w for w in macbeth if w.lower() not in sw]
fd = nltk.FreqDist (macbeth_filtered)
fd.most_common(10)
Out [ ]:
[(',', 1962),
 ('.', 1235),
 ("'", 637),
 (':', 477),
 ('?', 241),
 ('Macb', 137),
 ('haue', 117),
 ('-', 100),
 ('Enter', 80),
 ('thou', 63)]
```

Now that the first ten most common words are returned, you can see that the stopwords have been eliminated, but the result is still not satisfactory. In fact, punctuation is still present in the words. To eliminate all punctuation, you can change the previous code by inserting in the filter an array of punctuation containing the punctuation symbols. This punctuation array can be obtained by importing the `string` function.

```
import string
punctuation = set (string.punctuation)
macbeth_filtered2 = [w.lower () for w in macbeth if w.lower () not in sw and w.lower () not
in punctuation]
```

Now you can recalculate the frequency distribution of words.

```
fd = nltk.FreqDist (macbeth_filtered2)
fd.most_common(10)
Out [ ]:
[('macb', 137),
 ('haue', 122),
 ('thou', 90),
 ('enter', 81),
 ('shall', 68),
 ('macbeth', 62),
 ('vpon', 62),
 ('thee', 61),
 ('macd', 58),
 ('vs', 57)]
```

Finally, the result is what you were looking for.

Select Words from Text

Another form of processing and data analysis is the process of selecting words contained in a body of text based on particular characteristics. For example, you might be interested in extracting words based on their length.

To get all the longest words, for example words that are longer than 12 characters, you enter the following command.

```
long_words = [w for w in macbeth if len(w)> 12]
```

All words longer than 12 characters have now been entered in the `long_words` variable. You can list them in alphabetical order by using the `sort()` function.

```
sorted(long_words)
Out [ ]:
['Assassination',
 'Chamberlaines',
 'Distinguishes',
 'Gallowgrosses',
 'Metaphysicall',
 'Northumberland',
 'Voluptuousnesse',
```

```
'commendations',
'multitudinous',
'supernaturall',
'vnaccompanied']
```

As you can see, there are 11 words that meet this criteria.

Another example is to look for all the words that contain a certain sequence of characters, such as 'ious'. You only have to change the condition in the for in loop to get the desired selection.

```
ious_words = [w for w in macbeth if 'ious' in w]
ious_words = set(ious_words)
sorted(ious_words)
Out [ ]:
['Auaricious',
 'Gracious',
 'Industrious',
 'Iudicious',
 'Luxurious',
 'Malicious',
 'Obliuious',
 'Pious',
 'Rebellious',
 'compunctious',
 'furious',
 'gracious',
 'pernicious',
 'pernitious',
 'pious',
 'precious',
 'rebellious',
 'sacrilegious',
 'serious',
 'spacious',
 'tedious']
```

This example uses sort() to make a list casting, so that it did not contain duplicate words.

These two examples are just a starting point to show you the potential of this tool and the ease with which you can filter words.

Bigrams and Collocations

Another basic element of text analysis is to consider pairs of words (*bigrams*) instead of single words. The words "is" and "yellow" are for example a bigram, since their combination is possible and meaningful. So "is yellow" can be found in textual data. We all know that some of these bigrams are so common in our literature that they are almost always used together. Examples include "fast food," "pay attention," "good morning," and so on. These bigrams are called *collocations*.

Textual analysis can also involve the search for any bigrams within the text under examination. To find them, simply use the bigrams() function. In order to exclude stopwords and punctuation from the bigrams, you must use the set of words already filtered, such as macbeth_filtered2.

```
bgrms = nltk.FreqDist(nltk.bigrams(macbeth_filtered2))
bgrms.most_common(15)
Out [ ]:
[(('enter', 'macbeth'), 16),
 (('exeunt', 'scena'), 15),
 (('thane', 'cawdor'), 13),
 (('knock', 'knock'), 10),
 (('st', 'thou'), 9),
 (('thou', 'art'), 9),
 (('lord', 'macb'), 9),
 (('haue', 'done'), 8),
 (('macb', 'haue'), 8),
 (('good', 'lord'), 8),
 (('let', 'vs'), 7),
 (('enter', 'lady'), 7),
 (('wee', 'l'), 7),
 (('would', 'st'), 6),
 (('macbeth', 'macb'), 6)]
```

By displaying the most common bigrams in the text, linguistic locations can be found.

In addition to the bigrams, there can also be placements based on trigrams, which are combinations of three words. In this case, the trigrams() function is used.

```
tgrms = nltk.FreqDist(nltk.trigrams (macbeth_filtered2))
tgrms.most_common(10)
Out [ ]:
[(('knock', 'knock', 'knock'), 6),
 (('enter', 'macbeth', 'macb'), 5),
 (('enter', 'three', 'witches'), 4),
 (('exeunt', 'scena', 'secunda'), 4),
 (('good', 'lord', 'macb'), 4),
 (('three', 'witches', 'l'), 3),
 (('exeunt', 'scena', 'tertia'), 3),
 (('thunder', 'enter', 'three'), 3),
 (('exeunt', 'scena', 'quarta'), 3),
 (('scena', 'prima', 'enter'), 3)]
```

Preprocessing Steps

Text preprocessing is one of the most important and fundamental phases of text analysis. After collecting the text to be analyzed from various available sources, you will soon realize that in order to use the text in the various NLP techniques, it is necessary to clean it, transform it, and then prepare it specifically to be usable. This section looks at some of the more common preprocessing operations.

The *lower conversion* is perhaps the most frequent and common operation, not only in the preprocessing phase, but also in the following phases of the analysis. In fact, the text contains many words in which some characters are capitalized. Most parsing techniques require that all words be lowercased, so that "word" and "Word" are considered the same word.

In Python, the operation is very simple (there is no need to use the nltk library), since string variables have the lower() method that applies the conversion.

```
text = 'This is a Demo Sentence'
lower_text = text.lower()
lower_text
Out [ ]:
'this is a demo sentence'
```

Word tokenization is another very common operation in NLP. Text consists of words with spaces between them. Therefore, the operation of converting text into a list of words is fundamental in order to be able to process the text computationally. NLTK provides the word_tokenize() function for this purpose. But first you need to download the 'punkt' resource from nltk.

```
nltk.download('punkt')

text = 'This is a Demo Sentence'
tokens = nltk.word_tokenize(text)
tokens
Out [ ]:
['This', 'is', 'a', 'Demo', 'Sentence']
```

Tokenization can also be performed at a higher level, by separating the sentences that make up the text and converting them into elements of a list, instead of single words. In this case, the sent_tokenize() function is used.

```
text = 'This is a Demo Sentence. This is another sentence'
tokens = nltk.sent_tokenize(text)
tokens
Out [ ]:
['This is a Demo Sentence.', 'This is another sentence']
```

Another common preprocessing operation is *punctuation mark removal*. Often you have to submit comments taken from social networks or product reviews on the web for analysis. Many of these texts are rich in punctuation marks, which compromise correct tokenization of the words contained in them. For this operation, NLTK provides a particular tokenizer object called RegexpTokenizer, which allows you to define the tokenization criteria through regular expressions. The following example sets RegexpTokenizer to remove all punctuation marks present in the text.

```
from nltk.tokenize import RegexpTokenizer

text = 'This% is a #!!@ Sentence full of punctuation marks :-) '
regexpt = RegexpTokenizer(r'[a-zA-Z0-9]+')
tokens = regexpt.tokenize(text)
tokens
Out [ ]:
['This', 'is', 'a', 'Sentence', 'full', 'of', 'punctuation', 'marks']
```

The *stopword removal* operation is instead more complex. In fact, stopwords are not particular characters that can be discarded using regular expressions, but are real words that "do not provide information to the text." These words are clearly related to each individual language, and they vary in each language. In English, words like "the," "a," "on," and "in" are basically stopwords. To remove them from the text being analyzed, you can operate as follows.

First, you load the stopwords from nltk and then import them into the code. You do the normal word tokenization and then later remove the English stopwords.

```
nltk.download('stopwords')

from nltk.corpus import stopwords

text = 'This is a Demo Sentence. This is another sentence'
eng_sw = stopwords.words('english')
tokens = nltk.word_tokenize(text)
clean_tokens = [word for word in tokens if word not in eng_sw]
clean_tokens
Out [ ]:
['This', 'Demo', 'Sentence', '.', 'This', 'another', 'sentence']
```

Another type of preprocessing operation involves linguistics. Operations such as *stemming* and *lemmatization* operate on individual words by evaluating their linguistic root (in the first case) and lemma (in the second case). Stemming then groups all words having the same root, considering them the single, same word. Lemmatization instead looks for all the inflected forms of a word or a verb and groups them under the same lemma, considering them all a single word.

For stemming, you therefore have all the roots of the words contained in the text in the tokens. You import a stemmer available in NLTK as SnowballStemmer and set it to English. Then a classic tokenization is performed on the words. Only at this point are they cleaned up by converting them into their linguistic roots.

```
from nltk.stem import SnowballStemmer

text = 'This operation operates for the operator curiosity. A decisive decision'
stemmer = SnowballStemmer('english')
tokens = nltk.word_tokenize(text)
stemmed_tokens = [stemmer.stem(word) for word in tokens]
print(stemmed_tokens)
Out [ ]:
['this', 'oper', 'oper', 'for', 'the', 'oper', 'curios', '.', 'a', 'decis', 'decis']
```

As far as lemmatization is concerned, the operation is very similar. But first you need to download from nltk two components, like WordNet and Omw. A classic word tokenization is performed on the text and then a lemmatizer is defined. At this point this is applied to the tokens to perform the lemmatization of the single words. All inflected forms are lumped together, including singular and plural words and verb conjugations.

```
nltk.download('omw-1.4')
nltk.download('wordnet')

from nltk.stem import WordNetLemmatizer

text = 'A verb: I split, it splits. Splitted verbs.'
tokens = nltk.word_tokenize(text)
```

```
lmtzr = WordNetLemmatizer()
lemma_tokens = [lmtzr.lemmatize(word) for word in tokens]
print(lemma_tokens)
Out [ ]:
['A', 'verb', ':', 'I', 'split', ',', 'it', 'split', '.', 'Splitted', 'verb', '.']
```

Use Text on the Network

So far you have seen a series of examples that use ordered and included text (called a corpus) within the NLTK library as gutenberg. But in reality, you will need to access the Internet to extract the text and collect it as a corpus to be used for analysis with NLTK.

In this section, you see how simple this kind of operation is. First, you need to import a library that allows you to connect to the contents of web pages. The urllib library is an excellent candidate for this purpose, as it allows you to download the text content from the Internet, including HTML pages.

So first you import the request() function, which specializes in this kind of operation, from the urllib library.

```
from urllib import request
```

Then you have to write the URL of the page that contains the text to be extracted. Still referring to the gutenberg project, you can choose, for example, a book written by Dostoevsky (www.gutenberg.org). On the site, there is text in different formats; this example uses the one in the raw format (.txt).

```
url = "http://www.gutenberg.org/files/2554/2554-0.txt"
response = request.urlopen(url)
raw = response.read().decode('utf8')
```

Within the raw text is all the textual content of the book, downloaded from the Internet. Always check the contents of what you downloaded. To do this, the first 75 characters are enough.

```
raw[:75]
Out [ ]:
'\ufeffThe Project Gutenberg EBook of Crime and Punishment, by Fyodor Dostoevsky\r'
```

As you can see, these characters correspond to the title of the text. You can see that there is also an error in the first word of the text. In fact there is the Unicode character BOM \ufeff. This happened because this example used the utf8 decoding system, which is valid in most cases, but not in this case. The most suitable system in this case is utf-8-sig. Replace the incorrect value with the correct one.

```
raw = response.read().decode('utf8-sig')
raw[:75]
Out [ ]:
'The Project Gutenberg EBook of Crime and Punishment, by Fyodor Dostoevsky\r\n'
```

To be able to work on it, you have to convert it into a corpus compatible with NLTK. To do this, enter the following conversion commands.

```
tokens = nltk.word_tokenize (raw)
webtext = nltk.Text (tokens)
```

These commands do nothing more than split the character text into tokens (that is, words) using the nltk.word_tokenize() function and then convert the tokens into a textual body suitable for NLTK using nltk.Text().

You can see the title by entering this command:

```
webtext[:12]
Out [ ]:
['The',
 'Project',
 'Gutenberg',
 'EBook',
 'of',
 'Crime',
 'and',
 'Punishment',
 ',',
 'by',
 'Fyodor',
 'Dostoevsky']
```

Now you have a correct corpus on which to carry out your analysis.

Extract the Text from the HTML Pages

In the previous example, you created a NLTK corpus from text downloaded from the Internet. But most of the documentation on the Internet is in the form of HTML pages. In this section, you see how to extract text from HTML pages.

You always use the request() function of the urllib library to download the HTML content of a web page.

```
url = "https://news.bbc.co.uk/2/hi/health/2284783.stm"
html = request.urlopen(url).read().decode('utf8')
html[:120]
Out [ ]:
'<!doctype html public "-//W3C//DTD HTML 4.0 Transitional//EN" "http://www.w3.org/TR/REC-
html40/loose.dtd">\r\n<html>\r\n<hea'
```

Now, however, the conversion into NLTK corpus requires an additional library, bs4 (BeautifulSoup), which provides you with suitable parsers that can recognize HTML tags and extract the text contained in them.

```
from bs4 import BeautifulSoup
raw = BeautifulSoup(html, "lxml").get_text()
tokens = nltk.word_tokenize(raw)
text = nltk.Text(tokens)
```

Now you also have a corpus in this case, even if you often have to perform more complex cleaning operations than the previous case to eliminate the words that do not interest you.

Sentiment Analysis

Sentiment analysis is a new field of research that has developed very recently in order to evaluate people's opinions about a particular topic. This discipline is based on different techniques that use text analysis and its field of work in the world of social media and forums (*opinion mining*).

Thanks to comments and reviews by users, sentiment analysis algorithms can evaluate the degree of appreciation or evaluation based on certain keywords. This degree of appreciation is called *opinion* and has three possible values: positive, neutral, or negative. The assessment of this opinion thus becomes a form of classification.

So many sentiment analysis techniques are actually classification algorithms, similar to those you saw in previous chapters covering machine learning and deep learning (see Chapters 8 and 9).

As an example to better understand this methodology, I reference a classification tutorial using the Naïve Bayes algorithm on the official website (`www.nltk.org/book/ch06.html`), where it is possible to find many other useful examples to better understand this library.

As a training set, this example uses another corpus present in NLTK, which is very useful for these types of classification problems: `movie_reviews`. This corpus contains numerous film reviews in which there is text of a discrete length together with another field that specifies whether the critique is positive or negative. Therefore, it serves as great learning material.

The purpose of this tutorial is to find the words that recur most in negative documents, or words that recur more in positive ones, so as to focus on the keywords related to an opinion. This evaluation is carried out through a Naïve Bayes classification integrated into NLTK.

First of all, the corpus called `movie_reviews` is important.

```
nltk.download('movie_reviews')
Out [ ]:
[nltk_data] Downloading package movie_reviews to
[nltk_data]     C:\Users\nelli\AppData\Roaming\nltk_data...
[nltk_data]   Package movie_reviews is already up-to-date!
True
```

Then you build the training set from the corpus obtained, creating an array of element pairs called `documents`. This array contains in the first field the text of the single review, and in the second field the negative or positive evaluation. At the end, you mix all the elements of the array in random order.

```
import random
reviews = nltk.corpus.movie_reviews
documents = [(list(reviews.words(fileid)), category)
              for category in reviews.categories()
          for fileid in reviews.fileids(category)]
random.shuffle(documents)
```

To better understand this, take a look at the contents of the documents in detail. The first element contains two fields; the first is the review containing all the words used.

```
first_review = ' '.join(documents[0][0])
print(first_review)
Out [ ]:
topless women talk about their lives falls into that category that i mentioned in the
devil ' s advocate : movies that have a brilliant beginning but don ' t know how to end .
it begins by introducing us to a selection of characters who all know each other . there
is liz , who oversleeps and so is running late for her appointment , prue who is getting
married ,...
```

The second field instead contains the evaluation of the review:

```
documents[0][1]
Out [ ]:
'neg'
```

But the training set is not yet ready; in fact you have to create a frequency distribution of all the words in the corpus. This distribution is converted into a casting list with the `list()` function.

```
all_words = nltk.FreqDist(w.lower() for w in reviews.words())
word_features = list(all_words)
```

Then the next step is to define a function for the calculation of the features, that is, words that are important enough to establish the opinion of a review.

```
def document_features(document, word_features):
    document_words = set(document)
    features = {}
    for word in word_features:
        features ['{}'.format(word)] = (word in document_words)
    return features
```

Once you have defined the `document_features()` function, you can create feature sets from documents.

```
featuresets = [(document_features(d,word_features), c) for (d,c) in documents]
```

The aim is to create a set of all the words contained in the whole movie corpus, analyze whether they are present (True or False) in each single review, and see how much they contribute to the positive or negative judgment of the review. The more often a word is present in the negative reviews and the less often it's present in the positive ones, the more it's evaluated as a "bad" word. The opposite is true for a "good" word evaluation.

To determine how to subdivide this feature set for the training set and the testing set, you must first determine how many elements it contains.

```
len (featuresets)
Out [ ]:
2000
```

To evaluate the accuracy of the model, you use the first 1,500 elements of the set for the training set, and the last 500 items for the testing set.

```
train_set, test_set = featuresets[1500:], featuresets[:500]
```

Finally, you apply the Naïve Bayes classifier provided by the NLTK library to classify this problem. Then you calculate its accuracy, submitting the test set to the model.

```
classifier = nltk.NaiveBayesClassifier.train(train_set)
print (nltk.classify.accuracy(classifier, test_set))
Out [ ]:
0.85
```

The accuracy is not as high as in the examples from the previous chapters, but you are working with words contained in text, and therefore it is very difficult to create accurate models relative to numerical problems.

Now that you have completed the analysis, you can see which words have the most weight in evaluating the negative or positive opinion of a review.

```
classifier.show_most_informative_features(10)
Out [ ]:
Most Informative Features
                 compelling = True         pos : neg    =    11.9 : 1.0
                outstanding = True         pos : neg    =    11.2 : 1.0
                       lame = True         neg : pos    =    10.2 : 1.0
              extraordinary = True         pos : neg    =     9.7 : 1.0
                      lucas = True         pos : neg    =     9.7 : 1.0
                       bore = True         neg : pos    =     8.3 : 1.0
                      catch = True         neg : pos    =     8.3 : 1.0
                    journey = True         pos : neg    =     8.3 : 1.0
                 magnificent = True        pos : neg    =     8.3 : 1.0
                    triumph = True         pos : neg    =     8.3 : 1.0
```

Looking at the results, you will not be surprised to find that the word "badly" is a bad opinion word and that "finest" is a good opinion word. The interesting thing here is that "julie" is a bad opinion word.

Conclusions

In this chapter, you took a small glimpse of the text analysis world. In fact, there are many other techniques and examples that could be discussed. However, at the end of this chapter, you should be familiar with this branch of analysis and especially have begun to learn about the NLTK (Natural Language Toolkit) library, a powerful tool for text analysis.

■ ■ ■

## CHAPTER 14

Image Analysis and Computer Vision with OpenCV

In the previous chapters, the analysis of data was centered entirely on numerical and tabulated data, while in the previous chapter, you saw how to process and analyze data in textual form. This book rightfully closes by introducing the last aspect of data analysis: *image analysis*.

This chapter introduces topics such as computer vision and face recognition. You will see how the techniques of deep learning are at the base of this kind of analysis. Furthermore, another library is introduced, called openCV, which has always been the reference point for image analysis.

Image Analysis and Computer Vision

Throughout the book, you have seen how the purpose of the analysis is to extract new information, to draw new concepts and characteristics from a system under investigation. You did it with numerical and textual data, but the same can be done with images.

This branch of analysis is called *image analysis* and it is based on calculation techniques applied to images (called image filters), which you will see in the next sections.

In recent years, especially because of the development of deep learning, image analysis has experienced huge development in solving problems that were previously impossible, giving rise to a new discipline called *computer vision*.

In Chapter 9, you learned about artificial intelligence, which is the branch of calculation that deals with solving problems of pure "human relevance." Computer vision is part of this, since its purpose is to reproduce the way the human brain perceives images.

In fact, seeing is not just the acquisition of a two-dimensional image—above all it is the interpretation of the content of that area. The captured image is decomposed and elaborated into levels of representation that are gradually more abstract (contours, figures, objects, and words) and therefore recognizable by the human mind.

In the same way, computer vision intends to process a two-dimensional image and extract the same levels of representation from it. This is done through various operations that can be classified as follows:

- *Detection*: Detect shapes, objects, or other subjects of investigation in an image (for example, finding cars)

- *Recognition*: The identified subjects are then led back to generic classes (for example, subdividing cars by brands and types)

- *Identification*: An instance of the previous class is identified (for example, find my car)

OpenCV and Python

OpenCV (Open Source Computer Vision) is a library written in C ++ that is specialized for computer vision and image analysis (`https://opencv.org/`). This powerful library, designed by Gary Bradsky, was born as an Intel project. The first version was released in 2000. Then with the passage of time, it was released under an open source license, and since then has gradually becoming more widespread, reaching version 4.8 (June 2023). At this time, OpenCV supports many algorithms related to computer vision and machine learning and is expanding day by day.

Its usefulness and spread is due precisely to its antagonist: `matlab`. In fact, those who need to work with image analysis have two choices: purchase `matlab` packages or compile and install the open source version of OpenCV. Well, it is easy to see why many have opted for the second choice.

OpenCV and Deep Learning

There is a close relationship between computer vision and deep learning. Since 2017 was a significant year for the development of deep learning (read my article about it at `www.meccanismocomplesso.org/en/2017-the-year-of-deep-learning-frameworks/`), the release of the new version of OpenCV 3.3 has seen the enhancement of the library, with many new features of deep learning and neural networks in general. In fact, the library has a module called `dnn` (deep neural networks) dedicated to this aspect. This module has been specifically developed for use with many deep learning frameworks, including Caffe2, TensorFlow, and PyTorch (for information on these frameworks, see Chapter 9).

Installing OpenCV

You can install a OpenCV package on many operating systems (Windows, iOS, and Android) through the official website (`https://opencv.org/releases/`).

If you use Anaconda as a distribution medium, I recommend using this approach. The installation is very simple and clean.

```
conda install opencv
```

Unfortunately for Linux systems, there is no official PyPI package (with pip to be clear) to be installed. Manual installation is required and may vary depending on the distribution and version used. Many procedures are present on the Internet, some more or less valid.

First Approaches to Image Processing and Analysis

This section familiarizes you with the `opencv` library. First, you start to see how to upload and view images. Then you pass some simple operations to them, add and subtract two images, and see an example of image blending. All these operations are very useful, as they serve as a basis for many other image analysis operations.

Before Starting

Once the `opencv` library is installed, you can open an IPython session on the Jupyter QtConsole or in a Jupyter Notebook.

Before you start programming, you need to import the openCV library.

```
import numpy as np
import cv2
```

Load and Display an Image

First, mainly because OpenCV works on pictures, it is important to know how to load images in a program in Python, manipulate them again, and finally view them to see the results.

The first thing you need to do is read the file containing the image using the openCV library. You can do this using the imread() method. This method reads the file in a compressed format such as JPG and translates it into a data structure that's made of a numerical matrix corresponding to color gradations and position.

■ **Note**   You can find the images and files in the source code of this book.

```
img = cv2.imread('italy2018.jpg')
```

If you are interested in more details, you can see the content of an image directly. You will notice an array of arrays, each corresponding to a specific position of the image, and each characterized by numbers between 0 and 255.

In fact, if you look at the contents of the first element of the image, you get the following.

```
img[0]
Out [ ]:
array([[38, 43, 11],
       [37, 42, 10],
       [36, 41,  9],
       ...,
       [24, 37, 15],
       [22, 36, 12],
       [23, 36, 12]], dtype=uint8)
```

Continuing with the code, you now use the matplotlib library to show the image loaded in the img variable. First import what you need for plotting from the matplotlib library:

```
from matplotlib import pyplot as plt
```

Now you can use pyplot's imshow() method to display the loaded image. Because by default the color format in OpenCV is BGR (blue, green, red), it is important to specify the conversion to RGB format via cv2.COLOR_BGR2RGB. Also, since pyplot is set up for plotting, you need to remove the axes and ticks from the display, so disable them by setting plt.axis to off.

```
plt.axis('off')
plt.imshow(cv2.cvtColor(img, cv2.COLOR_BGR2RGB))
plt.show()
```

When you execute this command, a new window opens and shows the image in Figure 14-1.

Figure 14-1. *The photo of the Italian national football team during training*

Work with Images

Now that you've seen how to view existing images in your file system, you can proceed to the next step: processing an image by performing an operation on it and saving the result to a new file.

Continuing with the previous example, you will use the same code. This time, however, you will perform a simple image manipulation, for example, by decomposing the three RGB channels. Then you will exchange the channels to form a new image. This new image will have all altered colors.

After loading the image, decompose it into the three RGB channels. You can do this easily by using the split() method.

```
b,r,g = cv2.split(img)
```

Now reassemble the three channels, but change the order, for example by exchanging the red channel with the green channel. You can easily recombine the channels using the merge() method.

```
img2 = cv2.merge((b,g,r))
```

The new image is contained in the img2 variable. Display it along with the original in a new window.

```
plt.axis('off')
plt.imshow(cv2.cvtColor(img2, cv2.COLOR_BGR2RGB))
plt.show()
```

By running the program, a new window appears with altered colors (as shown in Figure 14-2).

Figure 14-2. *The processed image has altered colors*

Save the New Image

Finally you have to save your new image by saving the file system.

At the end of the program, add an `imwrite()` method with the name of the new file that you want to save, which can also be of another format, such as PNG.

```
cv2.imwrite('italy2018altered.png', img2)
Out [ ]:
True
```

Execute this command and you will notice a new `italy2018altered.png` file in the workspace.

Elementary Operations on Images

The most basic operation is the addition of two images. With the openCV library, this operation is very simple and you can do it using the `cv2.add()` function. The result obtained will be a combination of the two images.

But do not forget that the two images must have the same dimensions to be added together. In this case, the images are both 512x331 pixels.

The first thing you need to do is load a second image with the same dimensions, in this case `soccer.jpg` (you can find it in the source code).

```
img2 = cv2.imread('soccer.jpg')
plt.axis('off')
plt.imshow(img2)
plt.show()
```

By executing the code, you will get the image shown in Figure 14-3.

Figure 14-3. *A new image that's the same size (512x331 pixels)*

Now you add the two images using the add( ) function.

```
img3 = cv2.add(img,img2)
plt.axis('off')
plt.imshow(cv2.cvtColor(img3, cv2.COLOR_BGR2RGB))
plt.show()
```

By executing this code, you will receive a combination of the two images (as shown in Figure 14-4). Unfortunately, the effect is not very appealing.

Figure 14-4. *A new image obtained by adding the two images*

The result is not what you might expect. The prevalence of white is in fact the result of the simple arithmetic sum of the three RGB values, which is calculated for each individual pixel.

In fact, you know that each of the three RGB components takes values from 0 to 255. Therefore, if the sum of the values of a given pixel is greater than 255 (which is quite likely), the value will still be 255. Therefore, the simple task of adding the images does not lead to an image that's a merger of the two, but instead shifts gradually more and more toward white.

Later you learn how to add two images to create a new image that is half of the two (it is not the arithmetic sum).

You can do the same thing by subtracting two images. This operation can be performed with the `cv2.subtract()` function. This time you would expect an image that will tend more and more toward the black. Replace the `cv2.add()` function with the following.

```
img3 = cv2.subtract(img, img2)
plt.axis('off')
plt.imshow(cv2.cvtColor(img3, cv2.COLOR_BGR2RGB))
plt.show()
```

By running the program, you will get a much darker image (even if you do not see much), as shown in Figure 14-5.

Figure 14-5. *A new image obtained by subtracting one image from another*

Note that this effect is even worse if you do the reverse and subtract the second image from the first.

```
img3 = cv2.subtract(img2, img)
plt.axis('off')
plt.imshow(cv2.cvtColor(img3, cv2.COLOR_BGR2RGB))
plt.show()
```

You get a blackish image, as shown in Figure 14-6.

Figure 14-6. *A new image obtained by subtracting one image from another*

However, this is useful to know that the order of the operators is important for the result.

More concretely, you have already seen that an image created with the opencv library is nothing more than an array of arrays that responds perfectly to the canons of NumPy. Thus, you can use the operations between matrices provided by NumPy, such as the addition of matrices. But be careful, because the result will certainly not be the same.

```
img = img1 + img2
```

In fact, the cv2.add() and cv2.subtract() functions maintain the values between 0 and 255, regardless of the value of the operators. If the sum exceeds 255, the result is interpreted differently, thus creating a very strange color effect (maybe as a module of 255). The same thing happens when the removal produces a negative value; the result would be 0. Arithmetic operations do not have this feature.

However, you can try this directly.

```
img3 = img + img2
plt.axis('off')
plt.imshow(cv2.cvtColor(img3, cv2.COLOR_BGR2RGB))
plt.show()
```

Executing this code, you will get an image with a very strong color contrast (which are the points over 255), as shown in Figure 14-7.

Figure 14-7. *An image obtained by adding two images as two NumPy matrices*

Image Blending

In the previous example, you saw that the addition or subtraction of two images does not produce an intermediate image between the two, but instead changes the coloration toward white or black.

The correct operation is called *blending*. That is, you can consider the operation of superimposing the two images, one above the other, making the one placed above gradually more and more transparent. By adjusting the transparency gradually, you get a mixture of the two images, creating a new one that is the intermediate.

The blending operation does not correspond to a simple addition; the formula corresponds to the following equation.

$$img = \alpha \cdot img1 + (1 - \alpha) \cdot img2 \text{ with } 0 \geq \alpha \geq 1$$

As you can see from this equation, the two images have two numerical coefficients that take values between 0 and 1. With the growth of the α parameter, you will have a smooth transition from the first image to the second.

The opencv library provides the blending operation with the `cv2.addWeighted()` function.

Therefore, if you want to create an intermediate image between two source images, you can use the following code.

```
img3 = cv2.addWeighted(img, 0.3, img2, 0.7, 0)
plt.axis('off')
plt.imshow(cv2.cvtColor(img3, cv2.COLOR_BGR2RGB))
plt.show()
```

The result will be an image like the one shown in Figure 14-8.

Figure 14-8. *An image obtained with image blending*

Image Analysis

The purpose of the examples in the previous section was to understand that images are nothing but NumPy arrays. As such, these numerical matrices can be processed. Therefore, you can implement many mathematical functions that will process the numbers within these matrices to get new images. These new images, obtained from operations, will serve to provide new information.

This is the concept underlying *image analysis*. The mathematical operations carried out by a starting image (matrix) to a resultant image (matrix) are called *image filters* (see Figure 14-9). To understand this process, you will certainly have to deal with photo editing applications (like Photoshop). In any case, you have certainly seen that filters that can be applied to photos. These filters are nothing more than *algorithms* (sequences of mathematical operations) that modify the numerical values in the matrix of the starting image.

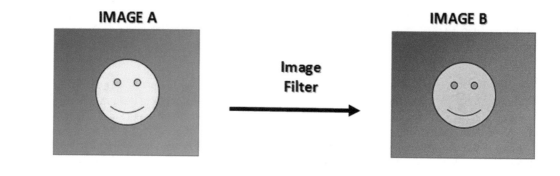

Figure 14-9. *A representation of the image filters that are the basis of image analysis*

Edge Detection and Image Gradient Analysis

In the previous sections, you learned how to perform some basic operations that are useful for image analysis. In this section, you start with a real case of image analysis, called *edge detection*.

Edge Detection

While analyzing an image, and especially during computer vision, one of the fundamental operations is to understand the content of the image, such as objects and people. It is first necessary to understand what possible forms are represented in the image. To understand the geometries represented, it is necessary to recognize the outlines that delimit an object from the background or from other objects. This is precisely the task of edge detection.

In edge detection, a great many algorithms and techniques have been developed and they exploit different principles in order to determine the contours of objects. Many of these techniques are based on the principle of color gradients, and they exploit the image gradient analysis process.

The Image Gradient Theory

Among the various operations that can be applied to images, there are the *convolutions* of an image in which certain filters are applied to edit the image in order to obtain information or some other utility. You have already seen that an image is represented as a large numerical matrix in which the colors of each pixel are represented by a number from 0 to 255 in the matrix. The convolutions process all these numerical values by applying a mathematical operation (*image filter*) to produce new values in a new matrix of the same size.

413

One of these operations is the *derivative*. In simple words, the derivative is a mathematical operation that allows you to get the numerical values indicating the speed at which a value changes (in space, time, etc.).

How could the derivative be important in the case of the images? It has to do with color variation, called a *gradient*.

Being able to calculate the gradient of a color is an excellent tool to calculate the edges of an image. In fact, your eye can distinguish the outlines of a figure present in an image, thanks to the jumps between one color to another. In addition, your eye can perceive the depths thanks to the various shades of color ranging from light to dark, which is the gradient.

From all this, it is quite clear that measuring a gradient in an image is crucial to being able to detect the edges of the image. It's done with a simple operation (a filter) that is carried out on the image.

To get a better look at this from a mathematical point of view, look at Figure 14-10.

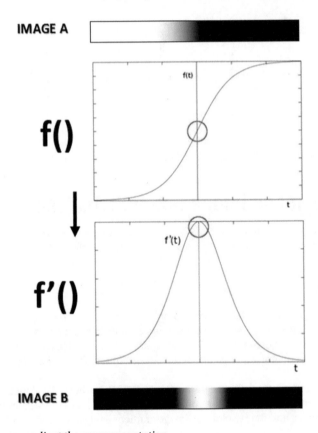

Figure 14-10. *The image gradient theory representation*

As you can see in Figure 14-10, an edge is no more than a quick transition from one hue to another. To simplify, 0 is black and 1 is white. All shades of gray are floating values between 0 and 1.

If you chart all corresponding values to the gradient values, you get the function f(). As you can see, there is a sudden transition from 0 to 1, which indicates the edge.

The derivative of the function f() results in the function f'(). As you can see, the maximum variation of the hue leads to values close to 1. So when converting colors, you will get a figure in which white will indicate the edge.

A Practical Example of Edge Detection with the Image Gradient Analysis

Moving on to the practical part, you will use two images created specifically to test the analysis of the contours, since they have several important characteristics in them.

The first image (shown in Figure 14-11) consists of two arrows in black and white and corresponds to the blackandwhite.jpg file. In this image, the color contrast is very strong and the contours of the arrows have all the possible orientations (horizontal, vertical, and diagonal). This test image will evaluate the effect of edge detection in a black-and-white system.

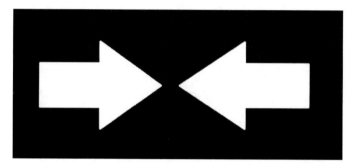

Figure 14-11. *A black-and-white image representing two arrows*

The second image, gradients.jpg, shows different gradients of gray, which, when placed next to each other, create rectangles whose edges have all the possible gradations and combinations of shades (as shown in Figure 14-12). This image is a good test to evaluate the true edge detection capabilities of the system.

Figure 14-12. *A set of gray gradients placed next to each other*

415

Now you can start to develop the code needed for edge detection. You will use `matplotlib` to display different images in the same window. In this test, you will use two different types of image filters provided by OpenCV: *sobel* and *laplacian*. In fact, their names correspond to the name of the mathematical operations performed on the matrices (images). The openCV library provides `cv2.Sobel()` and `cv2.Laplacian()` to apply these two calculations.

First it starts by analyzing the edge detection applied to the `blackandwhite.jpg` image.

```
from matplotlib import pyplot as plt
img = cv2.imread('blackandwhite.jpg',0)
laplacian = cv2.Laplacian(img, cv2.CV_64F)
sobelx = cv2.Sobel(img,cv2.CV_64F,1,0,ksize=5)
sobely = cv2.Sobel(img,cv2.CV_64F,0,1,ksize=5)
plt.subplot(2,2,1),plt.imshow(img,cmap = 'gray')
plt.title('Original'), plt.xticks([]), plt.yticks([])
plt.subplot(2,2,2),plt.imshow(laplacian,cmap = 'gray')
plt.title('Laplacian'), plt.xticks([]), plt.yticks([])
plt.subplot(2,2,3),plt.imshow(sobelx,cmap = 'gray')
plt.title('Sobel Y'), plt.xticks([]), plt.yticks([])
plt.subplot(2,2,4),plt.imshow(sobely,cmap = 'gray')
plt.title('Sobel Y'), plt.xticks([]), plt.yticks([])
plt.show()
```

When you run this code, you get a window with four boxes (as shown in Figure 14-13). The first box is the original image in black and white, while the other three boxes are the result of the three filters applied to the image.

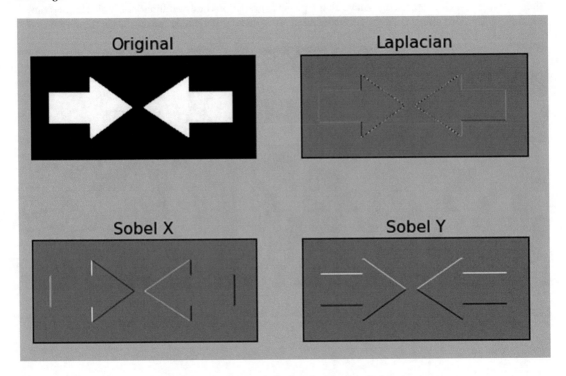

Figure 14-13. *The result from the edge detection applied to the blackandwhite.jpg image*

Regarding the Sobel filters, edge detection is perfect, even if limited horizontally or vertically. The diagonal lines are visible in both cases, since they have both horizontal and vertical components, but the horizontal edges in the Sobel X and those in the vertical Sobel Y are not detected in any way.

Combining the two filters (the calculation of two derivatives) to obtain the Laplacian filter, the determination of the edges is omnidirectional but has some loss of resolution. In fact, you can see that the ripples corresponding to the edges are more subdued.

The coloring in gray is very useful for detecting edges and gradients, but if you are interested in only detecting edges, you have to set as output an image file in cv2.CV_8U.

Therefore, you can change the type of output data from cv2.CV_64F to cv2.CV_8U in the filters function of the previous code. Replace the arguments passed to the two image filters as follows.

```
laplacian = cv2.Laplacian(img, cv2.CV_8U)
sobelx = cv2.Sobel(img,cv2.CV_8U,1,0,ksize=5)
sobely = cv2.Sobel(img,cv2.CV_8U,0,1,ksize=5)
```

By running the code, you will get similar results (as shown in Figure 14-14), but this time only in black and white, where the edges are displayed in white on a black background.

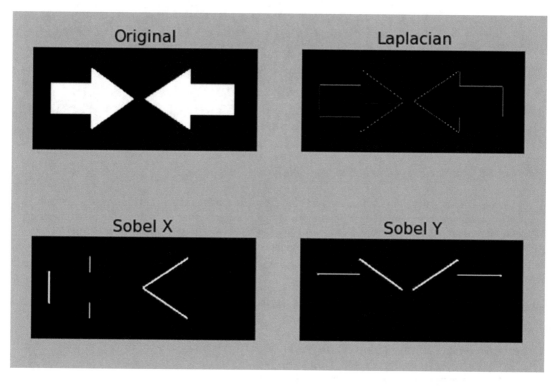

Figure 14-14. *The result from the edge detection applied to the blackandwhite.jpg image*

If you look carefully at the panels of the Sobel filter X and Y, you will notice right away that something is wrong. Where are the missing edges? Note this issue in Figure 14-15.

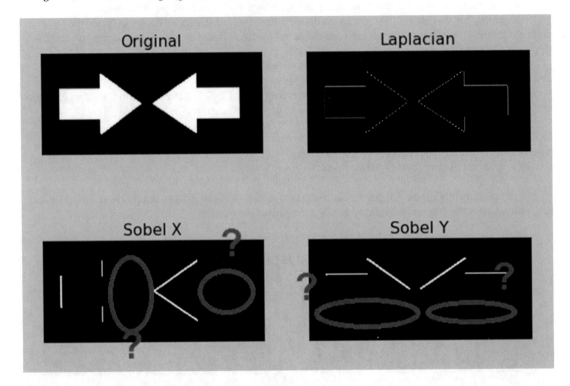

Figure 14-15. *Missing edges in the blackandwhite.jpg image*

In fact, there was a problem while converting the data. The gradients reported in the grayscale with cv2.CV_64F values are represented by positive values (positive slope) when changing from black to white. However, they are represented by negative values (negative slope) when switching from white to black. In the conversion from cv2.CV_64F to cv2.CV_8U, all negative slopes are reduced to 0, and then the information relating to those edges is lost. When the program displays the image, the edges from white to black are not shown.

To overcome this, you should keep the data in the output of the filter in cv2.CV_64F (instead of cv2. CV_8U), then calculate the absolute value, and finally do the conversion in cv2.CV_8U.

Make these changes to the code:

```
laplacian64 = cv2.Laplacian(img, cv2.CV_64F)
sobelx64 = cv2.Sobel(img,cv2.CV_64F,1,0,ksize=5)
sobely64 = cv2.Sobel(img,cv2.CV_64F,0,1,ksize=5)
laplacian = np.uint8(np.absolute(laplacian64))
sobelx = np.uint8(np.absolute(sobelx64))
sobely = np.uint8(np.absolute(sobely64))
plt.subplot(2,2,1),plt.imshow(img,cmap = 'gray')
plt.title('Original'), plt.xticks([]), plt.yticks([])
plt.subplot(2,2,2),plt.imshow(laplacian,cmap = 'gray')
plt.title('Laplacian'), plt.xticks([]), plt.yticks([])
plt.subplot(2,2,3),plt.imshow(sobelx,cmap = 'gray')
```

```
plt.title('Sobel Y'), plt.xticks([]), plt.yticks([])
plt.subplot(2,2,4),plt.imshow(sobely,cmap = 'gray')
plt.title('Sobel Y'), plt.xticks([]), plt.yticks([])
plt.show()
```

Now, if you execute it, you will get the right representation in white on the black edges of the arrows (as shown in Figure 14-16). As you can see, the edges do not appear in Sobel X and Sobel Y because they are parallel to the direction of detection (horizontal and vertical).

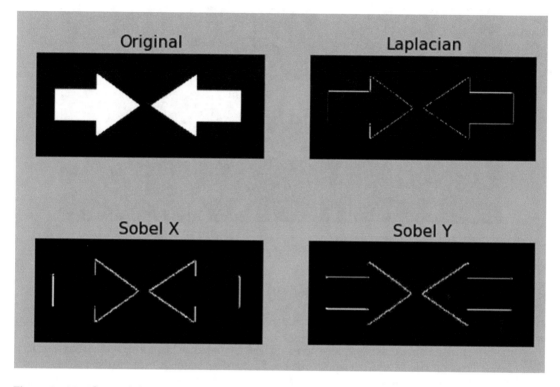

Figure 14-16. *The result from the edge detection applied to the blackandwhite.jpg image*

In addition to the edges, you see that the Laplacian and Sobel filters can also detect the level of gradients across a grayscale. Apply what you've seen to the gradient.jpg image. You have to make some changes to the previous code, leaving only one image (Laplacian) visible.

```
from matplotlib import pyplot as plt
img = cv2.imread('gradients.jpg',0)
laplacian = cv2.Laplacian(img, cv2.CV_64F)
sobelx = cv2.Sobel(img,cv2.CV_64F,1,0,ksize=5)
sobely = cv2.Sobel(img,cv2.CV_64F,0,1,ksize=5)
laplacian64 = cv2.Laplacian(img, cv2.CV_64F)
sobelx64 = cv2.Sobel(img,cv2.CV_64F,1,0,ksize=5)
sobely64 = cv2.Sobel(img,cv2.CV_64F,0,1,ksize=5)
laplacian = np.uint8(np.absolute(laplacian64))
sobelx = np.uint8(np.absolute(sobelx64))
sobely = np.uint8(np.absolute(sobely64))
```

```
plt.imshow(laplacian,cmap = 'gray')
plt.title('Laplacian'), plt.xticks([]), plt.yticks([])
plt.show()
```

By executing this code, you will get an image showing the white borders on a black background (as shown in Figure 14-17).

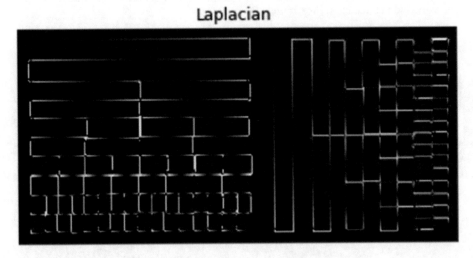

Figure 14-17. *The result from the edge detection applied to the gradients.jpg image*

A Deep Learning Example: Face Detection

In this last section of the chapter, you shift your attention to another highly studied and used case in computer vision, face detection.

This is a far more complex case than edge detection, and it is based on identifying human faces in an image. Given the complexity of the problem, face detection uses deep learning. In fact at the base of this technique, there are neural networks that are specially designed to recognize different subjects, including the faces of a person, in a photo. Object detection techniques also work very similarly. This example is very useful to fully understand the heart of computer vision, that of interpreting the subjects present in a photo.

In this example, you use an already learned neural network. In fact, educating a neural network for this kind of problem can be a complex operation and require a great deal of time and resources.

Fortunately, there are neural networks on the web already trained to perform these kinds of operations, and for this test you will use a model developed using the Caffe2 framework (see Chapter 9 for more information).

When you want to use a deep neural network module with Caffe models in the OpenCV environment, you need two types of files, as follows:

- A *prototxt* file, which defines the model architecture (i.e., the layers themselves). You will use a deploy.prototxt.txt file downloaded from the web (https://github.com/opencv/opencv/blob/master/samples/dnn/face_detector/deploy.prototxt).

- A *caffemodel* file, which contains the weights for the actual layers in the deep neural network. This file is the most important because it contains all the "learning" of that neural network to perform a given task. For your purposes, a caffemodel file is available at https://github.com/opencv/opencv_3rdparty/tree/dnn_samples_ face_detector_20170830.

■ **Note**   You can also find these files in the source code of this book.

Now that you have everything you need, start by uploading the neural network model and all the information about your learning.

The opencv library supports many deep learning frameworks, and it has many features in it that help you with this. In particular (mentioned at the beginning of the chapter), OpenCV has the dnn module, which specializes in these kinds of operations.

To load a learned neural network, you can use the dnn.readNetFromCaffe() function.

```
net = cv2.dnn.readNetFromCaffe('deploy.prototxt.txt', 'res10_300x300_ssd_iter_140000.
caffemodel')
```

As a test image, you can use the photo of the players of the Italian national team, italy2018.jpg. This image is a great example, as there are many faces inside.

```
image = cv2.imread('italy2018.jpg')
(h, w) = image.shape[:2]
```

Another function, called dnn.blobFromImage(), takes care of preprocessing the image to be adapted to neural networks. For example, resize the image to 300x300 pixels so that it can be used by the caffemodel file that has been trained for images of this size.

```
blob = cv2.dnn.blobFromImage(cv2.resize(image, (300, 300)), 1.0, (300, 300), (104.0,
177.0, 123.0))
```

Then define a confidence threshold with an optimal value of 0.5.

```
confidence_threshold = 0.5
```

And finally, perform the face detection test.

```
net.setInput(blob)
detections = net.forward()
for i in range(0, detections.shape[2]):
    confidence = detections[0, 0, i, 2]
    if confidence > confidence_threshold:
        box = detections[0, 0, i, 3:7] * np.array([w, h, w, h])
        (startX, startY, endX, endY) = box.astype("int")
        text = "{:.2f}%".format(confidence * 100)
        y = startY - 10 if startY - 10 > 10 else startY + 10
        cv2.rectangle(image, (startX, startY), (endX, endY),(0, 0, 255), 2)
        cv2.putText(image, text, (startX, y), cv2.FONT_HERSHEY_SIMPLEX, 0.45,
        (0, 0, 255), 2)
```

```
plt.axis('off')
plt.imshow(cv2.cvtColor(image, cv2.COLOR_BGR2RGB))
plt.show()
```

By executing the code, a window will appear with the results of processing the face detection (shown in Figure 14-18). The results are incredible, since the faces of all the players have been detected. You can see the faces surrounded by a red square that highlights them in the image with a percentage of confidence. Confidence percentages are all greater than 50 percent for the confidence_threshold parameter that you specified at the start of the test.

Figure 14-18. *The faces of the national football players have all been accurately recognized*

Conclusions

In this chapter, you saw some simple examples of techniques that form the basis of image analysis and in particular of computer vision. In fact, you saw how images are processed through image filters, and how some complex techniques can be built using edge detection. You also saw how computer vision works by using deep learning neural networks to recognize faces in an image (called face detection).

I hope this chapter has been a good starting point for your further insights on the subject.
If you are interested, you can find in-depth information on this topic on my website at https://meccanismocomplesso.org.

Writing Mathematical Expressions with LaTeX

LaTeX is extensively used in Python. In this appendix there are many examples that can be useful to represent LaTeX expressions inside Python implementations. This same information can be found at the link `https://matplotlib.org/2.0.2/users/mathtext.html`.

With matplotlib

You can enter the LaTeX expression directly as an argument of various functions that can accept it. For example, the `title()` function draws a chart title.

```
import matplotlib.pyplot as plt
%matplotlib inline
plt.title(r'$\alpha > \beta$')
```

With Jupyter Notebook in a Python Cell

To write expressions directly in LaTeX on Jupyter Notebook cells, import the `IPython.display` module.

```
from IPython.display import Latex
```

Then you can display the expressions with the `Latex()` function.

```
Latex('$\\frac{a}{b}$')
```

$$\frac{a}{b}$$

With Jupyter Notebook in a Markdown Cell

You can enter the LaTeX expression between two $$.

`$$c = \sqrt{a^2 + b^2}$$`

$$c = \sqrt{a^2 + b^2}$$

Subscripts and Superscripts

To create subscripts and superscripts, use the _ and ^ symbols:

`r'$\alpha_i > \beta_i$'`

$$\alpha_i > \beta_i$$

This can be very useful when you have to write summations:

`r'$\sum_{i=0}^\infty x_i$'`

$$\sum_{i=0}^{\infty} x_i$$

Fractions, Binomials, and Stacked Numbers

Fractions, binomials, and stacked numbers can be created with the `\frac{}{}`, `\binom{}{}`, and `\stackrel{}{}` commands, respectively:

`r'$\frac{3}{4} \binom{3}{4} \stackrel{3}{4}$'`

$$\frac{3}{4} \binom{3}{4} \stackrel{3}{4}$$

Fractions can be arbitrarily nested:

$$\frac{5 - \frac{1}{x}}{4}$$

Note that you need to take special care to place parentheses and brackets around fractions. You have to insert \left and \right preceding the bracket in order to inform the parser that those brackets encompass the entire object:

$$\left(\frac{5-\frac{1}{x}}{4}\right)$$

Radicals

Radicals can be produced with the \sqrt[]{} command.

r'$\sqrt{2}$'

$$\sqrt{2}$$

Fonts

The default font is italics for mathematical symbols. To change fonts, for example with trigonometric functions as sin:

$$s(t) = A\sin(2\omega t)$$

The choices available with all fonts are

```
from IPython.display import Math
display(Math(r'\mathrm{Roman}'))
display(Math(r'\mathit{Italic}'))
display(Math(r'\mathtt{Typewriter}'))
display(Math(r'\mathcal{CALLIGRAPHY}'))
```

$$\mathrm{Roman}$$

$$\mathit{Italic}$$

$$\mathtt{Typewriter}$$

$$\mathcal{CALLIGRAPHY}$$

Accents

An accent command may precede any symbol to add an accent above it. There are long and short forms for some of them.

\acute a or \'a	$\acute{a}$
\bar a	$\bar{a}$
\breve a	$\breve{a}$
\ddot a or \"a	$\ddot{a}$
\dot a or \.a	$\dot{a}$
\grave a or \`a	$\grave{a}$
\hat a or \^a	$\hat{a}$
\tilde a or \~a	$\tilde{a}$
\vec a	$\vec{a}$
\overline{abc}	$\overline{abc}$

Symbols

You can also use a large number of the TeX symbols.

Lowercase Greek

α \alpha	β \beta	χ \chi	δ \delta	F \digamma
ϵ \epsilon	η \eta	γ \gamma	ι \iota	κ \kappa
λ \lambda	μ \mu	ν \nu	ω \omega	ϕ \phi
π \pi	ψ \psi	ρ \rho	σ \sigma	τ \tau
θ \theta	υ \upsilon	ε \varepsilon	$\varkappa$ \varkappa	φ \varphi
ϖ \varpi	ϱ \varrho	ς \varsigma	ϑ \vartheta	ξ \xi
ζ \zeta				

Uppercase Greek

Δ \Delta	Γ \Gamma	Λ \Lambda	Ω \Omega	Φ \Phi	Π \Pi
Ψ \Psi	Σ \Sigma	Θ \Theta	Υ \Upsilon	Ξ \Xi	$\mho$ \mho
∇ \nabla					

Hebrew

$\aleph$ \aleph	$\beth$ \beth	$\daleth$ \daleth	$\gimel$ \gimel

Delimiters

/ /	[[	$\Downarrow$ \Downarrow	$\Uparrow$ \Uparrow	$\|$ \Vert	\ \backslash
$\downarrow$ \downarrow	\langle	\lceil	\lfloor	\llcorner	\lrcorner
\rangle	\rceil	\rfloor	\ulcorner	$\uparrow$ \uparrow	\urcorner
\vert	\{ \{	| |	\} \}	]]	|

Big Symbols

$\bigcap$ \bigcap	$\bigcup$ \bigcup	$\bigodot$ \bigodot	$\bigoplus$ \bigoplus	$\bigotimes$ \bigotimes
$\biguplus$ \biguplus	$\bigvee$ \bigvee	$\bigwedge$ \bigwedge	$\coprod$ \coprod	$\int$ \int
$\oint$ \oint	$\prod$ \prod	$\sum$ \sum		

Standard Function Names

$\Pr$ \Pr	$\arccos$ \arccos	$\arcsin$ \arcsin	$\arctan$ \arctan
$\arg$ \arg	$\cos$ \cos	$\cosh$ \cosh	$\cot$ \cot
$\coth$ \coth	$\csc$ \csc	$\deg$ \deg	$\det$ \det
$\dim$ \dim	$\exp$ \exp	$\gcd$ \gcd	$\hom$ \hom
$\inf$ \inf	$\ker$ \ker	$\lg$ \lg	$\lim$ \lim
$\liminf$ \liminf	$\limsup$ \limsup	$\ln$ \ln	$\log$ \log
$\max$ \max	$\min$ \min	$\sec$ \sec	$\sin$ \sin
$\sinh$ \sinh	$\sup$ \sup	$\tan$ \tan	$\tanh$ \tanh

Binary Operation and Relation Symbols

≎ \Bumpeq	⋓ \Cap	⋒ \Cup
≑ \Doteq	⋈ \Join	⋐ \Subset
⋑ \Supset	⊩ \Vdash	⊪ \Vvdash
≈ \approx	≊ \approxeq	∗ \ast
≍ \asymp	϶ \backepsilon	∽ \backsim
≃ \backsimeq	$\overline{wedge}$ \barwedge	∵ \because
⧖ \between	◯ \bigcirc	▽ \bigtriangledown
△ \bigtriangleup	◀ \blacktriangleleft	▶ \blacktriangleright
⊥ \bot	⋈ \bowtie	⊡ \boxdot
⊟ \boxminus	⊞ \boxplus	⊠ \boxtimes
● \bullet	≏ \bumpeq	∩ \cap
. \cdot	○ \circ	≗ \circeq
:= \coloneq	≅ \cong	∪ \cup
⋞ \curlyeqprec	⋟ \curlyeqsucc	⋎ \curlyvee

(continued)

⋏ \curlywedge	† \dag	⊣ \dashv
‡ \ddag	◇ \diamond	÷ \div
⋇ \divideontimes	*ėq* \doteq	*ėqdot* \doteqdot
plus \dotplus	⊼ \doublebarwedge	⧡ \eqcirc
=: \eqcolon	⩯ \eqsim	⩾ \eqslantgtr
⩽ \eqslantless	≡ \equiv	≒ \fallingdotseq
⌢ \frown	≥ \geq	≧ \geqq
⩾ \geqslant	≫ \gg	⋙ \ggg
⪆ \gnapprox	≩ \gneqq	⋧ \gnsim
⪊ \gtrapprox	⋗ \gtrdot	⋛ \gtreqless
⪌ \gtreqqless	≷ \gtrless	≳ \gtrsim
∈ \in	⊺ \intercal	⋌ \leftthreetimes
≤ \leq	≦ \leqq	⩽ \leqslant
⪅ \lessapprox	⋖ \lessdot	⋚ \lesseqgtr
⪋ \lesseqqgtr	≶ \lessgtr	≲ \lesssim
≪ \ll	⋘ \lll	⪉ \lnapprox
⪇ \lneqq	⋦ \lnsim	⋉ \ltimes
∣ \mid	⊧ \models	∓ \mp
⊯ \nVDash	⊮ \nVdash	≉ \napprox
≇ \ncong	≠ \ne	≠ \neq
≠ \neq	≢ \nequiv	≱ \ngeq
≯ \ngtr	∋ \ni	≰ \nleq
≮ \nless	∤ \nmid	∉ \notin

(*continued*)

Symbol	Command	Symbol	Command	Symbol	Command
⫴	\nparallel	⊀	\nprec	≁	\nsim
⊄	\nsubset	⊈	\nsubseteq	⊁	\nsucc
⊅	\nsupset	⊉	\nsupseteq	⋪	\ntriangleleft
⋬	\ntrianglelefteq	⋫	\ntriangleright	⋭	\ntrianglerighteq
⊭	\nvDash	⊬	\nvdash	⊙	\odot
⊖	\ominus	⊕	\oplus	⊘	\oslash
⊗	\otimes	‖	\parallel	⊥	\perp
⋔	\pitchfork	±	\pm	≺	\prec
⪷	\precapprox	≼	\preccurlyeq	⪯	\preceq
⪹	\precnapprox	⋨	\precnsim	≾	\precsim
∝	\propto	⋋	\rightthreetimes	≓	\risingdotseq
⋊	\rtimes	∼	\sim	≃	\simeq
╱	\slash	⌣	\smile	⊓	\sqcap
⊔	\sqcup	⊏	\sqsubset	⊏	\sqsubset
⊑	\sqsubseteq	⊐	\sqsupset	⊐	\sqsupset
⊒	\sqsupseteq	⋆	\star	⊂	\subset
⊆	\subseteq	⫅	\subseteqq	⊊	\subsetneq
⫋	\subsetneqq	≻	\succ	⪸	\succapprox
≽	\succcurlyeq	⪰	\succeq	⪺	\succnapprox
⋩	\succnsim	≿	\succsim	⊃	\supset
⊇	\supseteq	⫆	\supseteqq	⊋	\supsetneq
⫌	\supsetneqq	∴	\therefore	×	\times
⊤	\top	◁	\triangleleft	⊴	\trianglelefteq
≜	\triangleq	▷	\triangleright	⊵	\trianglerighteq
⊎	\uplus	⊨	\vDash	∝	\varpropto
◁	\vartriangleleft	▷	\vartriangleright	⊢	\vdash
∨	\vee	⊻	\veebar	∧	\wedge
≀	\wr				

Arrow Symbols

⇓	\Downarrow	⇐	\Leftarrow
⇔	\Leftrightarrow	⇚	\Lleftarrow
⇐	\Longleftarrow	⟺	\Longleftrightarrow
⟹	\Longrightarrow	↰	\Lsh
⇗	\Nearrow	⇖	\Nwarrow
⇒	\Rightarrow	⇛	\Rrightarrow
↱	\Rsh	⇘	\Searrow
⇙	\Swarrow	⇑	\Uparrow
⇕	\Updownarrow	↺	\circlearrowleft
↻	\circlearrowright	↶	\curvearrowleft
↷	\curvearrowright	⇠	\dashleftarrow
⇢	\dashrightarrow	↓	\downarrow
⇊	\downdownarrows	⇂	\downharpoonleft
⇃	\downharpoonright	↩	\hookleftarrow
↪	\hookrightarrow	⇝	\leadsto
←	\leftarrow	↢	\leftarrowtail
↽	\leftharpoondown	↼	\leftharpoonup
⇇	\leftleftarrows	↔	\leftrightarrow
⇆	\leftrightarrows	⇋	\leftrightharpoons
↭	\leftrightsquigarrow	↜	\leftsquigarrow
⟵	\longleftarrow	⟷	\longleftrightarrow
⟼	\longmapsto	⟶	\longrightarrow
↫	\looparrowleft	↬	\looparrowright
↦	\mapsto	⊸	\multimap
⇍	\nLeftarrow	⇎	\nLeftrightarrow
⇏	\nRightarrow	↗	\nearrow
↚	\nleftarrow	↮	\nleftrightarrow

(continued)

431

↛ \nrightarrow	↖ \nwarrow
→ \rightarrow	↣ \rightarrowtail
⇀ \rightharpoondown	⇁ \rightharpoonup
⇄ \rightleftarrows	⇄ \rightleftarrows
⇌ \rightleftharpoons	⇌ \rightleftharpoons
⇉ \rightrightarrows	⇉ \rightrightarrows
⇝ \rightsquigarrow	↘ \searrow
↙ \swarrow	→ \to
↞ \twoheadleftarrow	↠ \twoheadrightarrow
↑ \uparrow	↕ \updownarrow
↕ \updownarrow	↿ \upharpoonleft
↾ \upharpoonright	⇈ \upuparrows

Miscellaneous Symbols

$ \\$	Å \AA	⅃ \Finv
⅁ \Game	ℑ \Im	¶ \P
ℜ \Re	§ \S	∠ \angle
‵ \backprime	★ \bigstar	■ \blacksquare
▲ \blacktriangle	▼ \blacktriangledown	⋯ \cdots
✓ \checkmark	® \circledR	Ⓢ \circledS
♣ \clubsuit	∁ \complement	© \copyright
⃛ \ddots	◇ \diamondsuit	ℓ \ell
∅ \emptyset	ð \eth	∃ \exists
♭ \flat	∀ \forall	ℏ \hbar
♡ \heartsuit	ℏ \hslash	∭ \iiint
∬ \iint	∬ \iint	ı \imath

(*continued*)

∞ \infty	$\jmath$ \jmath	... \ldots
∡ \measuredangle	♮ \natural	¬ \neg
∄ \nexists	∰ \oiiint	∂ \partial
′ \prime	♯ \sharp	♠ \spadesuit
∢ \sphericalangle	β \ss	▽ \triangledown
∅ \varnothing	△ \vartriangle	⋮ \vdots
℘ \wp	¥ \yen	

APPENDIX B

Open Data Sources

Political and Government Data

Data.gov (`https://data.gov/`)—The resource for most government-related data.

    Socrata (`https://dev.socrata.com/data/Socrata`)—This is a good place to explore government-related data. Furthermore, it provides some visualization tools for exploring data.

    U.S. Census Bureau (`www.census.gov/data.html`)—This site provides information about U.S. citizens covering population data, geographic data, and education.

    UN3ta (`https://data.un.org/UNdata`)—This site is an Internet-based data service that includes UN statistical databases.

    European Union Open Data Portal (`https://data.europa.eu/en`)—This site provides a lot of data from European Union institutions.

    Data.gov.uk (`https://www.data.gov.uk/`)—This site of the UK Government includes the British National Bibliography, which has metadata on all UK books and publications published since 1950.

    The CIA World Factbook (`https://www.cia.gov/the-world-factbook/`)—This site of the Central Intelligence Agency provides a lot of information on history, population, economy, government, infrastructure, and military of 267 countries.

Health Data

Healthdata.gov (`https://www.healthdata.gov/`)—This site provides medical data about epidemiology and population statistics.

    NHS Health and Social Care Information Centre (`https://digital.nhs.uk/`)— This site contains health datasets from the UK National Health Service.

Social Data

Facebook Graph (`https://developers.facebook.com/docs/graph-api`)—Facebook provides this API, which allows you to query the huge amount of information that users are sharing with the world.

    Google Trends (`https://trends.google.com/trends/explore`)— This site includes statistics on search volume (as a proportion of total search) for any given term, since 2004.

© Fabio Nelli 2023
F. Nelli, *Python Data Analytics*, https://doi.org/10.1007/978-1-4842-9532-8

Miscellaneous and Public Datasets

Amazon Web Services public datasets (https://registry.opendata.aws/)—The public datasets on Amazon Web Services (AWS) provide a centralized repository of public datasets. An interesting dataset is the 1,000 Genome Project, an attempt to build the most comprehensive database of human genetic information. Also includes a NASA database of satellite imagery of Earth.

DBPedia (https://www.dbpedia.org/)—Wikipedia contains millions of pieces of data, structured and unstructured, on every subject. DBPedia is an ambitious project to catalogue and create a public, freely distributable database allowing anyone to analyze this data.

Gapminder (https://www.gapminder.org/data/)—This site provides data coming from the World Health Organization and World Bank covering economic, medical, and social statistics from around the world.

Financial Data

Google Finance (https://www.google.com/finance/)— This site includes forty years' worth of stock market data, updated in real time.

Climatic Data

National Climatic Data Center (https://www.ncei.noaa.gov/weather-climate-links#loc-clim)—This site is a huge collection of environmental, meteorological, and climate datasets from the U.S. National Climatic Data Center. The world's largest archive of weather data.

WeatherBase (https://www.weatherbase.com/)—This site provides climate averages, forecasts, and current conditions for over 40,000 cities worldwide.

Wunderground (https://www.wunderground.com/)—This site provides climatic data from satellites and weather stations, allowing you to get information about the temperature, wind, and other climatic measurements.

Sports Data

Pro-Football-Reference (https://www.pro-football-reference.com/)—This site provides data about football and several other sports.

Publications, Newspapers, and Books

The New York Times (https://archive.nytimes.com/www.nytimes.com/ref/membercenter/nytarchive.html)—This is a searchable, indexed archive of news articles going back to 1851.

Google Books Ngrams (https://storage.googleapis.com/books/ngrams/books/datasetsv2.html)—This source searches and analyzes the full text of any of the millions of books digitized as part of the Google Books project.

Musical Data

Million Song Dataset (https://aws.amazon.com/it/datasets/million-song-dataset/)—This site includes metadata on over a million songs and pieces of music. Part of Amazon Web Services (AWS).

Index

A

Accents, 426–433
Accuracy, 296, 297, 309, 315, 319, 377, 400
Activation function, 295–297
Adam optimization method, 318
add() function, 408
Adriatic Sea, 323–326
Aggregate functions, 55, 59, 60, 178
Algorithms, 3, 12, 259, 289–291, 321, 399, 412
Amazon Web Services (AWS), 436
Anaconda, 19, 20, 46, 74–78, 135, 144, 184, 245, 300, 404
Anaconda Navigator, 24, 34, 35, 75, 76, 185, 338
    button menu, 22
    conda command, 20
    environments panel, 22
    home panel, 21
    identical panel, 23
    learning panel, 23
    operating system environment, 21
annotate() function, 216, 217
apply() function, 103, 179, 180
arange() function, 50, 51, 194
Arithmetic operators, 52–54, 100, 101
Array manipulation, 47, (*see also* NumPy library)
    joining, 62, 63
    splitting, 63, 64
Artificial intelligence, 3, 289, 290
    ML, 290
    relationship, 290
Artificial neural networks, 293, 294, 298
    MLP, 296, 297
    SLP, 294–296
Assigning values, 81, 91–92
Attributes, 7, 47, 81, 196, 259, 277
autopct kwarg, 233
Axes, 164–166, 188, 196–198, 215, 241, 244
axis() function, 193, 224, 231
Axis labels, 186, 193, 198, 199

B

Back propagation algorithm, 297
Baldwin counties, 363
Bar charts
    horizontal, 222, 223
    matplotlib, 220
    multiserial, 223–225
    multiseries, 225–227
    representations, 230, 231
    stacked, 227–229
    in 3D, 240
    x-axis, 221
bar() function, 219, 221, 222, 224, 227, 237, 240
barh() function, 222, 224, 228
Barplot, 378–381, 383, 384
barplot() function, 257
bfill method, 97
Big Data, 292, 385
Bigrams, 393–394
bigrams() function, 394
Binary classification, 313
Binary files, 70
Binning, 166–169
Binomials, 424–425
Bins, 166–168, 218, 219, 345, 347
Biological neural networks, 298–299
Blending, 404, 411–412
Books, 134, 297, 367, 435, 436
Boolean operators, 60, 61
Bostock, 352, 360, 361
Broadcasting, 64–68

C

Caffe2, 292, 404, 420
caffemodel, 421
Cartesian axes, 215, 216, 247
Cartesian coordinates, 344
Cartesian plane, 305, 315–317

■ Q

■ R

Printed in the United States
by Baker & Taylor Publisher Services